Lecture Notes in Computer Science　9608

Commenced Publication in 1973
Founding and Former Series Editors:
Gerhard Goos, Juris Hartmanis, and Jan van Leeuwen

More information about this series at http://www.springer.com/series/7407

Ichiro Hasuo (Ed.)

Coalgebraic Methods in Computer Science

13th IFIP WG 1.3 International Workshop, CMCS 2016
Colocated with ETAPS 2016
Eindhoven, The Netherlands, April 2–3, 2016
Revised Selected Papers

 Springer

Editor
Ichiro Hasuo
University of Tokyo
Tokyo
Japan

ISSN 0302-9743   ISSN 1611-3349   (electronic)
Lecture Notes in Computer Science
ISBN 978-3-319-40369-4   ISBN 978-3-319-40370-0   (eBook)
DOI 10.1007/978-3-319-40370-0

Library of Congress Control Number: 2016941298

LNCS Sublibrary: SL1 – Theoretical Computer Science and General Issues

This Springer imprint is published by Springer Nature
The registered company is Springer International Publishing AG Switzerland

Preface

The 13th International Workshop on Coalgebraic Methods in Computer Science, CMCS 2016, was held during April 2–3, 2016, in Eindhoven, The Netherlands, as a satellite event of the Joint Conference on Theory and Practice of Software, ETAPS 2016. In more than a decade of research, it has been established that a wide variety of state-based dynamical systems, such as transition systems, automata (including weighted and probabilistic variants), Markov chains, and game-based systems, can be treated uniformly as coalgebras. Coalgebra has developed into a field of its own interest presenting a deep mathematical foundation, a growing field of applications, and interactions with various other fields such as reactive and interactive system theory, object-oriented and concurrent programming, formal system specification, modal and description logics, artificial intelligence, dynamical systems, control systems, category theory, algebra, analysis, etc. The aim of the workshop is to bring together researchers with a common interest in the theory of coalgebras, their logics, and their applications.

Previous workshops of the CMCS series have been organized in Lisbon (1998), Amsterdam (1999), Berlin (2000), Genoa (2001), Grenoble (2002), Warsaw (2003), Barcelona (2004), Vienna (2006), Budapest (2008), Paphos (2010), Tallinn (2012), and Grenoble (2014). Starting in 2004, CMCS has become a biennial workshop, alternating with the International Conference on Algebra and Coalgebra in Computer Science (CALCO), which, in odd-numbered years, has been formed by the union of CMCS with the International Workshop on Algebraic Development Techniques (WADT).

The CMCS 2016 program featured a keynote talk by Jiří Adámek (Technische Universität Braunschweig, Germany), an invited talk by Andreas Abel (University of Gothenburg, Sweden), and an invited talk by Filippo Bonchi (CNRS/ENS Lyon, France). In addition, a special session on weighted automata and coalgebras was held, featuring invited tutorials by Borja Balle (Lancaster University, UK) and Alexandra Silva (University College London, UK).

This volume contains revised regular contributions (10 accepted out of 13 submissions), an invited paper, and the abstracts of two keynote/invited talks. Special thanks go to all the authors for the high quality of their contributions, to the reviewers and Program Committee members for their help in improving the papers presented at CMCS 2016, and to all the participants for active discussions.

April 2016

Ichiro Hasuo

Organization

CMCS 2016 was organized as a satellite event of the Joint Conference on Theory and Practice of Software (ETAPS 2016).

Program Committee

Paolo Baldan	Università di Padova, Italy
Corina Cîrstea	University of Southampton, UK
Ugo Dal Lago	Università di Bologna, Italy
Ichiro Hasuo	University of Tokyo, Japan
Tom Hirschowitz	CNRS, Université de Savoie, France
Bart Jacobs	Radboud University Nijmegen, The Netherlands
Shin-ya Katsumata	Kyoto University, Japan
Bartek Klin	University of Warsaw, Poland
Barbara König	Universität Duisburg-Essen, Germany
Stefan Milius	FAU Erlangen-Nürnberg, Germany
Matteo Mio	CNRS/ENS Lyon, France
Larry Moss	Indiana University, USA
Rasmus Ejlers Møgelberg	IT University of Copenhagen, Denmark
Fredrik Nordvall Forsberg	University of Strathclyde, UK
Dirk Pattinson	The Australian National University
Daniela Petrisan	Université Paris Diderot—Paris 7, France
Jean-Eric Pin	LIAFA, CNRS and University Paris 7, France
John Power	University of Bath, UK
Jurriaan Rot	ENS Lyon, France
Jan Rutten	CWI, The Netherlands
Alexandra Silva	University College London, UK
Joost Winter	University of Warsaw, Poland
James Worrell	Oxford University, UK

Publicity Chair

Fabio Zanasi	Radboud University Nijmegen, The Netherlands

Additional Reviewers

Soichiro Fujii	Henning Kerstan	Toby Wilkinson
Helle Hvid Hansen	Lutz Schroeder	

Sponsoring Institutions

IFIP WG 1.3
Support Center for Advanced Telecommunications Technology Research (SCAT),
Tokyo, Japan

Contents

Fixed Points of Functors - A Short Abstract

Jiří Adámek[✉]

Institute for Theoretical Computer Science,
Technische Universität Braunschweig, Braunschweig, Germany
j.adamek@tu-braunschweig.de

Abstract. Fixed points of endofunctors play a central role in program semantics (initial algebras as recursive specification of domains), in coalgebraic theory of systems (terminal coalgebras and coinduction) and in a number of other connections such as iterative theories (rational fixed point). In this survey we present some older and new results on the structure of the three fixed points we have mentioned.

1 Initial Algebras

The classical example of Σ–algebras as algebras for the polynomial set functor F_Σ yields the initial algebra

$$\mu F_\Sigma = \text{all finite } \Sigma\text{-trees.}$$

This holds for all finitary signatures, whereas for the infinitary ones the initial algebra is formed by all well-founded trees. These are the trees in which every path is finite. The concept of well-foundedness is fundamental for initial algebras. This is a property of coalgebras studied by Osius [9] and Taylor [12].

In [4] we have proved that for all set functors F the initial algebra coincides with the terminal well-founded coalgebra. And every fixed point of F has a largest well–founded part which is the initial algebra of F, see [3].

The iterative construction of μF as $F^i(0)$ (for some ordinal i), introduced in [1], converges for every set functor with a fixed point, see [13]. Moreover the least such ordinal i is an infinite regular cardinal or at most 3, as proved in [6]. In contrast,

- on the category of many-sorted sets convergence can take place at any ordinal, and
- on the category of graphs endofunctors exist having an initial algebra although the iterative construction does not converge,

see [6].

Most of the above results extend from set functors to enfoductors of "reasonable" categories preserving monomorphisms. For example locally finitely presentable categories whose initial object is simple are "reasonable".

I. Hasuo (Ed.): CMCS 2016, LNCS 9608, pp. 1–4, 2016.
DOI: 10.1007/978-3-319-40370-0_1

2   Terminal Coalgebras

The importance of terminal coalgebras as systems was demonstrated by Rutten in his fundamental paper [10].

A number of classical examples are special cases of the following:

$$\nu F_\Sigma = \text{all } \Sigma\text{-trees.}$$

The case of automata with n inputs, $H_n X = X^n \times \{0,1\}$, is an example where Σ consists of two n–ary operations (and Σ–trees just represent languages in n^*). And streams, $FX = A \times X + B$, correspond to unary operations in A and constants in B.

In case of set functors, it is useful to work with pointed coalgebras. E.g., the classical automata are pointed coalgebras for H_n.

Definition 1. *A well–pointed coalgebra is a pointed coalgebra having no proper subobject and no proper quotient.*

Thus minimal automata are precisely the well-pointed coalgebras for H_n. Let T be the collection of all well–pointed coalgebras up to isomorphism. If F is a set functor preserving intersections, every pointed coalgebra has a canonical "minimization" to a well–pointed one. This yields a coalgebra structure on T. And in [4] we presented the following characterizations:

$$\nu F = \text{all well-pointed coalgebras(up to isomoprhism)}$$

and

$$\mu F = \text{all well-founded well-pointed coalgebras (up to isomoprhism).}$$

"The dual of the iterative initial-algebra construction, i.e. the cochain $F^i(1)$ as a construction of νF, was first explicitly considered by Barr [7]. In contrast to the initial chain the convergence does not always happen at a cardinal: Worrell proved that for the finite power-set functor that cochain converges in $\omega + \omega$ steps. And for all λ–accessible set functors (i.e., those preserving λ–filtered colimits), he proved that the construction converges in $\lambda + \lambda$ steps or sooner, see [14]. Until recently, for uncountable λ, no example has been known where the full $\lambda + \lambda$ steps are actually needed. Such an example is the functor of all λ–generated filters, see [2].

An important classical result concerns enriched endofunctors of the cartesian closed category CPO (of posets with directed joins and a bottom and strict continuous maps): Smyth and Plotkin proved in [11] that the terminal coalgebra equals the initial algebra, and both constructions converge in ω steps. This is based on the limit-colimit coincidence for embedding-projection pairs of Scott. In [3] we prove that for the (much wider) class of endofunctors that are just locally monotone the existence of a fixed point implies that the terminal coalgebra exists and equals the initial algebra.

3 Rational Fixed Point

This (much younger) fixed point arose in the study of iterative algebras. These are algebras A that have a unique solution of every finite guarded system of recursive equations (with parameters in A). In [5] we proved that every finitary set functor has an initial iterative algebra which is a fixed point, notation: ρF. And we called this the rational fixed point. Classical examples: for automata as coalgebras we have

$$\rho H_n = \text{all regular languages in } n^*,$$

and for general signatures

$$\rho F_\Sigma = \text{ all rational } \Sigma\text{-trees,}$$

that is, trees that have, up to isomorphism, only finitely many subtrees.

The rational fixed point can be described as the filtered colimit of all finite coalgebras. This has inspired Milius [8] to define locally finite coalgebras for a set functor as coalgebras that are directed unions of finite subcoalgebras. And then he proved that

$$\rho F = \text{ the terminal locally finite coalgebra.}$$

In a complete analogy to the above characterization of the initial and terminal fixed point we have that

$$\rho F = \text{all finite well-pointed coalgebras (up to isomorphism).}$$

All the above generalizes easily to finitary endofunctors of all finitely locally presentable categories.

Can we play the same game with countable sets in place of finite ones? No, only finite sets yield something new: the terminal locally countable coalgebra is all of νF, see [5].

References

1. Adámek, J.: Free algebras and automata realizations in the language of categories. Comment. Math. Univ. Carol. **15**, 589–602 (1974)
2. Adámek, J., Koubek, V., Palm, T.: Fixed points of set functors: how many iterations are needed? (submitted)
3. Adámek, J., Milius, S., Moss, L.: Initial algebras and terminal coalgebras (to appear)
4. Adámek, J., Milius, S., Moss, S., Sousa, L.: Well-pointed coalgebras. Log. Meth. Comput. Sci. **9**, 1–51 (2013)
5. Adámek, J., Milius, S., Velebil, J.: Iterative algebras at work. Math. Struct. Comput. Sci. **16**, 1085–1131 (2006)
6. Adámek, J., Trnková, V.: Initial algebras and terminal coalgebras in many-sorted sets. Mathem. Str. Comput. Sci. **21**, 481–509 (2011)

7. Barr, M.: Terminal coalgebras in well-founded set theory. Theoret. Comput. Sci. **124**, 182–192 (1994)
8. Milius, S.: A sound and complete calculus for finite stream circuits. In: Proceeding of 25th Annual Symposium on Logic in Computer Science (LICS 2010). IEEE Computer Society (2010)
9. Osius, G.: Categorical set theory: a characterization of the category of sets. J. Pure Appl. Algebra **4**, 79–119 (1974)
10. Rutten, J.J.M.M.: Universal coalgebra: a theory of systems. Theoret. Comput. Sci. **249**, 3–80 (2000)
11. Smyth, M., Plotkin, G.: Category-theoretical solution of recursive domain equations. SIAM Journ. Comput. **11**, 761–783 (1982)
12. Taylor, P.: Towards a unified treatment of induction i: the general recursion theorem. Preprint (1995–6). http://www.paultaylor.eu/ordinals/#towuti
13. Trnková, V., Adámek, J., Koubek, V., Reiterman, J.: Free algebras, input processes and free monads. Comment. Math. Univ. Carol. **16**, 339–351 (1975)
14. Worrell, J.: On the final sequence of a finitary set functor. Theoret. Comput. Sci. **338**, 184–199 (2005)

Compositional Coinduction with Sized Types

Andreas Abel[✉]

Department of Computer Science and Engineering, Gothenburg University,
Rännvägen 6, 41296 Göteborg, Sweden
andreas.abel@gu.se

Proofs by induction on some inductively defined structure, e.g., finitely-branching trees, may appeal to the induction hypothesis at any point in the proof, provided the induction hypothesis is only used for immediate substructures, e.g., the subtrees of the node we are currently considering in the proof. The basic principle of structural induction can be relaxed to course-of-value induction, which allows application of the induction hypothesis also to non-immediate substructures, like any proper subtree of the current tree. If course-of-value induction is not sufficient yet, we can resort to define a well-founded relation on the considered structure and use the induction hypothesis for any substructure which is strictly smaller with regard to the constructed relation. At a closer look, however, this well-founded induction is just *structural induction on the derivation* of being strictly smaller. This means that in a logical system that allows us to construct inductive predicate and relations, such as, e.g., Martin-Löf Type Theory (Nordström et al. 1990) or the Calculus of Inductive Constructions (Paulin-Mohring 1993), structural induction is complete for any kind of inductive proof.

In all these flavors of induction, validity of induction hypothesis application can be checked easily and locally, independent of its context. In proof assistants, in principle a *structural termination checker* (Giménez 1995; Abel 2000) suffices[1] to check such inductive proofs, which looks at the proof tree, extracts all calls to the induction hypotheses, and checks that they happen only on structurally smaller arguments. In practice, mutual induction is supported as well, based on a simple static call graph analysis (Abel and Altenkirch 2002; Barras 2010; Ben-Amram 2008; Hyvernat 2014).

Dually to structural induction, in a coinductive proof of a proposition defined as the greatest fixed-point of a set of rules, we may appeal to the coinduction hypothesis to fill the premises of the rule we have chosen to prove our goal. For instance, two infinite streams may be defined to be bisimilar if their heads are equal and their tails are bisimilar, coinductively. Our goal might be to show that bisimilarity is reflexive, i.e., any stream is bisimilar to itself. To establish bisimilarity, we use the sole rule with the first subgoal to show that the head of the stream is equal to itself. After this breath-taking enterprise, we are left

---

[1] Even for induction, type-based termination offers significant advantages for compositionality and robustness, as argued, e.g., by Barras and Sacchini (2013). However, at this point there is no mature implementation in proof assistants based on dependent types.

© IFIP International Federation for Information Processing 2016
Published by Springer International Publishing Switzerland 2016. All Rights Reserved
I. Hasuo (Ed.): CMCS 2016, LNCS 9608, pp. 5–10, 2016.
DOI: 10.1007/978-3-319-40370-0_2

with the second subgoal to show that the tail of the stream is bisimilar to itself, which we solve by appeal to the coinduction hypothesis. At this point, it is worth noting that not the stream got smaller (the tail of an infinite stream is still infinite), but the coinductive hypothesis is guarded by a rule application. This means that the coinductive proof can unfold into a possibly infinitely deep derivation without getting into a "busy loop", meaning the proof is *productive*.

In a similar way as for induction, we seek to relax the criterion for well-formed coinductive proofs, which states that only the immediate subgoals of the final rule application can be filled by the coinductive hypothesis. We can allow several rule applications until we reach the coinductive hypothesis from the root of the derivation. This is dual to course-of-value induction and could be called *guarded coinduction*.[2]

In contrast to induction, checking the validity of calls to the coinduction hypothesis requires us to look at the *context* of the calls rather than the call arguments. We have to check that the calls to the coinductive hypotheses happen in a *constructor* context, i. e., a context of coinductive rule applications only. This lack of locality also leads to a loss of compositionality of proofs by guarded coinduction. For instance, consider a coinductive proof of the bisimilarity of two streams through a bisimilarity chain, i. e., via some intermediate streams and the use of transitivity of bisimilarity. Transitivity is not a constructing rule for bisimilarity,[3] but an admissible rule proven by coinduction. As transitivity is not a constructor, we cannot use the coinduction hypothesis under transitivity nodes in the proof tree. In practice, often a severe restructuring of a natural informal proof is necessary to make it guarded and please a structural guardedness checker. The resulting proofs may be highly non-compositional and bloated, especially if proofs of previous lemmata have to be inlined and fused into the current proof.

To regain compositionality, we have to relax the contexts of coinductive hypothesis applications to include admissible rules and lemma invocation in general, without jeopardizing productivity. Such contexts need to produce one more rule constructor than they consume, which must be easily verifiable by the productivity checker. Sized types (Hughes et al. 1996; Amadio and Coupet-Grimal 1998; Barthe et al. 2004; Abel 2008; Sacchini 2013) offer the necessary technology. Coinductive types, propositions, and relations are parameterized by an ordinal $i \leq \omega$ which denotes the minimum *definedness depth* of their derivations. Semantically, this idea is already present in Mendler's work (Mendler et al. 1986; Mendler 1991), and it is implicit in the principle of ordinal iteration to construct

---

[2] Coquand (1994) calls it *guarded induction*, Giménez (1995) *guarded by constructors*.

[3] In fact, it is well-known that every coinductive relation with a transitivity rule is trivial, i. e., the total relation. The proof of relatedness of arbitrary objects is just the infinite tree all of whose nodes are applications of the transitivity rule. This problem can be overcome with mixed coinductive-inductive types (Abel 2007; Nakata and Uustalu 2010), to allow only finitely many applications of the transitivity rule in a row (Danielsson and Altenkirch 2010).

the greatest fixed point of a monotone operator F. We define the approximants $\nu^i F$ of the greatest fixed-point $\nu^\omega F$ by induction on ordinal i as follows:

$$\nu^0 \quad F = \top$$
$$\nu^{i+1} F = F\left(\nu^i F\right)$$
$$\nu^\omega \quad F = \sqcap_{i<\omega} \nu^i F$$

For monotone F, we obtain a descending chain $\nu^0 F \sqsupseteq \nu^1 F \sqsupseteq \cdots \sqsupseteq \nu^\omega F$. The greatest fixed point of F is reached at stage ω if F is continuous in the sense that $\sqcap_{i \in I} F(A_i) \sqsubseteq F(\sqcap_{i \in I} A_i)$. For instance, all strictly positive type transformers correspond to continuous operators (Abel 2003, Theorem 1).

An alternative construction of the greatest fixed-point uses deflationary iteration (Sprenger and Dam 2003; Abel 2012; Abel and Pientka 2013),

$$\nu^i F = \prod_{j<i} F\left(\nu^j F\right)$$

which gives a descending chain without the monotonicity of F. However, the same conditions on F are needed to reach the fixed point at stage ω.

Giving names to the approximants $\nu^i F$ of coinductive type $\nu^\omega F$, we can express through the type system when a term t, which acts as the context for the coinductive hypothesis, produces one more constructor than it consumes: it needs to have type $\forall i.\ \nu^i F \to \nu^{i+1} F$ polymorphic in "size" (depth) i. Such a context is called *guarding*. A weaker, but very common and useful property of a function t is to be *guardedness preserving*, i.e., having type $\forall i.\ \nu^i F \to \nu^i F$. For instance, consider bisimilarity on streams, which is defined using relation transformer $F(X)(x,y) = (\text{head}\, x \equiv \text{head}\, y) \times X\, (\text{tail}\, x)\, (\text{tail}\, y)$. The symmetry lemma of bisimilarity $\forall i.\, \nu^i F(x,y) \to \nu^i F(y,x)$ is guardedness preserving: to produce one constructor of the requested bisimilarity derivation, it only needs to inspect the last constructor of the given bisimilarity derivation. Analogously, transitivity of bisimilarity receives type $\forall i.\ \nu^i F(x,y) \to \nu^i F(y,z) \to \nu^i F(x,z)$. Here, to produce the last rule of the output derivations we only need to inspect the last rule of the two input derivations. This typing allows us to freely use transitivity in coinductive proofs without jeopardizing the validity of the coinductive hypothesis.

Tracking guardedness levels in the type systems through "sized" coinductive types gives us *compositional* coinduction, as we can freely abstract out, for instance, guardedness-preserving contexts without upsetting some structural productivity checker who can only deal with concretely given code.

Recently, guardedness-preserving functions have been rediscovered in the context of Isabelle/HOL as *friendly operations* (Blanchette et al. 2015). Prototypical implementations of type-based termination with sized types exist for Coq (Sacchini 2015) and Agda. The latter has been subjected to a larger case study: an implementation of normalization by evaluation using the coinductive delay monad (Abel and Chapman 2014). Compositional coinduction with sized types seems to be a promising alternative to parameterized coinduction (Hur et al. 2013) and up-to techniques (Pous and Sangiorgi 2012).

References

Abel, A.: Specification and verification of a formal system for structurally recursive functions. In: Coquand, T., Nordström, B., Dybjer, P., Smith, J. (eds.) TYPES 1999. LNCS, vol. 1956, pp. 1–20. Springer, Heidelberg (2000). http://dx.doi.org/10.1007/3-540-44557-9_1

Abel, A.: Termination and productivity checking with continuous types. In: Hofmann, M.O. (ed.) TLCA 2003. LNCS, vol. 2701, pp. 1–15. Springer, Heidelberg (2003). http://dx.doi.org/10.1007/3-540-44904-3_1

Abel, A.: Mixed inductive/coinductive types and strong normalization. In: Shao, Z. (ed.) APLAS 2007. LNCS, vol. 4807, pp. 286–301. Springer, Heidelberg (2007). http://dx.doi.org/10.1007/978-3-540-76637-7_19

Abel, A.: Semi-continuous sized types and termination. Log. Methods Comput. Sci. 4(2:3), 1–33 (2008). http://dx.doi.org/10.2168/LMCS-4(2:3)2008, CSL 2006 special issue

Abel, A.: Type-based termination, inflationary fixed-points, and mixed inductive-coinductive types. In: Miller, D., Ésik, Z. (eds.) Proceedings of 8th Workshop on Fixed Points in Computer Science (FICS 2012), Electronic Proceedings in Theoretical Computer Science, vol. 77, pp. 1–11 (2012). http://dx.doi.org/10.4204/EPTCS.77.1, invited talk

Abel, A., Altenkirch, T.: A predicative analysis of structural recursion. J. Funct. Program. 12(1), 1–41 (2002). http://dx.doi.org/10.1017/S0956796801004191

Abel, A., Chapman, J.: Normalization by evaluation in the delay monad: a case study for coinduction via copatterns and sized types. In: Levy, P., Krishnaswami, N. (eds.) Proceedings of 5th Workshop on Mathematically Structured Functional Programming, MSFP 2014, Grenoble, France, Electronic Proceedings in Theoretical Computer Science, vol. 153, pp. 51–67, 12 April 2014. http://dx.doi.org/10.4204/EPTCS.153.4

Abel, A., Pientka, B.: Wellfounded recursion with copatterns: a unified approach to termination and productivity. In: Morrisett, G., Uustalu, T. (eds.) Proceedings of 18th ACM SIGPLAN International Conference on Functional Programming, ICFP 2013, pp. 185–196. ACM Press, Boston, MA, USA, 25–27 September 2013. http://doi.acm.org/10.1145/2500365.2500591

Amadio, R.M., Coupet-Grimal, S.: Analysis of a guard condition in type theory (extended abstract). In: Nivat, M. (ed.) FOSSACS 1998. LNCS, vol. 1378, p. 48. Springer, Heidelberg (1998). http://dx.doi.org/10.1007/BFb0053541

Barras, B.: The syntactic guard condition of Coq. In: Talk at the Journée "égalité et Terminaison" du 2 février 2010 in Conjunction with JFLA 2010 (2010). http://coq.inria.fr/files/adt-2fev10-barras.pdf

Barras, B., Sacchini, J.L.: Type-based methods for termination and productivity in Coq. In: Mahboubi, A., Tassi, E. (eds.) The 5th Coq Workshop, A Satellite Workshop of ITP 2013, Rennes, 22 July 2013. https://coq.inria.fr/coq-workshop/2013#Sacchini

Barthe, G., Frade, M.J., Giménez, E., Pinto, L., Uustalu, T.: Type-based termination of recursive definitions. Math. Struct. Comput. Sci. 14(1), 97–141 (2004). http://dx.doi.org/10.1017/S0960129503004122

Ben-Amram, A.M.: Size-change termination with difference constraints. ACM Trans. Program. Lang. Syst. 30(3) (2008). http://doi.acm.org/10.1145/1353445.1353450

Blanchette, J.C., Popescu, A., Traytel, D.: Foundational extensible corecursion: a proof assistant perspective. In: Fisher, K., Reppy, J.H. (eds.) Proceedings of 20th ACM SIGPLAN International Conference on Functional Programming, ICFP 2015, pp. 192–204. ACM Press, Vancouver, BC, Canada, 1–3 September 2015. http://doi.acm.org/10.1145/2784731.2784732

Coquand, T.: Infinite objects in type theory. In: Barendregt, H., Nipkow, T. (eds.) TYPES 1993. LNCS, vol. 806, pp. 62–78. Springer, Heidelberg (1994). http://dx.doi.org/10.1007/3-540-58085-9_72

Danielsson, N.A., Altenkirch, T.: Subtyping, declaratively. In: Bolduc, C., Desharnais, J., Ktari, B. (eds.) MPC 2010. LNCS, vol. 6120, pp. 100–118. Springer, Heidelberg (2010). http://dx.doi.org/10.1007/978-3-642-13321-3_8

Giménez, E.: Codifying guarded definitions with recursive schemes. In: Dybjer, P., Nordström, B., Smith, J. (eds.) TYPES 1994. LNCS, vol. 996, pp. 39–59. Springer, Heidelberg (1994). http://dx.doi.org/10.1007/3-540-60579-7_3

Hughes, J., Pareto, L., Sabry, A.: Proving the correctness of reactive systems using sized types. In: Boehm, H.J., Jr., G.L.S. (eds.) Conference Record of POPL 1996: The 23rd ACM SIGPLAN-SIGACT Symposium on Principles of Programming Languages, pp. 410–423. ACM Press, St. Petersburg Beach, Florida, USA, 21–24 January 1996. http://doi.acm.org/10.1145/237721.240882

Hur, C., Neis, G., Dreyer, D., Vafeiadis, V.: The power of parameterization in coinductive proof. In: Giacobazzi, R., Cousot, R. (eds.) The 40th Annual ACM SIGPLAN-SIGACT Symposium on Principles of Programming Languages, POPL 2013, pp. 193–206. ACM Press, Rome, Italy, 23–25 January 2013. http://doi.acm.org/10.1145/2429069.2429093

Hyvernat, P.: The size-change termination principle for constructor based languages. Log. Methods Comput. Sci. 10(1) (2014). http://dx.doi.org/10.2168/LMCS-10(1:11)2014

Mendler, N.P., Panangaden, P., Constable, R.L.: Infinite objects in type theory. In: Proceedings, Symposium on Logic in Computer Science, pp. 249–255. IEEE Computer Society, Cambridge, Massachusetts, USA, 16–18 June 1986

Mendler, N.P.: Inductive types and type constraints in the second-order lambda calculus. Ann. Pure Appl. Log. 51(1–2), 159–172 (1991). http://dx.doi.org/10.1016/0168-0072(91)90069-X

Nakata, K., Uustalu, T.: Resumptions, weak bisimilarity and big-step semantics for while with interactive I/O: an exercise in mixed induction-coinduction. In: Aceto, L., Sobocinski, P. (eds.) Proceedings of 7th Workshop on Structural Operational Semantics, SOS 2010, Paris, France, Electronic Proceedings in Theoretical Computer Science, vol. 32, pp. 57–75, 30 August 2010. http://dx.doi.org/10.4204/EPTCS.32.5

Nordström, B., Petersson, K., Smith, J.M.: Programming in Martin Löf's Type Theory: An Introduction. Clarendon Press, Oxford (1990). http://www.cs.chalmers.se/Cs/Research/Logic/book/

Paulin-Mohring, C.: Inductive definitions in the system Coq - rules and properties. In: Bezem, M., Groote, J.F. (eds.) TLCA 1993. LNCS, vol. 664, pp. 328–345. Springer, Heidelberg (1993). http://dx.doi.org/10.1007/BFb0037116

Pous, D., Sangiorgi, D.: Enhancements of the bisimulation proof method. In: Sangiorgi, D., Rutten, J. (eds.) Advanced Topics in Bisimulation and Coinduction. Cambridge University Press, Cambridge (2012)

Sacchini, J.: Coq: Type-based termination in the Coq proof assistant (2015). project description, http://qatar.cmu.edu/sacchini/coq.html

Sacchini, J.L.: Type-based productivity of stream definitions in the calculus of constructions. In: 28th Annual ACM/IEEE Symposium on Logic in Computer Science, LICS 2013, pp. 233–242. IEEE Computer Society Press, New Orleans, LA, USA, 25–28 June 2013. http://dx.doi.org/10.1109/LICS.2013.29

Sprenger, C., Dam, M.: On the structure of inductive reasoning: circular and tree-shaped proofs in the μ-calculus. In: Gordon, A.D. (ed.) FOSSACS 2003 and ETAPS 2003. LNCS, vol. 2620, pp. 425–440. Springer, Heidelberg (2003). http://dx.doi.org/10.1007/3-540-36576-1_27

Lawvere Categories as Composed PROPs

Filippo Bonchi[1]([✉]), Pawel Sobocinski[2], and Fabio Zanasi[3]

[1] CNRS, École Normale Supérieure de Lyon, Lyon, France
`filippo.bonchi@ens-lyon.fr`
[2] University of Southampton, Southampton, UK
[3] Radboud University Nijmegen, Nijmegen, The Netherlands

Abstract. PROPs and Lawvere categories are related notions adapted to the study of algebraic structures borne by an object in a category, but whereas PROPs are symmetric monoidal, Lawvere categories are cartesian. This paper formulates the connection between the two notions using Lack's technique for composing PROPs via distributive laws. We show Lawvere categories can be seen as resulting from a distributive law of two PROPs — one expressing the algebraic structure in linear form and the other expressing the ability of copying and discarding variables.

1 Introduction

PROPs [28] are symmetric monoidal categories with objects the natural numbers. In the last two decades, they have become increasingly popular as an environment for the study of diagrammatic formalisms from diverse branches of science in a compositional, resource sensitive fashion. Focussing on computer science, they have recently featured in algebraic approaches to Petri nets [11,35], bigraphs [12], quantum circuits [15], and signal flow graphs [1,5,7,20].

PROPs describe both the *syntax* and the *semantics* of diagrams, with the interpretation expressed as a PROP morphism $[\![\cdot]\!]$: **Syntax** → **Semantics**. Typically, **Syntax** is freely generated by a signature Σ of operations with arbitrary arity/coarity and can be composed sequentially and in parallel. Thus diagram syntax—which we refer to as Σ-terms—is inherently 2-dimensional: the term structure is that of directed acyclic graphs, rather than trees, as in the familiar case of operations with coarity 1. A crucial aspect is *linearity*: variables in Σ-terms cannot be copied nor discarded.

It is often useful to axiomatise the equivalence induced by $[\![\cdot]\!]$ by means of a set of equations E, and then study the theory (Σ, E). For PROPs, completeness proofs typically provide a serious challenge, involving the retrieval of a normal form for Σ-terms modulo E. The difficulty can be drastically reduced by exploiting certain operations on PROPs: an example of this modular methodology is provided by [5,7], where the PROP operations of sum and composition are crucial for giving a sound and complete axiomatization of signal flow diagrams.

© IFIP International Federation for Information Processing 2016
Published by Springer International Publishing Switzerland 2016. All Rights Reserved
I. Hasuo (Ed.): CMCS 2016, LNCS 9608, pp. 11–32, 2016.
DOI: 10.1007/978-3-319-40370-0_3

Sum is just the coproduct in the category of PROPs. Whenever two PROPs T_1 and T_2 can be presented by the theories (Σ_1, E_1) and (Σ_2, E_2), then their sum $T_1 + T_2$ is presented by the disjoint union $(\Sigma_1 \uplus \Sigma_2, E_1 \uplus E_2)$.

Composition of PROPs is more subtle, as it requires certain compatibility conditions between the structure of T_1 and T_2. Lack [25] describes this operation formally by means of distributive laws, seeing PROPs as monads in the 2-categorical sense of Street [36]. In a nutshell, a distributive law $\lambda: T_1 \,;\, T_2 \to T_2 \,;\, T_1$ of PROPs is a recipe for moving arrows of T_1 past those of T_2. The resulting PROP $T_2 \,;\, T_1$ enjoys a factorisation property: every arrow in $T_2 \,;\, T_1$ decomposes as one of T_2 followed by one of T_1. The graph of λ can be seen as a set of directed equations $E_\lambda := (\xrightarrow{\in T_1}, \xrightarrow{T_2}) \approx (\xrightarrow{T_2}, \xrightarrow{T_1})$ and $T_2 \,;\, T_1$ is presented by the theory $(\Sigma_1 \uplus \Sigma_2, E_1 \uplus E_2 \uplus E_\lambda)$.

This work uses distributive laws of PROPs to characterise Lawvere categories[1], a well known class of structures adapted to the study of categorical universal algebra. The essential difference with PROPs is that Lawvere categories express *cartesian* theories (Σ, E), i.e. where Σ only features operations with coarity 1 and E may include non-linear equations. Our starting observation is that the Lawvere category $\mathcal{L}_\Sigma$ on a cartesian signature Σ exhibits a factorisation property analogous to the one of composed PROPs: arrows can always be decomposed as $\xrightarrow{\in \mathsf{Cm}}, \xrightarrow{\in T_\Sigma}$, where Cm is the PROP of commutative comonoids, generated by a copy $1 \to 2$ and a discard $1 \to 0$ operation, and T_Σ is the PROP freely generated by Σ. This factorisation represents cartesian Σ-terms by their syntactic tree — the T_Σ-part — with the possibility of explicitly indicating variable-sharing among sub-terms — the Cm-part. This simple observation leads us to the main result of the paper: for any cartesian signature Σ, there is a distributive law of PROPs $\lambda: T_\Sigma \,;\, \mathsf{Cm} \to \mathsf{Cm} \,;\, T_\Sigma$ which is presented by equations that express the naturality of copier and discarder; the resulting composed PROP $\mathsf{Cm} \,;\, T_\Sigma$ is the Lawvere category $\mathcal{L}_\Sigma$.

By a quotient construction on distributive laws, it follows immediately that the above theorem holds more generally for any cartesian theory (Σ, E) where the set of axioms E only contains linear equations. For instance the Lawvere category of commutative monoids $\mathcal{L}_{\mathrm{Mn}}$ can be obtained by means of PROP composition, while the one $\mathcal{L}_{\mathrm{Gr}}$ of abelian groups cannot, because of the non-linear axiom $x \times x^{-1} = 1$. Obviously, one can still formulate $\mathcal{L}_{\mathrm{Gr}}$ as the quotient of the composite $\mathcal{L}_{\mathrm{Mn}}$ by adding this equation, see Example 4.6 below.

As a side remark, we observe that, by taking the sum $\mathsf{Cm} + T_\Sigma$, rather than the composition, we are able to capture a different, well-known representation for cartesian Σ-terms, namely *term graphs*, which are acyclic graphs labeled over Σ. With respect to the standard tree representation, the benefit of term graphs is that the sharing of any common sub-term can be represented explicitly, making them particularly appealing for efficient rewriting algorithms, see e.g. [34] for a

---

[1] Usually called Lawvere *theories* in the literature: i.e. finite product categories with set of objects the natural numbers, where product on objects is addition [23, 27]. In order to keep the exposition uniform, we reserve the word *theory* for presentations and refer to the presentation (Σ, E) of a Lawvere theory as a *cartesian theory*.

survey on the subject. As shown in [16], Σ-term graphs are in 1-1 correspondence with the arrows of the free *gs-monoidal category* generated by Σ, a concept that actually amounts to forming the sum of PROPs $\mathsf{Cm} + \mathcal{T}_\Sigma$. Thus the only difference between term graphs and the representation of terms given by $\mathsf{Cm} \, ; \mathcal{T}_\Sigma$ is in the validity of naturality of copier and discarder. Intuitively, a term where a resource is explicitly copied is not identified with the term where two copies appear separately: in short, copying is not natural.

Related Works. The motto "*cartesian terms = linear terms + copying and discarding*" inspired several papers exploiting the role of Cm in Lawvere categories, see e.g. [14, 16, 26]. In our work, Lawvere categories feature as a distinguished example of a construction, PROP composition, that is increasingly important in many recent research threads [5, 6, 20, 32]. The significance of this exercise is two-fold. First, it gives a deeper understanding of the nature of Lawvere categories and how they formally relate to PROPs, by showing the provenance of the natural copy-discard structure. Second, our result provides a canonical means of defining a distributive law for freely generated PROPs, showing that the result of composition is a familiar algebraic notion and enjoys a *finite* axiomatisation.

The following construction, reported by Baez in [2], is close in spirit to our work. There is pseudo-adjunction between symmetric monoidal and categories with finite products

$$\mathsf{SMCat} \underset{R}{\overset{L}{\rightleftarrows}} \bot \; \mathsf{FPCat}$$

where R is the evident forgetful functor and L adds to any object of $\mathbb{C} \in \mathsf{SMCat}$ a natural copy-discard structure: natural diagonals and projections. Baez [2] states an equivalence between $RL(\mathbb{C})$ and $\mathbb{C} \otimes \mathsf{Cm}$, with the tensor $\otimes$ defined by $\mathsf{SMCat}[\mathbb{C}_1 \otimes \mathbb{C}_2, \mathbb{C}_3] \simeq \mathsf{SMCat}[\mathbb{C}_1, \mathsf{SMCat}[\mathbb{C}_2, \mathbb{C}_3]]$. Indeed, our main construction, as well as being a distributive law, is also an instance of a tensor or Kronecker product of symmetric monoidal theories; a concept that has been explored in some detail in the cartesian setting of Lawvere categories, see e.g. [22].

Our work restricts attention to PROPs $\mathcal{T}_\Sigma \in \mathsf{SMCat}$ freely generated by a cartesian signature Σ: in this case, it is enough to add a copy-discard structure for the object 1 and $RL(\mathcal{T}_\Sigma)$ coincides with PROP composition $\mathsf{Cm} \, ; \mathcal{T}_\Sigma$. Our perspective exploiting distributive laws has the advantage of providing a *finite* presentation in terms of the naturality axioms.

It is also worth mentioning that the relationship between symmetric monoidal and cartesian structures is central in the categorical semantics of linear logic; in this perspective, the presence of Cm allows to interpret the structural rules of contraction and weakening — see e.g. [24, 30].

Prerequisites and Notation. We assume familiarity with the basics of category theory (see e.g. [10, 29]), the definition of symmetric strict monoidal category [29, 33] (often abbreviated as SMC) and of bicategory [3, 10]. We write $f \, ; g \colon a \to c$ for composition of $f \colon a \to b$ and $g \colon b \to c$ in a category $\mathbb{C}$, and $\mathbb{C}[a, b]$ for the

$$(t_1 \,;\, t_3) \oplus (t_2 \,;\, t_4) = (t_1 \oplus t_2) \,;\, (t_3 \oplus t_4)$$

$$(t_1 \,;\, t_2) \,;\, t_3 = t_1 \,;\, (t_2 \,;\, t_3) \quad id_n \,;\, c = c = c \,;\, id_m$$
$$(t_1 \oplus t_2) \oplus t_3 = t_1 \oplus (t_2 \oplus t_3) \quad id_0 \oplus t = t = t \oplus id_0$$
$$\sigma_{1,1} \,;\, \sigma_{1,1} = id_2 \quad (t \oplus id_z) \,;\, \sigma_{m,z} = \sigma_{n,z} \,;\, (id_z \oplus t)$$

Fig. 1. Axioms of symmetric strict monoidal categories for a PROP $\mathcal{T}$.

hom-set of arrows $a \to b$. It will be sometimes convenient to indicate an arrow $f \colon a \to b$ of $\mathbb{C}$ as $x \xrightarrow{f \in \mathbb{C}} y$ or also $\xrightarrow{\in \mathbb{C}}$, if names are immaterial. For $\mathbb{C}$ an SMC, $\oplus$ is its monoidal product, with unit object I, and $\sigma_{a,b} \colon a \oplus b \to b \oplus a$ is the symmetry associated with $a, b \in \mathbb{C}$.

2   PROPs

Our exposition is founded on *symmetric monoidal theories*: specifications for algebraic structures borne by objects in a symmetric monoidal category.

Definition 2.1. *A (one-sorted) symmetric monoidal theory (SMT) is a pair (Σ, E) consisting of a signature Σ and a set of equations E. The signature Σ is a set of generators $o \colon n \to m$ with arity n and coarity m. The set of Σ-terms is obtained by composing generators in Σ, the unit $id \colon 1 \to 1$ and the symmetry $\sigma_{1,1} \colon 2 \to 2$ with $;$ and $\oplus$. This is a purely formal process: given Σ-terms $t \colon k \to l,\ u \colon l \to m,\ v \colon m \to n$, one constructs new Σ-terms $t \,;\, u \colon k \to m$ and $t \oplus v \colon k + n \to l + n$. The set E of equations contains pairs $(t, t' \colon n \to m)$ of Σ-terms with the same arity and coarity.*

The categorical concept associated with symmetric monoidal theories is the notion of PROP (**pro**duct and **p**ermutation category [28]).

Definition 2.2. *A PROP is a symmetric strict monoidal category with objects the natural numbers, where $\oplus$ on objects is addition. Morphisms between PROPs are strict symmetric identity-on-objects monoidal functors: PROPs and their morphisms form the category* **PROP**. *We call a* sub-PROP *a sub-category of a PROP which is also a PROP; i.e. the inclusion functor is a PROP morphism.*

The PROP $\mathcal{T}$ freely generated by an SMT (Σ, E) has as its set of arrows $n \to m$ the set of Σ-terms $n \to m$ taken modulo the laws of symmetric strict monoidal categories — Fig. 1 — and the smallest congruence (with respect to $;$ and $\oplus$) containing the equations $t = t'$ for any $(t, t') \in E$.

There is a natural graphical representation for arrows of a PROP as string diagrams, which we now sketch, referring to [33] for the details. A Σ-term $n \to m$ is pictured as a box with n ports on the left and m ports on the right. Composition via $;$ and $\oplus$ are rendered graphically by horizontal and vertical juxtaposition of boxes, respectively.

$t\,;s$ is drawn $\boxed{t}\ \boxed{s}$ $t\oplus s$ is drawn $\dfrac{\boxed{t}}{\boxed{s}}$. (1)

In any SMT there are specific Σ-terms generating the underlying symmetric monoidal structure: these are $id_1\colon 1\to 1$, represented as $\boxminus$, the symmetry $\sigma_{1,1}\colon 1+1\to 1+1$, represented as $\bowtie$, and the unit object for $\oplus$, that is, $id_0\colon 0\to 0$, whose representation is an empty diagram $\square$. Graphical representation for arbitrary identities id_n and symmetries $\sigma_{n,m}$ are generated according to the pasting rules in (1).

Example 2.3.

(a) We write (Σ_M, E_M) for the SMT of *commutative monoids*. The signature Σ_M contains a *multiplication* $\rhd\!\!-\colon 2\to 1$ and a *unit* $\circ\!\!-\colon 0\to 1$.. Equations E_M assert associativity (A1), commutativity (A2) and unitality (A3).

$$\boxed{\rhd\!\!-} = \boxed{\rhd\!\!-}\quad\text{(A1)}\qquad \boxed{\rhd\!\!-} = \boxed{\rhd\!\!\circ\!\!-}\quad\text{(A2)}\qquad \boxed{\circ\!\!\rhd} = \boxed{}\quad\text{(A3)}$$

We call Mn the PROP freely generated by the SMT (Σ_M, E_M).

(b) We also introduce the SMT (Σ_C, E_C) of *cocommutative comonoids*. The signature Σ_C consists of a *comultiplication* $-\!\!\lhd\colon 1\to 2$ and a *counit* $-\!\!\bullet\colon 1\to 0$.. E_C is the following set of equations.

$$\boxed{-\!\!\lhd} = \boxed{-\!\!\lhd}\quad\text{(A4)}\qquad \boxed{-\!\!\lhd} = \boxed{-\!\!\circ\!\!\lhd}\quad\text{(A5)}\qquad \boxed{-\!\!\lhd\!\!\bullet} = \boxed{}\quad\text{(A6)}$$

We call Cm the PROP freely generated by (Σ_C, E_C). Modulo the white vs. black colouring, string diagrams of Cm can be seen as those of Mn "reflected about the y-axis". This observation yields $\mathsf{Cm}\cong\mathsf{Mn}^{op}$.

(c) The PROP B of (commutative/cocommutative) *bialgebras* is generated by the theory $(\Sigma_M\uplus\Sigma_C, E_M\uplus E_C\uplus B)$, where B is the following set of equations.

$$\boxed{\rhd\!\!\bullet} = \boxed{\bullet\ \bullet}\quad\text{(A7)}\qquad\qquad \boxed{\rhd\!\!\lhd} = \boxed{\bowtie}\quad\text{(A8)}$$

$$\boxed{\circ\!\!\lhd} = \boxed{\circ\ \circ}\quad\text{(A9)}\qquad\qquad \boxed{\circ\!\!\bullet} = \boxed{}\quad\text{(A10)}$$

Remark 2.4 (Models of a PROP). The assertion that (Σ_M, E_M) *is the SMT of commutative monoids*—and similarly for other SMTs appearing in our exposition—can be made precise using the notion of *model* (sometimes also called algebra) of a PROP. Given a strict symmetric monoidal category $\mathbb{C}$, a model of a PROP $\mathcal{T}$ in $\mathbb{C}$ is a symmetric strict monoidal functor $\mathcal{F}\colon\mathcal{T}\to\mathbb{C}$. Then $\mathsf{LinMod}(\mathcal{T},\mathbb{C})$ is the category of models of $\mathcal{T}$ in $\mathbb{C}$ and natural transformations between them.

Turning to commutative monoids, there is a category $\mathsf{Monoid}(\mathbb{C})$ whose objects are the commutative monoids in $\mathbb{C}$, i.e., objects $x\in\mathbb{C}$ equipped with arrows $x \oplus x \to x$ and $I \to x$, satisfying the usual equations. Given any model $\mathcal{F}\colon \mathsf{Mn} \to \mathbb{C}$, it follows that $\mathcal{F}(1)$ is a commutative monoid in $\mathbb{C}$: this yields a functor $\mathsf{LinMod}(\mathsf{Mn}, \mathbb{C}) \to \mathsf{Monoid}(\mathbb{C})$. Saying that (Σ_M, E_M) is the SMT of commutative monoids means that this functor is an equivalence natural in $\mathbb{C}$. We shall not focus on models as they are not central in our developments and refer the reader to [19,25] for more information.

Example 2.3 only shows PROPs freely generated from an algebraic specification. However, one can also define PROPs in a more direct manner, without relying on SMTs. Two basic examples will be useful for our exposition:

- the PROP F whose arrows $n \to m$ are functions from $\overline{n}$ to $\overline{m}$, where $\overline{n} = \{0, 1, \ldots, n-1\}$.
- the PROP P whose arrows $n \to m$ exist only if $n = m$, in which case they are the permutations on $\overline{n}$.

This kind of definition is often useful as a different, more concrete perspective on PROPs that arise from symmetric monoidal theories. For instance, F is *presented* by the theory of commutative monoids, in the sense that F and Mn are isomorphic PROPs: once can consider a string diagram $t \in \mathsf{Mn}[n, m]$ as the graph of a function of type $\{1, \ldots, n\} \to \{1, \ldots, m\}$. For instance, $\boxed{\triangleright}\!\!-\!\oplus\boxed{\circ}\!\!-\colon 2 \to 2$ describes the function $f\colon \{1, 2\} \to \{1, 2\}$ mapping both elements to 1. By duality, $\mathsf{Cm} \cong \mathsf{F}^{op}$, that is, F^{op} is presented by the theory of commutative comonoids.

Similarly, P provides a concrete description of the theory $(\emptyset, \emptyset)$ with empty signature and no equations. It is the initial object in the category **PROP**.

3   PROP Composition

A basic operation on SMTs (Σ, E) and (Σ', E') is to take their *sum* $(\Sigma \uplus \Sigma', E \uplus E')$. In **PROP**, the PROP generated by $(\Sigma \uplus \Sigma', E \uplus E')$ is the *coproduct* $\mathcal{T} + \mathsf{S}$ of the PROP $\mathcal{T}$ generated by (Σ, E) and S, generated by (Σ', E').

The sum $\mathcal{T} + \mathsf{S}$ is the least interesting way of combining theories, because there are no equations that express compatibility between the algebraic structures in $\mathcal{T}$ and S. This is a standard pattern in algebra: e.g. a ring is given by a monoid and an abelian group, subject to equations that ensure that the former distributes over the latter. Similarly, the equations of bialgebras (Example 2.3) describe the interplay of a monoid and a comonoid. Ordinary functions, which can always be decomposed as a surjection followed by an injection, are another example.

In [25] Lack shows how these phenomena can be uniformly described as the operation of composing PROPs. The conceptual switch is to understand PROPs as monads, and their composition as a distributive law. These monads live in a certain bicategory [3], as in the classical work by Street [36][2].

Definition 3.1. *A* monad *on an object x of a bicategory $\mathfrak{B}$ is a 1-cell $\mathcal{F}: x \to x$ with 2-cells $\eta^{\mathcal{F}}: id_x \to \mathcal{F}$ and $\mu^{\mathcal{F}}: \mathcal{F}; \mathcal{F} \to \mathcal{F}$ (called the* unit *and the* multiplication *respectively) making the following diagrams–in which we suppress the associativity isomorphisms—commute.*

$$\mathcal{F} \xrightarrow{\mathcal{F}\eta^{\mathcal{F}}} \mathcal{F}; \mathcal{F} \xleftarrow{\eta^{\mathcal{F}}\mathcal{F}} \mathcal{F}$$
$$id \searrow \quad \downarrow \mu^{\mathcal{F}} \quad \swarrow id$$
$$\mathcal{F}$$

(2)

$$\mathcal{F}; \mathcal{F}; \mathcal{F} \xrightarrow{\mathcal{F}\mu^{\mathcal{F}}} \mathcal{F}; \mathcal{F}$$
$$\mu^{\mathcal{F}}\mathcal{F} \downarrow \qquad \downarrow \mu^{\mathcal{F}}$$
$$\mathcal{F}; \mathcal{F} \xrightarrow{\mu^{\mathcal{F}}} \mathcal{F}$$

(3)

A morphism between monads $x \xrightarrow{\mathcal{F}} x$ *and* $x \xrightarrow{\mathcal{G}} x$ *is a 2-cell* $\theta: \mathcal{F} \to \mathcal{G}$ *making the following diagrams commute*[3].

$$id_x$$
$$\eta^{\mathcal{F}} \downarrow \quad \searrow \eta^{\mathcal{G}}$$
$$\mathcal{F} \xrightarrow{\theta} \mathcal{G}$$

(4)

$$\mathcal{F}; \mathcal{F} \xrightarrow{\theta\theta} \mathcal{G}; \mathcal{G}$$
$$\mu^{\mathcal{F}} \downarrow \qquad \downarrow \mu^{\mathcal{G}}$$
$$\mathcal{F} \xrightarrow{\theta} \mathcal{G}$$

(5)

An epimorphic monad morphism is called a monad quotient.

For $\mathfrak{B} = \mathbf{Cat}$, the above definition yields the standard notion of monad as an endofunctor with a pair of natural transformations. Something interesting happens for the case of the bicategory $\mathfrak{B} = \mathsf{Span}(\mathbf{Set})$ whose objects are sets, 1-cells are spans of functions (with composition defined by pullback) and 2-cells are span morphisms: monads in $\mathsf{Span}(\mathbf{Set})$ are precisely the small categories. Indeed, a monad $(\mathcal{F}, \eta, \mu)$ there consists of a span $Ob \xleftarrow{dom} Ar \xrightarrow{cod} Ob$, which yields a set Ob of objects, one Ar of arrows and domain/codomain maps $Ar \rightrightarrows Ob$. The unit $\eta: id \to \mathcal{F}$ is a span morphism associating an identity arrow to each object (below left). The multiplication $\mu: \mathcal{F}; \mathcal{F} \to \mathcal{F}$ is a span morphism defining composition for any two arrows $a \xrightarrow{f} b \xrightarrow{g} c$ in Ar (below right).

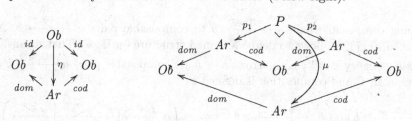

By thinking of categories as monads, one can define the composition of categories with the same set of objects as monad composition by a *distributive law* in $\mathsf{Span}(\mathbf{Set})$. This phenomenon is studied in [31].

Definition 3.2. *Let* $(\mathcal{F}, \eta^{\mathcal{F}}, \mu^{\mathcal{F}})$, $(\mathcal{G}, \eta^{\mathcal{G}}, \mu^{\mathcal{G}})$ *be monads in a bicategory $\mathfrak{B}$ on the same object. A* distributive law *of $\mathcal{F}$ over $\mathcal{G}$ is a 2-cell* $\lambda: \mathcal{F}; \mathcal{G} \to \mathcal{G}; \mathcal{F}$ *in $\mathfrak{B}$ making the following diagrams—in which we again omit associativity—commute.*

---

[3] A notion of morphism can be defined also between monads on different objects, like in [36]. We will not need that level of generality here.

$$
\begin{array}{ccc}
 & \mathcal{F} & \\
\mathcal{F}\eta^{\mathcal{G}}\Big\downarrow & \searrow{\scriptstyle \eta^{\mathcal{G}}\mathcal{F}} & \\
\mathcal{F};\mathcal{G} \xrightarrow{\ \lambda\ } \mathcal{G};\mathcal{F} & & \\
\eta^{\mathcal{F}}\mathcal{G}\Big\uparrow & \nearrow{\scriptstyle \mathcal{G}\eta^{\mathcal{F}}} & \\
 & \mathcal{G} &
\end{array}
\qquad
\begin{array}{ccc}
\mathcal{F};\mathcal{G};\mathcal{G} \xrightarrow{\ \lambda\mathcal{G}\ } \mathcal{G};\mathcal{F};\mathcal{G} \xrightarrow{\ \mathcal{G}\lambda\ } \mathcal{G};\mathcal{G};\mathcal{F} \\
\mathcal{F}\mu^{\mathcal{G}}\Big\downarrow \hspace{6cm} \Big\downarrow\mu^{\mathcal{G}}\mathcal{F} \\
\mathcal{F};\mathcal{G} \xrightarrow{\hspace{4.5cm}\lambda\hspace{4.5cm}} \mathcal{G};\mathcal{F} \\
\mu^{\mathcal{F}}\mathcal{G}\Big\uparrow \hspace{6cm} \Big\uparrow\mathcal{G}\mu^{\mathcal{F}} \\
\mathcal{F};\mathcal{F};\mathcal{G} \xrightarrow[\ \mathcal{F}\lambda\ ]{} \mathcal{F};\mathcal{G};\mathcal{F} \xrightarrow[\ \lambda\mathcal{F}\ ]{} \mathcal{G};\mathcal{F};\mathcal{F}
\end{array}
\qquad (6)
$$

A *distributive law* $\lambda\colon \mathcal{F};\mathcal{G} \to \mathcal{G};\mathcal{F}$ yields a monad $\mathcal{G};\mathcal{F}$ with the following unit and multiplication:

$$
\eta^{\mathcal{G};\mathcal{F}} :\ id \xrightarrow{\ \eta^{\mathcal{F}}\ } \mathcal{F} \xrightarrow{\ \eta^{\mathcal{G}}\mathcal{F}\ } \mathcal{G};\mathcal{F}
$$
$$
\mu^{\mathcal{G};\mathcal{F}} :\ \mathcal{G};\mathcal{F};\mathcal{G};\mathcal{F} \xrightarrow{\ \mathcal{G}\lambda\mathcal{F}\ } \mathcal{G};\mathcal{G};\mathcal{F};\mathcal{F} \xrightarrow{\ \mu^{\mathcal{G}}\mathcal{F}\mathcal{F}\ } \mathcal{G};\mathcal{F};\mathcal{F} \xrightarrow{\ \mathcal{G}\mu^{\mathcal{F}}\ } \mathcal{G};\mathcal{F}
$$

$$(7)$$

Let us verify how the abstract definition works for the case of categories. Pick categories $\mathbb{C}$ and $\mathbb{D}$ with the same set Ob of objects, seen as monads $Ob \xleftarrow{\ dom_{\mathbb{C}}\ } Ar_{\mathbb{C}} \xrightarrow{\ cod_{\mathbb{C}}\ } Ob$ and $Ob \xleftarrow{\ dom_{\mathbb{D}}\ } Ar_{\mathbb{D}} \xrightarrow{\ cod_{\mathbb{D}}\ } Ob$ in $\mathsf{Span}(\mathbf{Set})$. A distributive law $\lambda\colon \mathbb{C};\mathbb{D} \to \mathbb{D};\mathbb{C}$ is a span morphism

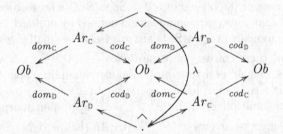

mapping composable pairs $a \xrightarrow{\ \in\mathbb{C}\ } \xrightarrow{\ \in\mathbb{D}\ } b$ to composable pairs $a \xrightarrow{\ \in\mathbb{D}\ } \xrightarrow{\ \in\mathbb{C}\ } b$. As described in (7), λ allows to define a monad structure on $\mathbb{D};\mathbb{C}$. That means, λ yields a category $\mathbb{D};\mathbb{C}$ whose arrows $a \to b$ are composable pairs $a \xrightarrow{\ \in\mathbb{D}\ } \xrightarrow{\ \in\mathbb{C}\ } b$ of arrows of $\mathbb{D}, \mathbb{C}$ and composition is defined as

$$
\left(a \xrightarrow{\ f\in\mathbb{D}\ } \xrightarrow{\ g\in\mathbb{C}\ } b\right) ; \left(b \xrightarrow{\ f'\in\mathbb{D}\ } \xrightarrow{\ g'\in\mathbb{C}\ } c\right) \ :=\ \left(a \xrightarrow{\ f\in\mathbb{D}\ } \lambda(\xrightarrow{\ g\in\mathbb{C}\ } \xrightarrow{\ f'\in\mathbb{D}\ }) \xrightarrow{\ g'\in\mathbb{C}\ } c\right).
$$

PROPs are understood as monads in the same sense that small categories are. The difference is that one needs to refine the bicategory of interest, in order for the composition of PROPs-monads to yield another PROP-monad and not an arbitrary small category. These refinements are in two steps. First, one takes the bicategory $\mathsf{Span}(\mathbf{Mon})$ whose 1-cells are spans in the category of *monoids*, instead of sets. This accounts for the monoidal structure — in fact, monads in $\mathsf{Span}(\mathbf{Mon})$ are small strict monoidal categories. The second refinement has the purpose of correctly account for the symmetry structure of PROPs: one takes the bicategory Bimod(Span(**Mon**)) whose objects are small strict monoidal categories and 1-cells are the *bimodules* in Span(**Mon**). PROPs are then monads on the object P of Bimod(Span(**Mon**)).

We shall gloss over further details about the exact formalisation of this observation, as it is out of the scope of this paper — we refer to [25] and [37, § 2.4] for the detailed definitions. The simpler setting of composition of mere categories should provide enough guidance to follow the rest of our exposition.

It is important for our purposes to remark how composition works for PROPs T_1, T_2 generated by SMTs, say (Σ_1, E_1) and (Σ_2, E_2). The PROP $T_1 \,;\, T_2$ induced by a distributive law $\lambda \colon T_2 \,;\, T_1 \to T_1 \,;\, T_2$ will also enjoy a presentation by generators and equations, consisting of the sum $(\Sigma_1 \uplus \Sigma_2, E_1 \uplus E_2)$ plus the equations E_λ *arising* from the the distributive law. The set E_λ is simply the graph of λ, now seen as a set of directed equations $(\xrightarrow{\in T_2}\xrightarrow{\in T_1}) \approx (\xrightarrow{\in T_1}\xrightarrow{\in T_2})$ telling how Σ_2-terms modulo E_2 distribute over Σ_1-terms modulo E_1. In fortunate cases, it is possible to present E_λ by a simpler, or even finite, set of equations, thus giving a sensible axiomatisation of the compatibility conditions expressed by λ. This is the case for both examples considered below.

Example 3.3.

(a) The PROP F of functions can be described as the composite of PROPs for surjections and injections. Let In be the PROP whose arrows $n \to m$ are injective functions from $\overline{n}$ to $\overline{m}$. The PROP Su of surjective functions is defined analogously. There is a distributive law $\lambda \colon$ In $;$ Su $\to$ Su $;$ In defined by epi-mono factorisation: it maps a composable pair $\xrightarrow{\in \mathrm{In}}\xrightarrow{\in \mathrm{Su}}$ to a composable pair $\xrightarrow{\in \mathrm{Su}}\xrightarrow{\in \mathrm{In}}$ [25]. The resulting PROP Su $;$ In is isomorphic to F because any function in F can be uniquely factorised (up-to permutation) as a surjection followed by an injection. In more syntactic terms, using the isomorphism F $\cong$ Mn, this result says that Mn is the composite Mu $;$ Un, where Mu $\cong$ Su is the PROP freely generated by the SMT $(\{\;\boxed{\text{>-}}\;\}, \{(A1), (A2)\})$ and Un $\cong$ In by the SMT $(\{\;\boxed{\text{o-}}\;\}, \emptyset)$. The distributive law $\lambda \colon$ In $;$ Su $\to$ Su $;$ In is then presented by the remaining equation (A3) of Mn, which indeed describes how the generator $\boxed{\text{o-}}$ of Un can be moved past the one $\boxed{\text{>-}}$ of Mu.

(b) The composition of Cm and Mn yields the PROP B of commutative bialgebras. First, because Mn $\cong$ F and Cm $\cong$ F^{op}, we can express a distributive law $\lambda \colon$ Mn $;$ Cm $\to$ Cm $;$ Mn as having the type F $;$ F$^{op} \to$ F$^{op} ;$ F. This amounts to saying that λ maps *cospans* $n \xrightarrow{f \in \mathsf{F}} \xleftarrow{g \in \mathsf{F}} m$ to *spans* $n \xleftarrow{p \in \mathsf{F}} \xrightarrow{q \in \mathsf{F}} m$. Defining this mapping via (chosen) pullback in F satisfies the conditions of distributive laws [25]. One can now read the equations arising by the distributive law from pullback squares in F. For instance:

where the second diagram is obtained from the pullback by applying the isomorphisms $F \cong Mn$ and $F^{op} \cong Cm$. In fact, Lack [25] shows that the equations presenting Cm ; Mn arise from (those of $Cm+Mn$ and) just four pullback squares, yielding equations (A7)–(A10). Therefore, Cm ; Mn is isomorphic to the PROP B of bialgebras encountered in Example 2.3. Furthermore, these PROPs have a "concrete" descriptions as F^{op} ; F. In the terminology of [3], one can see F^{op} ; F as the *classifying category* of the bicategory $Span(F)$, obtained by identifying the isomorphic 1-cells and forgetting the 2-cells.

There is a tight relationship between distributive laws and factorisation systems. Distributive laws of small categories are in 1-1 correspondence with so-called *strict* factorisation systems [31], in which factorisations must be specified uniquely on the nose, rather than merely up-to isomorphism. Distributive laws of PROPs correspond instead to a more liberal kind of factorisation system, for which decompositions are up-to permutation. As this perspective will be useful later, we recall the following result from Lack [25].

Proposition 3.4 ([25], **Theorem 4.6**). *Let* S *be a PROP and* T_1, T_2 *be sub-PROPs of* S. *Suppose that each arrow* $n \xrightarrow{f \in S} m$ *can be factorised as* $n \xrightarrow{g_1 \in T_1} \xrightarrow{g_2 \in T_2} m$ *uniquely up-to permutation, that is, for any other decomposition* $n \xrightarrow{h_1 \in T_1} \xrightarrow{h_2 \in T_2} m$ *of* f, *there exists permutation* $\xrightarrow{p \in P}$ *such that the following diagram commutes.*

$$
\begin{array}{c}
\xrightarrow{g_1} \quad \uparrow p \quad \xrightarrow{g_2} \\
\xrightarrow{h_1} \quad \xrightarrow{h_2}
\end{array}
$$

Then there exists a distributive law $\lambda\colon T_2$; $T_1 \to T_1$; T_2, *defined by associating to a composable pair* $\xrightarrow{f \in T_2} \xrightarrow{g \in T_1}$ *the factorisation of* f ; g, *yielding* $S \cong T_1$; T_2.

Quotient of a Distributive Law. Definition 3.1 introduced the notion of *quotient* $\theta\colon \mathcal{F} \to \mathcal{G}$ of a monad $\mathcal{F}$: the idea is that the monad $\mathcal{G}$ is obtained by imposing additional equations on the algebraic theory described by $\mathcal{F}$. As one may expect, distributive laws are compatible with monad quotients, provided that the law preserves the newly added equations. This folklore result appears in various forms in the literature: [9] gives it for distributive laws of endofunctors over monads and [4,8] for distributive laws of monads. All these references concern distributive laws in **Cat**. For our purposes, it is useful to state the result for arbitrary bicategories.

Proposition 3.5. *Suppose that* $\lambda\colon \mathcal{F}$; $\mathcal{H} \to \mathcal{H}$; $\mathcal{F}$ *is a distributive law in a bicategory* $\mathfrak{B}$, $\theta\colon \mathcal{F} \to \mathcal{G}$ *a monad quotient and* $\lambda'\colon \mathcal{G}$; $\mathcal{H} \to \mathcal{H}$; $\mathcal{G}$ *another 2-cell of* $\mathfrak{B}$ *making the following diagram commute.*

$$
\begin{array}{ccc}
\mathcal{F} ; \mathcal{H} & \xrightarrow{\theta\mathcal{H}} & \mathcal{G} ; \mathcal{H} \\
\lambda \downarrow & & \downarrow \lambda' \\
\mathcal{H} ; \mathcal{F} & \xrightarrow{\mathcal{H}\theta} & \mathcal{H} ; \mathcal{G}
\end{array}
\tag{8}
$$

Then λ' is a distributive law of monads.

Proof. The diagrams for compatibility of λ' with unit and multiplication of $\mathcal{G}$ commute because θ is a monad morphism and (8) commutes. For compatibility of λ' with unit and multiplication of $\mathcal{H}$, one needs to use commutativity of (8) and the fact that θ is epi.

We remark that Proposition 3.5 holds also in the version in which one quotients the monad $\mathcal{H}$ instead of $\mathcal{F}$. It is now useful to instantiate the result to the case of distributive laws of PROPs.

Proposition 3.6. *Let $\mathcal{T}$ be the PROP freely generated by (Σ, E) and $E' \supseteq E$ be another set of equations on Σ-terms. Suppose that there exists a distributive law $\lambda \colon \mathcal{T} \mathbin{;} \mathsf{S} \to \mathsf{S} \mathbin{;} \mathcal{T}$ such that, if E' implies $c = d$, then $\lambda(\xrightarrow{c \in \mathcal{T}} \xrightarrow{e \in \mathsf{S}}) = \lambda(\xrightarrow{d \in \mathcal{T}} \xrightarrow{e \in \mathsf{S}})$. Then there exists a distributive law $\lambda' \colon \mathcal{T}' \mathbin{;} \mathsf{S} \to \mathsf{S} \mathbin{;} \mathcal{T}'$ presented by the same equations as λ, i.e., $E_{\lambda'} = E_\lambda$.*

Proof. There is a PROP morphism $\theta \colon \mathcal{T} \to \mathcal{T}'$ defined by quotienting string diagrams in $\mathcal{T}$ by E'. This is a monad quotient in the bicategory $\mathsf{Prof}(\mathbf{Mon})$ where PROPs are monads. We now define another 2-cell $\lambda' \colon \mathcal{T}' \mathbin{;} \mathsf{Cm} \to \mathsf{Cm} \mathbin{;} \mathcal{T}'$ as follows: given $\xrightarrow{e \in \mathcal{T}'} \xrightarrow{c \in \mathsf{Cm}}$, pick any $\xrightarrow{d \in \mathcal{T}}$ such that $\theta(d) = e$ and let $\xrightarrow{c' \in \mathsf{Cm}} \xrightarrow{d' \in \mathcal{T}}$ be $\lambda(\xrightarrow{d \in \mathcal{T}} \xrightarrow{c \in \mathsf{Cm}})$. Define $\lambda'(\xrightarrow{e \in \mathcal{T}'} \xrightarrow{c \in \mathsf{Cm}})$ as $\xrightarrow{c' \in \mathsf{Cm}} \xrightarrow{\theta(d') \in \mathcal{T}'}$. λ' is well-defined because, by assumption, if $\theta(d_1) = \theta(d_2)$ then E' implies that $d_1 = d_2$ and thus $\lambda(\xrightarrow{d_1} \xrightarrow{c}) = \lambda(\xrightarrow{d_2} \xrightarrow{c})$.

Now, λ, λ' and θ satisfy the assumptions of Proposition 3.5. In particular, (8) commutes by definition of λ' in terms of λ and θ. The conclusion of Proposition 3.5 guarantees that λ' is a distributive law. By construction, λ' is presented by the same equations as λ.

We will see an application of Proposition 3.6 in the next section (Lemma 4.8).

4   Lawvere Categories as Composed PROPs

This section introduces and characterises Lawvere categories via a certain class of distributive laws of PROPs. As mentioned in the introduction, Lawvere categories are closely related to PROPs: the essential difference is that, whereas a Lawvere category is required to be a category with finite products — henceforth called a *cartesian* category, a PROP may carry any symmetric monoidal structure, not necessarily cartesian.

Just as PROPs, Lawvere categories can be also freely obtained by generators and equations. By analogy with symmetric monoidal theories introduced in Sect. 2, we organise this data as a *cartesian theory*: it simply amounts to the notion of equational theory that one typically finds in universal algebra, see e.g. [13].

Definition 4.1. *A (one-sorted) cartesian theory* (Σ, E) *consists of a signature* $\Sigma = \{o_1 \colon n_1 \to 1, \ldots, o_k \colon n_k \to 1\}$ *and a set E of equations between* cartesian Σ-terms, *which are defined as follows:*

- *for each $i \in \mathbb{N}$, the variable x_i is a cartesian term;*
- *suppose $o \colon n \to 1$ is a generator in Σ and $t_1, \ldots, t_n$ are cartesian terms. Then* $o(t_1, \ldots, t_n)$ *is a cartesian term.*

The Lawvere category $\mathcal{L}_{(\Sigma, E)}$ *freely generated by* (Σ, E) *is the category whose objects are the natural numbers and arrows $n \to m$ are lists $\langle t_1, \ldots, t_m \rangle$ of cartesian Σ-terms quotiented by E, such that, for each t_i, only variables among $x_1, \ldots, x_n$ appear in t_i. Composition is by substitution:*

$$\left(n \xrightarrow{\langle t_1, \ldots, t_m \rangle} m \right) ; \left(m \xrightarrow{\langle s_1, \ldots, s_z \rangle} z \right) = n \xrightarrow{\langle s_1[t_i/x_i \mid 1 \leq i \leq m], \ldots, s_z[t_i/x_i \mid 1 \leq i \leq m] \rangle} z$$

where $t[t'/x]$ denotes the cartesian term t in which all occurrences of the variable x have been replaced with t'.

$\mathcal{L}_{(\Sigma, E)}$ *is equipped with a product $\times$ which is defined on objects by addition and on arrows by list concatenation and suitable renaming of variables:*

$$\left(n \xrightarrow{\langle t_1, \ldots, t_m \rangle} m \right) \times \left(z \xrightarrow{\langle s_1, \ldots, s_l \rangle} l \right)$$
$$= n + z \xrightarrow{\langle t_1, \ldots, t_m, s_1[x_{i+m}/x_i \mid 1 \leq i \leq l], \ldots, s_l[x_{i+m}/x_i \mid 1 \leq i \leq l] \rangle} m + l.$$

We use notation $ovar(t)$ for the list of occurrences of variables appearing (from left to right) in t and, more generally, $ovar(t_1, \ldots, t_m)$ for the list $ovar(t_1)$ :: $\cdots$ :: $ovar(t_m)$. Also, $|l| \in \mathbb{N}$ denotes the length of a list l. We say that a list $\langle t_1, \ldots, t_m \rangle \colon n \to m$ is *linear* if each variable among $x_1, \ldots, x_n$ appears exactly once in $ovar(t_1, \ldots, t_m)$.

Our first observation is that Lawvere categories are PROPs.

Proposition 4.2. $\mathcal{L}_{(\Sigma, E)}$ *is a PROP.*

Proof. Let $\times$ act as the monoidal product, 0 as its unit and define the symmetry $n + m \to m + n$ as the list $\langle x_{n+1}, \ldots, x_{n+m}, x_1, \ldots, x_n \rangle$. It follows that $\mathcal{L}_{(\Sigma, E)}$ equipped with this structure satisfies the laws of symmetric strict monoidal categories, thus it is a PROP.

As a side observation, note that the unique PROP morphism $\mathsf{P} \to \mathcal{L}_{(\Sigma, E)}$ given by initiality of P in **PROP** sends $p \colon n \to n$ to $\langle x_{p^{-1}(1)}, \ldots, x_{p^{-1}(n)} \rangle$.

Remark 4.3. In spite of Proposition 4.2, cartesian theories are *not* a subclass of symmetric monoidal theories: in fact, the two concepts are incomparable. On the one hand, a symmetric monoidal theory (Σ, E) is cartesian if and only if all generators in Σ have coarity 1 and, for all equations $t = s$ in E, t and s are Σ-terms with coarity 1. Under these conditions, there is a canonical way to interpret any

Σ-term $n \to m$ as a list of m cartesian Σ-terms on variables $x_1, \ldots, x_n$. Below, o ranges over Σ:

$$\boxminus : 1 \to 1 \;\mapsto\; \langle x_1 \rangle \qquad\qquad \bowtie : 2 \to 2 \mapsto \langle x_2, x_1 \rangle$$

$$\boxed{O}\!\!- : n \to 1 \;\mapsto\; \langle o(x_1, \ldots, x_n) \rangle.$$

The inductive cases are defined using the operations ; and $\oplus$ on lists given in Definition 4.1. Note that Σ-terms always denote (lists of) *linear* cartesian terms. This explains why, conversely, not all the cartesian theories are symmetric monoidal: their equations possibly involve non-linear Σ-terms, which are not expressible with (symmetric monoidal) Σ-terms. The subtlety here is that, in a sense, we can still *simulate* a cartesian theory on signature Σ with a symmetric monoidal theory, which however will be based on a larger signature Σ', recovering the possibility of copying and discarding variables by the use of additional generators. This point will become more clear below, where we will see how copier and discharger, i.e., the cartesian structure, can be mimicked with the use of the PROP Cm.

Example 4.4. The SMT (Σ_M, E_M) of commutative monoids is cartesian. It generates the Lawvere category $\mathcal{L}_{\Sigma_M, E_M}$ whose arrows $n \to m$ are lists $\langle t_1, \ldots, t_m \rangle$ of elements of the free commutative monoid on $\{x_1, \ldots, x_n\}$.

The example of commutative monoids is particularly instructive for sketching our approach to Lawvere categories as composed PROPs. First, note that the Lawvere category $\mathcal{L}_{\Sigma_M, E_M}$ *includes* the PROP Mn freely generated by (Σ_M, E_M). Indeed, any string diagram of Mn can be interpreted as a list of terms following the recipe of Remark 4.3. For instance,

$$\boxed{\text{(diagram)}} : 4 \to 3 \text{ is interpreted as } \langle \,\rhd\!\!-(x_2, \rhd\!\!-(x_1, x_3)), x_4, \multimap \rangle : 4 \to 3$$

As we observed above, string diagrams of Mn can only express *linear* terms. What makes $\mathcal{L}_{\Sigma_M, E_M}$ more general than Mn is the ability to *copy* and *discard* variables. Indeed, just as any monoidal category in which $\oplus$ is the cartesian product, $\mathcal{L}_{\Sigma_M, E_M}$ comes equipped with canonical choices of a "copy" and "discard" operation

$$cpy(n) := \langle x_1, \ldots, x_n, x_1, \ldots, x_n \rangle : n \to 2n \quad dsc(n) := \langle \rangle : n \to 0 \quad n \in \mathbb{N}$$

natural in n, which satisfy some expected equations: copying is commutative and associative; copying and then discarding is the same as not doing anything — see e.g. [14,18].

How can we mimic the copy and discard structure in the language of PROPs? First, for each $n > 1$ one can define $cpy(n)$ and $dsc(n)$ in terms of $cpy(1)$ and $dsc(1)$, which can therefore be regarded as the only fundamental operations[4].

---

[4] For $n = 0$, both operations are equal to the identity on 0.

Also, the equations that they satisfy can be synthesised as saying that $cpy(1)$ acts as the comultiplication and $dsc(1)$ as the counit of a commutative comonoid on 1. Therefore, they are none other than the generators of the PROP Cm:

$$\text{■◁} : 1 \to 2 \qquad\qquad \text{■●} : 1 \to 0.$$

Our approach suggests that a copy of Mn and of Cm "live" inside $\mathcal{L}_{\Sigma_M, E_M}$. We claim that these two PROPs provide a *complete* description of $\mathcal{L}_{\Sigma_M, E_M}$, that means, any arrow of $\mathcal{L}_{\Sigma_M, E_M}$ can be presented diagrammatically by using Mn and Cm. For instance,

$$\langle \text{▷}(x_2, \text{▷}(x_1, x_4)), x_1, \text{○}\rangle : 4 \to 3 \text{ corresponds to } \overset{\in \text{Cm} \quad \in \text{Mn}}{\boxed{}} : 4 \to 3$$

Observe that the diagram is of the factorised form $\xrightarrow{\in \text{Cm}} \xrightarrow{\in \text{Mn}}$. Intuitively, Cm is deputed to model the interplay of variables — in this case, the fact that x_1 is copied and x_3 is deleted — and Mn describes the syntactic tree of the terms. Of course, to claim that this factorisation is always possible, we need additional equations to model composition of factorised diagrams. For instance:

$$\langle \text{▷}(x_1, x_2), x_1\rangle \,;\, \langle x_1, \text{▷}(x_1, x_2)\rangle \;=\; \langle \text{▷}(x_1, x_2), \text{▷}(\text{▷}(x_1, x_2), x_1)\rangle.$$

The second equality holds if we assume the equation (A8) of the SMT of *bialgebras*. Thus the example suggests that composition by substitution in $\mathcal{L}_{\Sigma_M, E_M}$ can be mimicked at the diagrammatic level by allowing the use of bialgebra equations, which as we know from Example 3.3(b) present the composite PROP Cm ; Mn. Therefore, the conclusive conjecture of our analysis is that $\mathcal{L}_{\Sigma_M, E_M}$ must be isomorphic to Cm ; Mn and can be presented by equations (A1)–(A10).

We now generalise and make formal the above approach. Our main result is the following.

Theorem 4.5. *Suppose that (Σ, E) is an SMT which is also cartesian and let $\mathcal{T}_{(\Sigma, E)}$ be its freely generated PROP. Then $\mathcal{L}_{(\Sigma, E)} \cong$ Cm ; $\mathcal{T}_{(\Sigma, E)}$, where distributive law $\mathcal{T}_{(\Sigma, E)}$; Cm $\to$ Cm ; $\mathcal{T}_{(\Sigma, E)}$ yielding $\mathcal{L}_{(\Sigma, E)}$ is presented by equations*

for each $o \in \Sigma$.

Before moving to the proof of Theorem 4.5, we show its significance by revisiting some well-known theories in terms of our result.

Example 4.6.

(a) If we instantiate Theorem 4.5 to the theory (Σ_M, E_M) of commutative monoids (Example 2.3), then (Lw1)–(Lw2) are the bialgebra equations (A7)–(A10). The result that $B \cong Cm; Mn$ (Example 3.3(b)) is now an immediate consequence of Theorem 4.5 and tells us that the Lawvere category of commutative monoids can be considered as the PROP of bialgebras.

(b) In the case of monoids, the resulting Lawvere category is precisely a composite PROP, because all the equations only affect the linear part of the theory, that means, the generating cartesian theory is also an SMT. As observed in Remark 4.3, this is not true in general: for instance, the cartesian theory (Σ_G, E_G) of abelian groups extends the one (Σ_M, E_M) of commutative monoids with an inverse operation ▬ : $1 \to 1$ and a non-linear equation ⊳(x, ▬(x)) = ∘⊦ . In such cases, Theorem 4.5 still yields useful information about the structure of the resulting Lawvere category. For instance, it means that $\mathcal{L}_{(\Sigma_G, E_G)}$ is isomorphic to the PROP $Cm; \mathcal{T}_{\Sigma_G, E_M}$ quotiented by the above non-linear equation, which is rendered in string diagrams as:

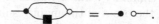

Interestingly, the result of this quotient is isomorphic to the PROP of integer matrices, see e.g. [37, §3.5] and its models in a symmetric monoidal category are the *Hopf algebras* [17], with ▬ playing the role of the antipode.

(c) In [21] Fritz presents the category of finite sets and probabilistic maps using generators and equations. The resulting Lawvere category $\mathcal{L}_{\text{Prob}}$ can be decomposed following the scheme of Theorem 4.5: there is a linear part (Σ_P, E_P) of the theory — given by binary convex combinations $\Sigma_P = \{ \underline{\lambda} : 2 \to 1 \mid \lambda \in [0,1]\}$ and suitable associativity and commutativity laws in E_P, a commutative comonoid structure, and the two interact according to (Lw1)–(Lw2). This interaction yields a composite PROP $Cm; \mathcal{T}_{(\Sigma_P, E_P)}$ which, quotiented by non-linear equations $\boxed{0} = \bullet$ and $\bullet\!\!\underline{\lambda} = $ ——— , yields $\mathcal{L}_{\text{Prob}}$.[5]

Remark 4.7. It is instructive to observe how Theorem 4.5 translates to models of theories. We recalled what is a model for a PROP in Remark 2.4; there is an analogous notion for Lawvere categories. A model for a Lawvere category $\mathcal{L}_{(\Sigma, E)}$ is a cartesian category $\mathbb{C}$ together with a cartesian (i.e., finite-products preserving) functor $\mathcal{L}_{(\Sigma, E)} \to \mathbb{C}$: models of $\mathcal{L}_{(\Sigma, E)}$ in $\mathbb{C}$ and natural transformations between them form a category $\text{CartMod}(\mathcal{L}_{(\Sigma, E)}, \mathbb{C})$. Now, for (Σ, E) and $\mathcal{T}_{(\Sigma, E)}$ as in Theorem 4.5, we have that models of $\mathcal{L}_{(\Sigma, E)}$ in $\mathbb{C}$ cartesian are the same as models of $\mathcal{T}_{(\Sigma, E)}$ in $\mathbb{C}$, now seen more abstractly as a symmetric monoidal category. That means, there is an equivalence $\text{LinMod}(\mathcal{T}_{(\Sigma, E)}, \mathbb{C}) \simeq \text{CartMod}(\mathcal{L}_{(\Sigma, E)}, \mathbb{C})$.

---

[5] In fact, the Lawvere category in [21] has finite coproducts, while our $\mathcal{L}_{\text{Prob}}$ is based on finite products. This is just a matter of co-/contra-variant presentation of the same data: one can switch between the two by "vertical rotation" of diagrams.

The rest of the section is devoted to proving Theorem 4.5. First we observe that, by the following lemma, it actually suffices to check our statement for SMTs with no equations. This reduction has just the purpose of making computations in $\mathcal{L}_{(\Sigma,E)}$ easier, by working with terms instead of equivalence classes.

Lemma 4.8. *If the statement of* Theorem 4.5 *holds in the case* $E = \varnothing$, *then it holds for any cartesian SMT* (Σ, E).

Proof. Let (Σ, E) be a cartesian SMT and $\mathcal{T}_\Sigma$, $\mathcal{T}_{(\Sigma,E)}$ be the PROPs freely generated, respectively, by $(\Sigma, \varnothing)$ and (Σ, E). By assumption, Theorem 4.5 holds for $(\Sigma, \varnothing)$, yielding a distributive law $\lambda \colon \mathcal{T}_\Sigma \, ; \mathsf{Cm} \to \mathsf{Cm} \, ; \mathcal{T}_\Sigma$. A routine check shows that λ preserves the equations of E, whence Proposition 3.6 gives a distributive law $\lambda' \colon \mathcal{T}_{(\Sigma,E)} \, ; \mathsf{Cm} \to \mathsf{Cm} \, ; \mathcal{T}_{(\Sigma,E)}$ with the required properties.

In the sequel, let us abbreviate $\mathcal{L}_{\Sigma,\emptyset}$ as $\mathcal{L}_\Sigma$. By virtue of Lemma 4.8, we shall prove Theorem 4.5 for $\mathcal{L}_\Sigma$ and by letting $\mathcal{T}_\Sigma$ be the PROP freely generated by $(\Sigma, \emptyset)$. We shall obtain the distributive law $\mathcal{T}_\Sigma \, ; \mathsf{Cm} \to \mathsf{Cm} \, ; \mathcal{T}_\Sigma$ from the recipe of Proposition 3.4, by showing that any arrow of $\mathcal{L}_\Sigma$ decomposes as $\xrightarrow{\;\in\mathsf{Cm}\;\;\in\mathcal{T}_\Sigma\;}$.

We now give some preliminary lemmas that are instrumental for the definition of the factorisation and the proof of the main result. We begin by showing how string diagrams of Cm and $\mathcal{T}_\Sigma$ are formally interpreted as arrows of $\mathcal{L}_\Sigma$.

Lemma 4.9.

- Cm *is the sub-PROP of* $\mathcal{L}_\Sigma$ *whose arrows are lists of variables. The inclusion of* Cm *in* $\mathcal{L}_\Sigma$ *is the morphism* $\Phi \colon \mathsf{Cm} \to \mathcal{L}_\Sigma$ *defined on generators of* Cm *by*

$$\boxed{\blacktriangleleft} \; \mapsto \; \langle x_1, x_1 \rangle \colon 1 \to 2 \qquad \boxed{\bullet} \; \mapsto \; \langle\,\rangle \colon 0 \to 1.$$

- $\mathcal{T}_\Sigma$ *is the sub-PROP of* $\mathcal{L}_\Sigma$ *whose arrows are linear terms. The inclusion of* $\mathcal{T}_\Sigma$ *in* $\mathcal{L}_\Sigma$ *is the morphism* $\Psi \colon \mathcal{T}_\Sigma \to \mathcal{L}_\Sigma$ *defined on generators of* $\mathcal{T}_\Sigma$ *by*

$$\boxed{o} \vdash \; \mapsto \; \langle o(x_1, \ldots, x_n) \rangle \colon n \to 1 \qquad (o \colon n \to 1) \in \Sigma.$$

Proof. First, it is immediate to verify that lists of variables are closed under composition, monoidal product and include all the symmetries of $\mathcal{L}_\Sigma$: therefore, they form a sub-PROP. The same holds for linear terms.

We now consider the first statement of the lemma. There is a 1-1 correspondence between arrows $n \xrightarrow{f \in \mathcal{L}_\Sigma} m$ that are lists of variables and functions $\overline{m} \to \overline{n}$: the function for f maps k, for $1 \le k \le m$, to the index l of the variable x_l appearing in position k in f. This correspondence yields an isomorphism between the sub-PROP of $\mathcal{L}_\Sigma$ whose arrows are lists of variables and F^{op}. Composing this isomorphism with $\mathsf{Cm} \xrightarrow{\cong} \mathsf{F}^{op}$ yields Φ as in the statement of the lemma.

For the second statement, faithfulness is immediate by the fact that arrows of $\mathcal{T}_\Sigma$ are Σ-terms modulo the laws of SMCs, with no additional equations. One can easily verify that $\Psi \colon \mathcal{T}_\Sigma \to \mathcal{L}_\Sigma$ identifies the linear terms in $\mathcal{L}_\Sigma$ following the observations in Remark 4.3.

Henceforth, for the sake of readability we shall not distinguish between Cm and the isomorphic sub-PROP of $\mathcal{L}_\Sigma$ identified by the image of Φ, and similarly for $\mathcal{T}_\Sigma$ and Ψ. Lemma 4.9 allows us to use $\mathcal{L}_\Sigma$ as an environment where Cm and $\mathcal{T}_\Sigma$ interact. The following statement guarantees the soundness of the interaction described by (LW1)–(Lw2).

Lemma 4.10. *Equations (LW1) and (Lw2) are sound in* $\mathcal{L}_\Sigma$.

Proof. We first focus on (Lw1). Following the isomorphisms of Lemma 4.9, $\boxed{O}$ $\in \mathcal{T}_\Sigma[n,1]$ is interpreted as the arrow $\langle o(x_1,\ldots,x_n)\rangle \in \mathcal{L}_\Sigma[n,1]$ and $\multimap \in$ Cm$[0,1]$ as as $\langle\rangle \in \mathcal{L}_\Sigma[1,0]$. The left-hand side of (Lw1) is then the composite $\langle o(x_1,\ldots,x_n)\rangle ; \langle\rangle \in \mathcal{L}_\Sigma[n,0]$, which is equal by definition to $\langle\rangle \in \mathcal{L}_\Sigma[n,0]$. Therefore, the left- and right-hand side of (Lw1) are the same arrow of $\mathcal{L}_\Sigma$.

It remains to show soundness of (Lw2). Following Lemma 4.9, the left-hand side $\boxed{O}$; $\prec$ is interpreted in $\mathcal{L}_\Sigma$ as $\langle o(x_1,\ldots,x_n)\rangle ; \langle x_1,x_1\rangle$ and the right-hand side as $\langle x_1,\ldots,x_n,x_1,\ldots,x_n\rangle ; \langle o(x_1,\ldots,x_n),o(x_{n+1},\ldots,x_{n+n})\rangle$. By definition, both composites are equal to $\langle o(x_1,\ldots,x_n),o(x_1,\ldots,x_n)\rangle$ in $\mathcal{L}_\Sigma$. Therefore, (Lw2) is also sound in $\mathcal{L}_\Sigma$.

It is useful to observe that (Lw1)–(Lw2) allows us to copy and discard not only the generators but arbitrary string diagrams of $\mathcal{T}_\Sigma$.

Lemma 4.11. *Suppose* $\boxed{d}$ *is a string diagram of* $\mathcal{T}_\Sigma$. *Then the following holds in* $\mathcal{T}_\Sigma +$ Cm *quotiented by (Lw1)–(Lw2).*

$$\boxed{d} \; = \; \genfrac{}{}{0pt}{}{\boxed{d}}{\boxed{d}} \qquad \text{(Lw3)} \qquad \boxed{d}\, \bullet \; = \; \bullet \qquad \text{(Lw4)}$$

Proof. The proof is by induction on $\boxed{d}$ . . For (Lw3), the base cases of $-\!\square$ and $\bowtie$ follow by the laws of SMCs (Fig. 1). The base case of $\boxed{O}$, for o a generator in Σ, is given by (Lw2). The inductive cases of composition by ; and $\oplus$ immediately follow by induction hypothesis. The proof of (Lw4) is analogous.

We can now show the factorisation lemma.

Lemma 4.12. *Any arrow* $n \xrightarrow{f \in \mathcal{L}_\Sigma} m$ *has a factorisation* $n \xrightarrow{\hat{c} \in \mathsf{Cm}} \xrightarrow{\hat{d} \in \mathcal{T}_\Sigma} m$ *which is unique up-to permutation.*

Proof. Since the cartesian theory generating $\mathcal{L}_\Sigma$ has no equations, $n \xrightarrow{f} m$ is just a list of cartesian Σ-terms $\langle t_1,\ldots,t_m\rangle$. The factorisation consists in replacing all variables appearing in $\langle t_1,\ldots,t_m\rangle$ with fresh ones $x_1,\ldots,x_z$, so that no repetition occurs: this gives us the second component of the decomposition as a list of linear terms $z \xrightarrow{\hat{d} \in \mathcal{T}_\Sigma} m$. The first component $\hat{c}$ will be the list $n \xrightarrow{ovar(t_1,\ldots,t_m) \in \mathsf{Cm}} z$ of variables originally occurring in f, so that post-composition with $\hat{d}$ yields $\langle t_1,\ldots,t_m\rangle$. It is simple to verify uniqueness up-to permutation of this factorisation.

We now have all the ingredients to conclude the proof of our main statement.

Proof (Theorem 4.5). Using the conclusion of Lemma 4.12, Proposition 3.4 gives us a distributive law $\lambda\colon \mathcal{T}_\Sigma\,;\mathsf{Cm} \to \mathsf{Cm}\,;\mathcal{T}_\Sigma$ such that $\mathcal{L}_\Sigma \cong \mathsf{Cm}\,;\mathcal{T}_\Sigma$. It remains to show that (Lw1)–(Lw2) allow to prove all the equations arising from λ. By Proposition 3.4, λ maps a composable pair $n \xrightarrow{d\in\mathcal{T}_\Sigma} \xrightarrow{c\in\mathsf{Cm}} m$ to the factorisation $n \xrightarrow{c'\in\mathsf{Cm}} \xrightarrow{d'\in\mathcal{T}_\Sigma} m$ of $d\,;c$ in $\mathcal{L}_\Sigma$, calculated according to Lemma 4.12. The corresponding equation generated by λ is $d\,;c = c'\,;d'$, with d,c,c',d' now seen as string diagrams of $\mathcal{T}_\Sigma + \mathsf{Cm}$. The equational theory of $\mathcal{L}_\Sigma \cong \mathsf{Cm}\,;\mathcal{T}_\Sigma$ consists of all the equations arising in this way plus those of $\mathcal{T}_\Sigma + \mathsf{Cm}$. What we need to show is that

> the string diagrams $d\,;c$ and $c'\,;d'$ are equal modulo the   (†)
> equations of $\mathcal{T}_\Sigma + \mathsf{Cm}$ and (Lw1)–(Lw2).

Since our factorisation is unique up-to permutation, it actually suffices to show a weaker statement, namely that

> *there exists* a factorisation $n \xrightarrow{c''\in\mathsf{Cm}} \xrightarrow{d''\in\mathcal{T}_\Sigma} m$ of $d\,;c$ (‡)
> in $\mathcal{L}_\Sigma$ such that the string diagrams $d\,;c$ and $c''\,;d''$ are
> equal modulo the equations of $\mathcal{T}_\Sigma + \mathsf{Cm}$ and (Lw1)–(Lw2).

Statement (‡) implies (†) because, by uniqueness of the factorisation $c'\,;d'$ up-to permutation, there exists $\xrightarrow{p\in\mathsf{P}}$ such that $d' = p\,;d''$ and $c'' = c'\,;p$ in $\mathcal{L}_\Sigma$. Since p is an arrow of both sub-PROPs $\mathcal{T}_\Sigma$ and Cm, the first equality also holds in $\mathcal{T}_\Sigma$ and the second in Cm. So $c'\,;d' = c'\,;p\,;d'' = c''\,;d''$ in $\mathcal{T}_\Sigma + \mathsf{Cm}$.

Therefore, we turn to a proof of (‡). We describe a procedure to transform the string diagram $\xrightarrow{d\in\mathcal{T}_\Sigma} \xrightarrow{c\in\mathsf{Cm}}$ into the form $\xrightarrow{c''\in\mathsf{Cm}} \xrightarrow{d''\in\mathcal{T}_\Sigma}$ by only using the equations in $\mathcal{T}_\Sigma + \mathsf{Cm}$ plus (Lw1)-(Lw2). Lemmas 4.9 and 4.10 guarantee that $d\,;c = c''\,;d''$ as arrows of $\mathcal{L}_\Sigma$.

1. First, there is a preparatory step in which we move all symmetries to the outmost part of the string diagram $d\,;c$, to ease the application of (Lw1)–(Lw2). By definition, d only contains components of the kind ⊟O⊢$\colon k \to 1$, , $o \in \Sigma$, ⋈$\colon 2 \to 2$ and ⊟$\colon 1 \to 1$. Using naturality (Fig. 1), we can move all symmetries ⋈ to the left of components ⊟O⊢ .

The result is a string diagram $p\,;\bar{d}\,;c'$, where p only contains components ⋈ and ⊟ — i.e., it is a string diagram of P — and $\bar{d}$ is a string diagram of $\mathcal{T}_\Sigma$ where ⋈ does not appear.

We then perform a symmetric transformation on the string diagram c. By definition, c contains components ⟨: $1 \to 2$, •⟨: $0 \to 1$, ⋈: $2 \to 2$ and ▭: $1 \to 1$. By naturality, we can move all symmetries ⋈ to the right of any component ⟨ and •⟨ .

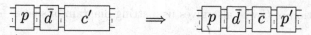

The result is a string diagram p; $\bar{d}$; $\bar{c}$; p', where $\bar{c}$ is a string diagram of Cm in which ⋈ does not appear and p' is a string diagram of P.

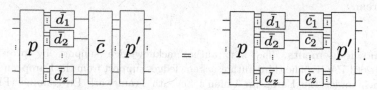

2. We now make $\bar{d}$ and $\bar{c}$ interact. First note that, since $\bar{d}$ does not contain ⋈ and all generators $o \in \Sigma$ have coarity 1, $\bar{d}$ must the $\oplus$-product $\bar{d}_1 \oplus \ldots \oplus \bar{d}_z$ of string diagrams $\bar{d}_i \colon k_i \to 1$ of $\mathcal{T}_\Sigma$.

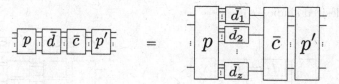

For analogous reasons, $\bar{c}$ is also a $\oplus$-product $\bar{c}_1 \oplus \ldots \oplus \bar{c}_z$ where, for $1 \leq i \leq z$,

$$-\bar{c}_i- \quad = \quad \text{(diagram)} \qquad \text{or} \qquad -\bar{c}_i- \quad = \quad \bullet. \qquad (9)$$

We thus can present $\bar{c}$ as follows:

(diagram equality)

We are now in position to distribute each $\bar{d}_i$ over the corresponding $\bar{c}_i$. Suppose first $\bar{c}_i$ satisfies the left-hand equality in (9). By assumption, all the equations of $\mathcal{T}_\Sigma + \text{Cm}$, Lw1 and (Lw2) hold. Thus, by Lemma 4.11, also (Lw3) holds. Starting from $\bar{d}_i$; $\bar{c}_i$, we can iteratively apply (Lw3) to obtain a string diagram of shape $\xrightarrow{\in \text{Cm}} \xrightarrow{\in \mathcal{T}_\Sigma}$:

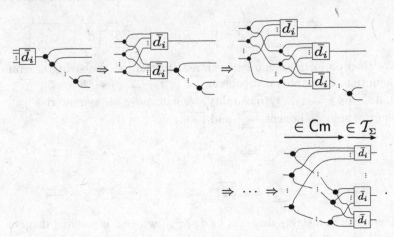

In the remaining case, $\bar{c}_i$ satisfies the right-hand equality in (9). Then, one application of (Lw1) also gives us a string diagram of shape $\xrightarrow{\in \mathsf{Cm}} \xrightarrow{\in \mathcal{T}_\Sigma}$.

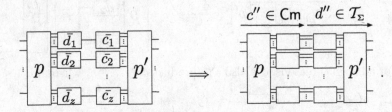

Applying the above transformations for each $\bar{d}_i$; $\bar{c}_i$ yields a string diagram of the desired shape $\xrightarrow{c'' \in \mathsf{Cm}} \xrightarrow{d'' \in \mathcal{T}_\Sigma}$.

Observe that all the transformations that we described only used equations in $\mathcal{T}_\Sigma + \mathsf{Cm}$, (Lw1) and (Lw2). This concludes the proof of (‡) and thus of the main theorem.

Acknowledgements. The first author acknowledge support by project ANR 12IS02001 PACE. The third author acknowledges support from the European Research Council under the European Union's Seventh Framework Programme (FP7/2007-2013) / ERC grant agreement n° 320571.

References

1. Baez, J.C., Erbele, J.: Categories in control (2014). CoRR abs/1405.6881

2. Baez, J.C.: Universal algebra and diagrammatic reasoning. Lecture (2006). http://math.ucr.edu/home/baez/universal/universal_hyper.pdf
3. Studer, C.: Introduction. In: Studer, C. (ed.) Numerics of Unilateral Contacts and Friction. LNACM, vol. 47, pp. 1–8. Springer, Heidelberg (2009)
4. Bonchi, F., Milius, S., Silva, A., Zanasi, F.: Killing epsilons with a dagger – a coalgebraic study of systems with algebraic label structure. Theoret. Comput. Sci. **604**, 102–126 (2015)
5. Bonchi, F., Sobociński, P., Zanasi, F.: A categorical semantics of signal flow graphs. In: Baldan, P., Gorla, D. (eds.) CONCUR 2014. LNCS, vol. 8704, pp. 435–450. Springer, Heidelberg (2014)
6. Bonchi, F., Sobociński, P., Zanasi, F.: Interacting bialgebras are Frobenius. In: Muscholl, A. (ed.) FOSSACS 2014 (ETAPS). LNCS, vol. 8412, pp. 351–365. Springer, Heidelberg (2014)
7. Bonchi, F., Sobocinski, P., Zanasi, F.: Full abstraction for signal flow graphs. In: Proceedings of the 42nd Annual ACM SIGPLAN-SIGACT Symposium on Principles of Programming Languages, POPL 2015, pp. 515–526 (2015)
8. Bonchi, F., Zanasi, F.: Bialgebraic semantics for logic programming. Log. Methods Comput. Sci. **11**(1:14), 1–47 (2015)
9. Bonsangue, M.M., Hansen, H.H., Kurz, A., Rot, J.: Presenting distributive laws. In: Heckel, R., Milius, S. (eds.) CALCO 2013. LNCS, vol. 8089, pp. 95–109. Springer, Heidelberg (2013)
10. Borceux, F.: Handbook of Categorical Algebra 1 - Basic Category Theory. Cambridge Univ. Press, Cambridge (1994)
11. Bruni, R., Melgratti, H., Montanari, U.: A connector algebra for P/T nets interactions. In: Katoen, J.-P., König, B. (eds.) CONCUR 2011. LNCS, vol. 6901, pp. 312–326. Springer, Heidelberg (2011)
12. Bruni, R., Montanari, U., Plotkin, G.D., Terreni, D.: On hierarchical graphs: reconciling bigraphs, gs-monoidal theories and gs-graphs. Fundam. Inform. **134**(3–4), 287–317 (2014)
13. Burris, S., Sankappanavar, H.P.: A Course in Universal Algebra. Springer,break Heidelberg (1981)
14. Burroni, A.: Higher dimensional word problems with applications to equational logic. Theor. Comput. Sci. **115**, 43–62 (1993)
15. Coecke, B., Duncan, R.: Interacting quantum observables: categorical algebra and diagrammatics. New J. Phys. **13**(4), 043016 (2011)
16. Corradini, A., Gadducci, F.: An algebraic presentation of term graphs, via gs-monoidal categories. Appl. Categorical Struct. **7**(4), 299–331 (1999)
17. Dǎscǎlescu, S., Nastasescu, C., Raianu, S.: Hopf Algebras, Pure and Applied Mathematics, vol. 235. Marcel Dekker Inc., New York (2001). An Introduction
18. Eilenberg, S., Kelly, G.M.: Closed categories. In: Eilenberg, S., Harrison, D.K., MacLane, S., Röhri, H. (eds.) Proceedings of the Conference on Categorical Algebra, pp. 421–562. Springer, Heidelberg (1966)
19. Fiore, M., Devesas Campos, M.: The algebra of directed acyclic graphs. In: Coecke, B., Ong, L., Panangaden, P. (eds.) Computation, Logic, Games and Quantum Foundations. LNCS, vol. 7860, pp. 37–51. Springer, Heidelberg (2013)
20. Fong, B., Rapisarda, P., Sobociński, P.: A categorical approach to open and interconnected dynamical systems (2015). CoRR/1510.05076
21. Fritz, T.: A presentation of the category of stochastic matrices (2009). CoRR abs/0902.2554. http://arxiv.org/abs/0902.2554
22. Hyland, M., Plotkin, G., Power, J.: Combining effects: sum and tensor. Theor. Comput. Sci. **357**(1–3), 70–99 (2006)

23. Hyland, M., Power, J.: The category theoretic understanding of universal algebra: Lawvere theories and monads. Electron. Notes Theor. Comput. Sci. **172**, 437–458 (2007)
24. Jacobs, B.: Semantics of weakening and contraction. Ann. Pure Appl. Logic **69**(1), 73–106 (1994)
25. Lack, S.: Composing PROPs. Theor. App. Categories **13**(9), 147–163 (2004)
26. Lafont, Y.: Equational reasoning with 2-dimensional diagrams. Term Rewriting. LNCS, vol. 909, pp. 170–195. Springer, Heidelberg (1995)
27. Lawvere, W.F.: Functorial semantics of algebraic theories. Ph.D. thesis (2004)
28. Mac Lane, S.: Categorical algebra. Bull. Am. Math. Soc. **71**, 40–106 (1965)
29. Mac Lane, S.: Categories for the Working Mathematician. Springer, Heidelberg (1998)
30. Melliès, P.A.: Categorical semantics of linear logic. In: Interactive Models of Computation and Program Behaviour, Panoramas et Synthèses 27, Société Mathématique de France 1196 (2009)
31. Rosebrugh, R., Wood, R.: Distributive laws and factorization. J. Pure Appl. Algebra **175**(13), 327–353 (2002). Special Volume celebrating the 70th birthday of Professor Max Kelly
32. Ross Duncan, K.D.: Interacting Frobenius algebras are Hopf. CoRR abs/1601.04964 (2016). http://arxiv.org/abs/1601.04964
33. Selinger, P.: A survey of graphical languages for monoidal categories. Springer Lect. Notes Phys. **13**(813), 289–355 (2011)
34. Sleep, M.R., Plasmeijer, M.J., van Eekelen, M.C.J.D. (eds.): Term Graph Rewriting: Theory and Practice. Wiley, Chichester (1993)
35. Sobociński, P., Stephens, O.: A programming language for spatial distribution of net systems. In: Petri Nets 2014 (2014)
36. Street, R.: The formal theory of monads. J. Pure Appl. Algebra **2**(1), 243–265 (2002)
37. Zanasi, F.: Interacting Hopf algebras: the theory of linear systems. Ph.D. thesis, Ecole Normale Supérieure de Lyon (2015)

Transitivity and Difunctionality of Bisimulations

Mehdi Zarrad[⊠] and H. Peter Gumm

Philipps-Universität Marburg, Marburg, Germany
zarrad@mathematik.uni-marburg.de

Abstract. Bisimilarity and observational equivalence are notions that agree in many classical models of coalgebras, such as e.g. Kripke structures. In the general category Set_F of $F-$coalgebras these notions may, however, diverge. In many cases, observational equivalence, being transitive, turns out to be more useful.

In this paper, we shall investigate the role of transitivity for the largest bisimulation of a coalgebra. Passing to relations between two coalgebras, we choose difunctionality as generalization of transitivity. Since in Set_F bisimulations are known to coincide with $\bar{F}-$simulations, we are led to study the notion of $L-$similarity, where L is a relation lifting.

1 Introduction

In the 1990s, J. Rutten developed a universal theory for state-based systems. Such systems were represented as coalgebras in [19]. A coalgebra of type F is a pair consisting of a base set A and a structure map $\alpha : A \longrightarrow F(A)$ where F is any Set-endofunctor. Many of the early structural results, we just exemplarily mention [16], assumed that the type functor F should preserve weak pullbacks. In [5], one of the present authors began studying the role of weak pullback preservation. In [4,14] he extended many results from [19] to the case of arbitrary functors, adding new coalgebraic constructions, e.g. for building terminal coalgebras. A systematic study relating preservation properties of functors to structural (coalgebraic) properties of their coalgebras was given in [8]. In particular, preservation properties were related to properties of bisimulations and congruences. For instance, it was shown that bisimulations restrict to subcoalgebras, if and only if the functor F preserves preimages, i.e. pullbacks along regular monos.

Bisimulations, as compatible relations have been introduced by Aczel and Mendler in [1]. An alternative definition, equivalent for Set-coalgebras was given by Hermida and Jacobs [13].

A concept, competing with bisimulations is the notion of a congruence relation θ as a kernel $\ker \varphi$ of a homomorphism φ. Just as there is always a largest bisimulation $\sim_{\mathcal{A}}$, there is also a largest congruence $\nabla_{\mathcal{A}}$ on every coalgebra $\mathcal{A}$. The disadvantage of $\sim_{\mathcal{A}}$ versus $\nabla_{\mathcal{A}}$ is, that even though $\sim_{\mathcal{A}}$ is reflexive and symmetric, it need not be transitive, hence it is not able to reflect logical equivalence.

© IFIP International Federation for Information Processing 2016
Published by Springer International Publishing Switzerland 2016. All Rights Reserved
I. Hasuo (Ed.): CMCS 2016, LNCS 9608, pp. 33–52, 2016.
DOI: 10.1007/978-3-319-40370-0_4

Since the equivalence hull of any bisimulation is always a congruence relation [1], we have the implications

$$\sim_{\mathcal{A}} \subseteq \sim_{\mathcal{A}}^{*} \subseteq \nabla_{\mathcal{A}}$$

where each of the inclusions may be strict.

Already in [1], bisimulations were defined as relations between two (possible different) coalgebras $\mathcal{A}$ and $\mathcal{B}$. The notion of congruence can as easily be extended to the notion of 2-congruence, defined as the kernel of a sink of two homomorphisms $\varphi : \mathcal{A} \to \mathcal{C}$, and $\psi : \mathcal{B} \to \mathcal{C}$ i.e. as $\ker(\varphi, \psi) := \{(a, b) \mid \varphi(a) = \psi(b)\}$. (2-congruences were studied by S. Staton [21] under the name of kernel-bisimulations. However they are not necessarily bisimulations in the original sense, so we prefer to call them 2-simulations). Again, for the case of two coalgebras $\mathcal{A}$ and $\mathcal{B}$ it is easy to see that there is again a largest bisimulation $\sim_{\mathcal{A},\mathcal{B}}$ as well as a largest 2-congruence $\nabla_{\mathcal{A},\mathcal{B}}$. The above inequalities do not immediately generalize, since transitivity makes no sense for relations between different sets. An alternative notion, however, is *difunctionality* as introduced by Riguet [18]. The inequality corresponding to the above is then

$$\sim_{\mathcal{A},\mathcal{B}} \subseteq \sim_{\mathcal{A},\mathcal{B}}^{d} \subseteq \nabla_{\mathcal{A},\mathcal{B}}$$

where R^d denotes the difunctional closure of a relation R.

The definition of bisimulation given by Hermida and Jacobs used the fact that a relation $R \subseteq A \times B$ can easily be lifted to a relation $\bar{F}R \subseteq F(A) \times F(B)$, often called the Barr extension of R. Generalizing this to an arbitrary relation lifting L leads to the notion of $L-$simulation as studied by Thijs [22] and in a series of papers by Venema et al., see e.g. [15].

The congruences which can be obtained as transitive closure of bisimulations are exactly the regular congruences i.e. kernels of coequalizers [8]. In the same paper it was shown that a functor F preserves weak kernel pairs if and only every congruence on a single coalgebra in Set_F is a bisimulation. In the present paper, which subsumes some results obtained in the first author's Master thesis [24], we want to explain further the relationship between bisimulations and $2-$congruences and, assuming a monotonic relation lifting L, we study the difunctional closure of $L-$similarity. A quite useful tool is provided by a result, giving conditions for a bisimulation to restrict to subcoalgebras without assuming anything about the functor F.

2 Basic Notions

For a product $A \times B$ we denote the projections to the components by π_A and π_B. Given a relation $R \subseteq A \times B$, the restrictions to R of the projections are $\pi_A^R := \pi_A \circ \iota_R$ where $\iota_R : R \hookrightarrow A \times B$ is set inclusion. Given a second relation S with $R \subseteq S \subseteq A \times B$ we note that $\pi_A^R = \pi_A^S \circ \iota_R^S$, where $\iota_R^S : R \hookrightarrow S$ is, again, the inclusion map.

The converse of a relation R is $R^- := \{(b, a) \mid (a, b) \in R\}$. For a subset $U \subseteq A$, we let $R[U] := \{b \in B \mid \exists u \in U.(u, b) \in R\}$. A relation $R \subseteq A \times A$ is transitive if for all $x, y, z \in A$ we have $(x, y), (y, z) \in R \implies (x, z) \in R$. Equivalently, R is transitive, if $R \circ R \subseteq R$ where $\circ$ is relational composition. The reflexive transitive closure of R is called $R^\star$.

Transitivity makes no sense for relations $R \subseteq A \times B$ between different sets, so *difunctionality* [18] can be considered as a possible generalization of transitivity:

Definition 1. *A relation $R \subseteq A \times B$ is called* difunctional, *if for all $a_1, a_2 \in A$ and for all $b_1, b_2 \in B$ it satisfies:*

$$(a_1, b_1), (a_2, b_1), (a_2, b_2) \in R \implies (a_1, b_2) \in R.$$

This notion can be illustrated by the following figure, which explains why "difunctional" is sometimes called "z-closed":

It is elementary to see that the difunctional closure of a relation R is

$$R^d := R \circ (R^- \circ R)^\star = (R \circ R^-)^\star \circ R$$

where R^- is the converse relation to R. The difunctional closure can also be obtained as the pullback of the pushout of π_A^R with π_B^R [9, 18].

2.1 Categorical Notions

We assume only elementary categorical notions and we use the terminology of [2]. *Regular monos* are equalizers of a parallel pair of morphisms. Analogously, *regular epis* are coequalizers. Split epis (split monos) i.e. the right-(left-)invertible morphisms are regular epi (regular mono). We denote the sum of a family $(A_i)_{i \in I}$ of objects and with the canonical injections by $e_{A_i} : A_i \to \Sigma_{i \in I} A_i$. Given a sink $(q_i : A_i \longrightarrow Q)_{i \in I}$ we denote the induced morphisms from the sum $\sum A_i \to Q$ by $[(q_i)_{i \in I}]$. The following lemma will be needed later. It is obviously true in every category with sums:

Lemma 1. *For all morphisms $f : A \longrightarrow B$ the map $[f, id_B] : A + B \to B$ is split epi.*

Definition 2. *A weak limit of a diagram D is a cone over D such that for every other (competing) cone there is* **at least one** *morphism making the relevant triangles commutative.*

If we replace "at least one" in the above definition with "exactly one", this is the definition of limit.

Set-Endofunctors. In the following, F will always be a *Set*-endofunctor. F preserves epis, and F preserves monos with nonempty domain. Next, let D a diagram.

Definition 3. *F weakly preserves D-limits if F maps every D-limit into a weak D-limit.*

It is well known that F weakly preserves D-limits if and only if it *preserves weak D-limits*, see [5].

An important property of *Set*-endofunctors is that they preserve finite nonempty intersections [23]. In order to preserve all finite intersections, it might be necessary to redefine F on the empty set and on empty mappings to obtain a (marginally modified) new functor preserving all finite intersections. Therefore, we are safe to only consider functors which preserve all finite intersections.

2.2 F-coalgebras

Definition 4. *Let $F : Set \to Set$ be a Set-endofunctor. An F-coalgebra $\mathcal{A} = (A, \alpha)$ consists of a set A and a structure map $\alpha : A \to F(A)$. A map $\varphi : A \to B$ between two coalgebras $\mathcal{A} = (A, \alpha)$ and $\mathcal{B} = (B, \beta)$ is called a* homomorphism, *if $\beta \circ \varphi = F\varphi \circ \alpha$. A subcoalgebra of a coalgebra $\mathcal{A}$ is a subset $U \subseteq A$ with a structure map α_U such that the inclusion map $\iota_U^X : U \hookrightarrow X$ is a homomorphism.*

The class of all F−coalgebras with coalgebra homomorphisms forms the category Set_F. In [3] it is proved that Set_F is cocomplete i.e. every colimit exists. Epis are exactly the surjective homomorphisms [19]. Monomorphisms in Set_F need not be injective. Regular monos are exactly the injective homomorphisms [6].

Bisimulations

Definition 5 (Aczel and Mendler [1]). *A bisimulation between two coalgebras $\mathcal{A}$ and $\mathcal{B}$ is a relation $R \subseteq A \times B$ for which there exists a coalgebra structure $\rho : R \to F(R)$ such that the projections $\pi_A^R : R \to A$ and $\pi_B^R : R \to B$ are homomorphisms. A bisimulation R on a coalgebra $\mathcal{A}$ is a bisimulation between $\mathcal{A}$ and itself.*

The union of bisimulations is a bisimulation and $\emptyset$ is always a bisimulation, so that the bisimulations between $\mathcal{A}$ and $\mathcal{B}$ form a complete lattice with largest element called $\sim_{\mathcal{A},\mathcal{B}}$. For the same reason, every relation R between coalgebras $\mathcal{A}$ and $\mathcal{B}$ contains a largest bisimulation, which we denote by $[R]$. It is the union of all bisimulations contained in R.

Congruences and 2−congruences

Definition 6. *A congruence θ on a coalgebra $\mathcal{A}$ is the kernel of a homomorphism $\varphi : \mathcal{A} \to \mathcal{C}$, i.e $\theta = ker\varphi = \{(a, a') \in A \times A \mid \varphi(a) = \varphi(a')\}$.*

If θ is a congruence on $\mathcal{A}$, then there is a structure map on the factor set $A/\theta := \{[a]\theta \,|\, a \in A\}$, such that $\pi_\theta : \mathcal{A} \longrightarrow A/\theta$ with $\pi_\theta(a) := [a]\theta := \{a' \in A \,|\, a\theta a'\}$ becomes a homomorphism. The set of all congruences on a coalgebra $\mathcal{A}$ forms a complete lattice [7] with largest element called $\nabla_\mathcal{A}$ and smallest element $\Delta_\mathcal{A} = \{(a,a) \,|\, a \in A\}$. The supremum of a family of congruences is obtained as the transitive closure of their union. We denote the lattice of all congruences by $Con(\mathcal{A})$.

Given a bisimulation R on $\mathcal{A}$, its equivalential hull, that is the smallest equivalence relation containing R is a congruence relation [1]. It follows immediately, that $\sim_\mathcal{A} \subseteq \nabla_\mathcal{A}$. Congruences arising as equivalential hulls of a bisimulation are called *regular*. This notion is suggested by the following result from [8]:

Lemma 2. *A morphism $\varphi : \mathcal{A} \to \mathcal{B}$ is mono iff $[ker\varphi] = \Delta_\mathcal{A}$ and regular mono, iff it is injective.*

An epi $\varphi : \mathcal{A} \twoheadrightarrow \mathcal{B}$ is regular epi (in Set_F) iff its kernel is a regular congruence iff $ker\varphi = [ker\varphi]^\star$.

For technical reasons, we call a homomorphism φ *strictly regular*, if its kernel is a bisimulation, i.e. if $ker\varphi = [ker\varphi]$. Notice that this notion is not an abstract categorical one, but a coalgebraic notion.

Definition 7. *A 2−congruence between two coalgebras $\mathcal{A}$ and $\mathcal{B}$ is the kernel (pullback in the category Set) of two homomorphisms $\varphi : \mathcal{A} \to \mathcal{C}$ and $\psi : \mathcal{B} \to \mathcal{C}$, i.e.*

$$\theta = ker(\varphi, \psi) := \{(a,b) \in A \times B \,|\, \varphi(a) = \psi(b)\}.$$

Just as congruences are transitive, 2−congruences are difunctional. The largest 2−congruence *between $\mathcal{A}$ and $\mathcal{B}$ is called observational equivalence* and written $\nabla_{\mathcal{A},\mathcal{B}}$. For each bisimulation R *between* $\mathcal{A}$ and $\mathcal{B}$, its difunctional hull R^d, being the kernel of the pushout of the components $\pi_\mathcal{A}^R$ and $\pi_\mathcal{B}^R$ is a 2-congruence, so $R^d \subseteq \nabla_{\mathcal{A},\mathcal{B}}$.

3   Observational Equivalence and Bisimilarity

When $\mathcal{A}$ and $\mathcal{B}$ are coalgebras, elements $a \in \mathcal{A}$ and $b \in \mathcal{B}$ are called *bisimilar*, if there exists a bisimulation R between $\mathcal{A}$ and $\mathcal{B}$ containing (a,b). This is the same as saying that for some coalgebra $\mathcal{R}$ and homomorphisms $\varphi_\mathcal{A} : \mathcal{R} \to \mathcal{A}$ and $\varphi_\mathcal{B} : \mathcal{R} \to \mathcal{B}$ there is some $r \in \mathcal{R}$ with $\varphi_\mathcal{A}(r) = a$ and $\varphi_\mathcal{B}(r) = b$. In short, a and b are bisimilar, if they have a common ancestor.

Dually, a and b are called *observationally equivalent*, iff they have a common offspring, meaning that there exists a coalgebra $\mathcal{C}$ and homomorphisms $\psi_\mathcal{A} : \mathcal{A} \to \mathcal{C}$ and $\psi_\mathcal{B} : \mathcal{B} \to \mathcal{C}$ with $\psi_\mathcal{A}(a) = \psi_\mathcal{B}(b)$. We first study the situation for the case $\mathcal{A} = \mathcal{B}$. Here a, a' are bisimilar iff $(a, a') \in \sim_\mathcal{A}$ and observationally equivalent iff $(a, a') \in \nabla_\mathcal{A}$.

3.1 Nabla and Simple Coalgebras

Definition 8. *A coalgebra $\mathcal{A}$ is called* simple, *if $\nabla_\mathcal{A} = \Delta_\mathcal{A}$ and* extensional, *if $\sim_\mathcal{A} = \Delta_\mathcal{A}$.*

It is well known that a coalgebra $\mathcal{A}$ is simple iff each morphism starting in $\mathcal{A}$ is injective [4]. If a terminal coalgebra $\mathcal{T}$ exists, simple coalgebras are precisely the subcoalgebras of $\mathcal{T}$ [4]. For every coalgebra $\mathcal{A}$, the factor coalgebra $\mathcal{A}/\nabla_\mathcal{A}$ is simple. The notions of simplicity and extensionality differ [7]. The coalgebra of the following example is extensional but not simple.

Example 1.

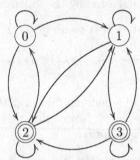

Consider the above $2 \times \mathbb{P}_{\leq 3}$–coalgebra $\mathcal{A} := (\{0,1,2,3\}, \alpha)$ with $\alpha(0) := (0, \{0,1,2\}), \alpha(2) := (1, \{0,1,2\}), \alpha(1) = (0, \{1,2,3\}), \alpha(3) = (1, \{1,2,3\})$. It is easy to check that $\nabla = \{(0,1),(1,0),(2,3),(3,2)\} \cup \Delta_\mathcal{A}$. Suppose that $(0,1) \in \sim$. Then there is $(0,M) \in 2 \times \mathbb{P}_{\leq 3} \sim$ with $\pi_1[M] = (0,1,2)$ and $\pi_2[M] = (1,2,3)$. Hence $M = \{(0,1),(1,1),(2,2),(2,3)\}$. This is a contradiction, because $|M| = 4$. Similarly $(2,3) \notin \sim$. We obtain $\sim = \Delta \neq \nabla$.

The following lemma helps us to verify whether a congruence θ is the largest congruence. It suffices to check that the codomain of a factor is simple.

Lemma 3. *Subcoalgebras of simple coalgebras are simple. More generally, if $\varphi : \mathcal{A} \to \mathcal{C}$ is a homomorphism and $\mathcal{C}$ is simple, then $\ker \varphi = \nabla_\mathcal{A}$.*

Proof. Let $\iota : \mathcal{U} \hookrightarrow \mathcal{C}$ be the inclusion morphism, and assume that θ is a congruence on $\mathcal{U}$. The pushout of ι with π_θ yields a homomorphism $\psi : \mathcal{C} \to \mathcal{P}$. Since $\mathcal{C}$ is simple, ψ is injective. It follows that π_θ is injective. This means that $\nabla_\mathcal{U} = \Delta_\mathcal{U}$. Next, let $\varphi : \mathcal{A} \to \mathcal{C}$ be a homomorphism and $\mathcal{C}$ simple. The image of $\mathcal{A}$ under φ is a subcoalgebra of $\mathcal{C}$, which, as we have just seen, must be simple. Thus we may as well assume that φ is epi. Obviously, $\ker \varphi \subseteq \nabla_\mathcal{A}$, so there is a homomorphism $\psi : \mathcal{C} \to \mathcal{A}/\nabla_\mathcal{A}$ with $\pi_{\nabla_\mathcal{A}} = \psi \circ \varphi$. So ψ is surjective, and injective, since $\mathcal{C}$ is simple. It follows that ψ is an isomorphism, whence $\ker \varphi = \ker \pi_{\nabla_\mathcal{A}} = \nabla_\mathcal{A}$.

The following lemma relates observational equivalence *between* two coalgebras to observational equivalence *on* their sum:

Lemma 4. [14] $(a,b) \in \nabla_{\mathcal{A},\mathcal{B}} \iff (e_\mathcal{A}(a), e_\mathcal{B}(b)) \in \nabla_{\mathcal{A}+\mathcal{B}}$.

Proof.

$$(a, b) \in \nabla_{A,B} \iff \exists C, \varphi : A \to C, \psi : B \to C. \varphi(a) = \psi(b)$$
$$\iff \exists C, \chi : A + B \to C, \chi(e_A(a)) = \chi(e_B(b))$$
$$\iff (e_A(a), e_B(b)) \in \nabla_{A+B}.$$

We need to extend Lemma 3 to the case of a cospan of two homomorphisms:

Lemma 5. *If $\varphi : A \to C$ and $\psi : B \to C$ are homomorphism and C is simple, then $\ker(\varphi, \psi) = \nabla_{A,B}$.*

Proof. Consider $[\varphi, \psi] : A + B \to C$ with $\varphi = [\varphi, \psi] \circ e_A$ and $\psi = [\varphi, \psi] \circ e_B$. By Lemma 3, we obtain $\ker[\varphi, \psi] = \nabla_{A+B}$, so with this and Lemma 4 we calculate:

$$\varphi(a) = \psi(b) \iff [\varphi, \psi] \circ e_A(a) = [\varphi, \psi] \circ e_B(b)$$
$$\iff (e_A(a), e_B(b)) \in \ker[\varphi, \psi]$$
$$\iff (e_A(a), e_B(b)) \in \nabla_{A+B}$$
$$\iff (a, b) \in \nabla_{A,B}.$$

The following construction will be used in various places. It can be used to change the structure map α of a coalgebra while at the same time preserving other important properties as seen in the ensuing lemma:

Definition 9. *Given a coalgebra $A = (A, \alpha)$, element $x_0 \in A$ and subset $U \subseteq A$. We define a new coalgebra $A_{x_0}^U := (A, \bar{\alpha})$ on the same base set by constantly mapping all elements of U to $\alpha(x_0)$ and retaining α on all other elements, i.e.:*

$$\bar{\alpha}(x) := \begin{cases} \alpha(x_0) & \text{if } x \in U \\ \alpha(x) & \text{else.} \end{cases}$$

Lemma 6. *Let $\varphi : A \to C$ be a homomorphism and $U \subseteq [x_0] \ker \varphi$ for some $x_0 \in A$. Then the map $\varphi : A \to C$ is also a homomorphism $\varphi : \bar{A} \to C$ where $\bar{A} := A_{x_0}^U$. Moreover, $\nabla_A = \nabla_{\bar{A}}$.*

Proof. If $x \in U$, then $(F\varphi \circ \bar{\alpha})(x) = (F\varphi)(\alpha_A(x_0)) = \alpha_C(\varphi(x_0)) = \alpha_C(\varphi(x))$. If $x \notin U$, nothing has changed, so φ keeps being a homomorphism. With $C = A/\nabla_A$ and Lemma 3, we obtain $\nabla_A = \ker \varphi = \nabla_{\bar{A}}$.

3.2   Restricting Bisimulations

For the following we assume that $R \subseteq A_1 \times A_2$ is a bisimulation between coalgebras $A_1 = (A, \alpha_1)$ and $A_2 = (A, \alpha_2)$, and $U_i \leq A_i$ are subcoalgebras for $i = 1, 2$.

Definition 10. *R restricts to $U := U_1 \times U_2$, if $R \upharpoonright U := R \cap (U_1 \times U_2)$ is a bisimulation between U_1 and U_2.*

Without any further assumption, bisimulations will not necessarily restrict to subcoalgebras. In fact in [6] it was shown that all bisimulations restrict to all subcoalgebras, globally throughout Set_F, if and only if F preserves preimages. In spite of this, here we can identify conditions on R and on the U_i that guarantee that R restricts to $U_1 \times U_2$ without any condition on F. The main result of this section is the following theorem and its applications:

Theorem 1. *If there exist* $\kappa_i : A_i \to U_i$ *left inverses to the inclusion maps, satisfying for all* $a_1 \in A_1, a_2 \in A_2$:

$$(a_1, a_2) \in R \implies (\kappa_1 a_1, \kappa_2 a_2) \in R$$

then R *restricts to* U.

Instantiating the existential quantifier in this theorem with particularly natural left inverses to the inclusion maps, we shall obtain the following, easy-to-apply criterion (Here $R[U]$ denotes $\{y \mid \exists x \in U.(x,y) \in R\}$ and R^- is the converse relation to R):

Theorem 2. *If* $R[U_1] \subseteq U_2$ *and* $R^-[U_2] \subseteq U_1$ *then* R *restricts to a bisimulation between* $\mathcal{U}_1$ *and* $\mathcal{U}_2$.

The following further specialization with $\mathcal{U} \leq \mathcal{A}$ and R a bisimulation on $\mathcal{A}$ will likely be the most useful one:

Corollary 1. *A bisimulation* R *on* $\mathcal{A}$ *restricts to the subcoalgebra* $\mathcal{U} \leq \mathcal{A}$, *provided that* $R[U] \subseteq U$ *and* $R^-[U] \subseteq U$.

Given an epimorphism $\varphi : \mathcal{A} \twoheadrightarrow \mathcal{B}$, the largest bisimulation contained in its kernel reveals, whether φ is a regular epi in the category Set_F or not. We start with $[\ker \varphi]$, the largest bisimulation contained in the kernel of φ. The criterion found in [8] is, that the transitive hull of $[\ker \varphi]$ should be all of $\ker \varphi$. Expressed as a formula this is: $[\ker \varphi]^* = \ker \varphi$. Studying this further, we show the following result, which turns out to be another corollary:

Corollary 2. *From* $\varphi_i : \mathcal{A}_i \to \mathcal{B}_i$ *construct* $\varphi_1 + \varphi_2 : \mathcal{A} \to \mathcal{B}$ *with* $\mathcal{A} := \mathcal{A}_1 + \mathcal{A}_2$ *and* $\mathcal{B} := \mathcal{B}_1 + \mathcal{B}_2$. *Consider* $\ker \varphi_i$ *as subsets of* A, *then* $[\ker(\varphi_1 + \varphi_2)] = [\ker\varphi_1] \cup [\ker\varphi_2]$.

To see why this follows from Corollary 1, choose $\mathcal{U}_i := \mathcal{A}_i \leq \mathcal{A}$: If $R \subseteq \ker(\varphi_1 + \varphi_2)$ is a bisimulation, then $R[A_i] \subseteq \ker(\varphi_1 + \varphi_2)[A_i] = A_i$ and symmetrically $R^-[A_i] \subseteq A_i$. Thus R restricts to $A_i \subseteq A_1 + A_2$. This proves that $[\ker(\varphi_1 + \varphi_2)] \subseteq [\ker\varphi_1] \cup [\ker\varphi_2]$. The other direction is trivial, since $\ker\varphi_i \subseteq \ker(\varphi_1 + \varphi_2)$, hence $[\ker\varphi_i] \subseteq [\ker(\varphi_1 + \varphi_2)]$.

We now turn to the proof of Theorem 1:

Proof (of Theorems 1 and 2). Consider the following diagram, where we use ι, possibly with indices, to denote inclusion maps and similarly π_i or $\bar{\pi}_i$ to denote projection maps to the i-th component, for $i = 1, 2$. The κ_i are the mentioned left inverses to the ι_i and $\kappa := \kappa_1 \times \kappa_2$. The condition of the theorem guarantees that $\kappa's$ codomain is indeed $R \restriction U$, and by definition, we have:

$$\bar{\pi}_i \circ \kappa = \kappa_i \circ \pi_i.$$

Simply define

$$\bar{\rho} := F(\kappa) \circ \rho \circ \iota$$

and chase the following diagram:

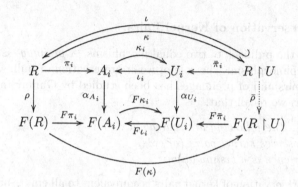

$$
\begin{aligned}
F\bar{\pi}_i \circ \bar{\rho} &= F\bar{\pi}_i \circ F\kappa \circ \rho \circ \iota \\
&= F\kappa_i \circ F\pi_i \circ \rho \circ \iota \\
&= F\kappa_i \circ \alpha_{A_i} \circ \pi_i \circ \iota \\
&= F\kappa_i \circ \alpha_{A_i} \circ \iota_i \circ \bar{\pi}_i \\
&= F\kappa_i \circ F\iota_i \circ \alpha_{U_i} \circ \bar{\pi}_i \\
&= F id_{A_i} \circ \alpha_{U_i} \circ \bar{\pi}_i \\
&= \alpha_{U_i} \circ \bar{\pi}_i
\end{aligned}
$$

For the proof of Theorem 2, we note that the case where $R \restriction U$ is empty becomes trivial. Otherwise fix any pair $(u_1, u_2) \in R \restriction U$ and define the following inverses κ_i to the inclusions $\iota_i : U_i \to A_i$:

$$\kappa_i(a) := if\,(a \in U_i)\,then\,a\,else\,u_i.$$

Given $(a_1, a_2) \in R$, the conditions $R[U_1] \subseteq U_2$ and $R^{-}[U_2] \subseteq U_1$ guarantee that $a_1 \in U_1 \iff a_2 \in U_2$. By definition of κ_i then, either $(\kappa_1 a_1, \kappa_2 a_2) = (a_1, a_2)$ or $(\kappa_1 a_1, \kappa_2 a_2) = (u_1, u_2)$.

4 Relationships Between Bisimilarity and Observational Equivalence

Bisimilarity and observational equivalence need not be distinguished in many classical systems, such as e.g. in Kripke structures. In particular, the famous

Hennessy-Milner theorem [12] relates bisimilarity with logical equivalence. However, the generalization of this result to coalgebras, as given by Pattinson [17], demonstrates that logical equivalence should rather be related to observational equivalence. In hindsight, this appears obvious, as bisimilarity need not be transitive, whereas observational equivalence always is transitive, and clearly so is logical equivalence. Thus in the case of the Hennessy-Milner theorem, "luckily" both notions agreed. The reason for this is, that the type functor for Kripke-structures, is rather well behaved, in that it weakly preserves kernel pairs, even arbitrary pullbacks.

4.1   Weak Preservation of Kernel Pairs

A *kernel pair* is the pullback of two equal morphisms. A *preimage* is a pullback with a regular epimorphism. The role of weak preservation of pullbacks in general, of kernel pairs and of preimages has been studied by Gumm and Schröder [8]. In particular, we recall that:

Theorem 3 [8]. *The following are equivalent :*
(1) F weakly preserves kernel pairs (of epis).
(2) Every congruence is a bisimulation.

Thus weak preservation of kernel pairs is equivalent to all epis φ being *strictly regular*, meaning that $[ker\ \varphi] = ker\varphi$. We shall show below that this is equivalent to all epis just being regular, i.e. $[ker\ \varphi]^* = ker\ \varphi$.

A joint result of the second author and C. Henkel, to be found in the latter's Master thesis [11], also turns out to be useful:

Theorem 4 [11]. *The following are equivalent:*

1. *F weakly preserves kernel pairs*
2. *F weakly preserves pullbacks of epimorphisms.*

In [8], furthermore, the implications (1) $\implies$ (2) $\implies$ (3) have been shown for the following properties:

1. F weakly preserves kernel pairs
2. every epi is regular epi
3. every mono is regular mono.

In [20] it was further claimed that (3) $\Rightarrow$ (1), rendering all three properties equivalent. Unfortunately, the proof contained a gap, so until today, (3) $\Rightarrow$ (1) remains open. Nevertheless, we are able to close the loop at (2) $\Rightarrow$ (1) in the following Theorem 5, so (1) and (2) indeed turn out to be equivalent. The following innocuous lemma holds a key for the proof.

Lemma 7. *Let φ be the coequalizer of $\psi_1, \psi_2 : Q \to A$ and let $x \in A$. If there exists $y \neq x$ with $\varphi(x) = \varphi(y)$ then for some q in Q either $\psi_1(q) = x \neq \psi_2(q)$ or $\psi_2(q) = x \neq \psi_1(q)$.*

Proof. In the category *Set*, the coequalizer φ of ψ_1 and ψ_2 is obtained by factoring A by the equivalence relation generated by the relation $R = \{(\psi_1(q), \psi_2(q)) \mid q \in Q\}$. Thus $\ker \varphi = Eq(R) = (\Delta_A \cup R \cup R^-)^*$. If $x \ker y$ for some $y \neq x$, there must be therefore be at least some $y' \neq x$ with xRy' or xR^-y'.

The following theorem was obtained in the master thesis of the first author:

Theorem 5 [24]. *F weakly preserves kernel pairs iff every epi is regular epi.*

Proof "$\Rightarrow$" is from [8].

"$\Leftarrow$": By Theorem 3 it suffices to show, that F preserves weak kernel pairs of epis. Let $\varphi : A \twoheadrightarrow C$ be a surjective map, $\tilde{a}, \tilde{b} \in F(A)$ and $\tilde{c} \in F(C)$ with $(F\varphi)\tilde{a} = \tilde{c} = (F\varphi)\tilde{b}$. We need to find $\tilde{p} \in F(\ker\varphi)$ with $(F\pi_1)\tilde{p} = \tilde{a}$ and $(F\pi_2)\tilde{p} = \tilde{b}$. In case φ is injective, then the pullback of φ is an intersection. We have assumed in this paper that functors preserve intersections. Otherwise there are $x, y \in A$ with $x \neq y$ and $\varphi(x) = \varphi(y)$. We define structure maps $\alpha_A : A \longrightarrow FA$ and $\alpha_C : C \longrightarrow FC$ as follows:

$$\alpha_A(z) := \begin{cases} \tilde{a} & if\ z = x \\ \tilde{b} & otherwise \end{cases}$$

and

$$\alpha_C(z) := \tilde{c}.$$

Clearly φ is a surjective homomorphisms, since $(\alpha_C \circ \varphi)(z) = \alpha_C(\varphi(z)) = \tilde{c} = (F\varphi)(\alpha_A(z)) = ((F\varphi) \circ \alpha_A)(z)$ for every $z \in A$. By assumption φ is regular epi, so it is the coequalizer of two homomorphisms ψ_1 and $\psi_2 \colon Q \longrightarrow A$. By Lemma 7 there is some $q \in Q$ with $(\psi_1(q) = x$ and $\psi_2(q) \neq x)$ or $(\psi_1(q) \neq x$ and $\psi_2(q) = x)$. Since φ is the coequalizer of ψ_1 and ψ_2, (Q, ψ_1, ψ_2) becomes a competitor for the pullback $(\ker\varphi, \pi_1, \pi_2)$ in *Set*. Thus there is a map m with $\pi_1 \circ m = \psi_1$ and $\pi_2 \circ m = \psi_2$.

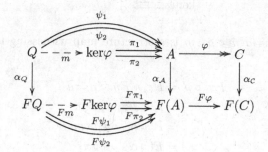

Since $\psi_1(q) = x$ and $\psi_2(q) \neq x$ it follows that $\alpha_A(\psi_1(q)) = \tilde{a}$ and $\alpha_A(\psi_2(q)) = \tilde{b}$. Then $F\psi_1 \circ \alpha_Q(q) = \tilde{a}$ and $F\psi_2 \circ \alpha_Q(q) = \tilde{b}$ because ψ_1 and ψ_2 are homomorphisms. From property of functor this implies: $F\pi_1 \circ Fm \circ \alpha_Q(q) = \tilde{a}$ and $F\pi_2 \circ Fm \circ \alpha_Q(q) = \tilde{b}$. Then $F\pi_1(Fm \circ \alpha_Q(q)) = \tilde{a}$ and $F\pi_2(Fm \circ \alpha_Q(q)) = \tilde{b}$. This shows the existence of some $\tilde{p} \in F\ker\varphi$ with $(F\pi_1)\tilde{p} = \tilde{a}$ and $(F\pi_2)\tilde{p} = \tilde{b}$.

Lemma 8. *If an epimorphism $\varphi : \mathcal{U} \twoheadrightarrow \mathcal{B}$ can be factored as $\varphi = \psi \circ e$ with $e : \mathcal{U} \hookrightarrow \mathcal{A}$ regular mono and $\psi : \mathcal{A} \twoheadrightarrow \mathcal{B}$ a strictly regular epi, then φ is strictly regular epi.*

Proof. If $U = \emptyset$ then $\ker \varphi = \emptyset$ is a bisimulation, so we may assume that $U \neq \emptyset$. Since φ is epi, then φ is surjective and the axiom of choice yields a right inverse map r for φ. With its help we obtain a map $l : A \longrightarrow U$ as

$$l(x) := \begin{cases} u & \text{if } e(u) = x \\ r(\psi(x)) & \text{otherwise.} \end{cases}$$

It is easy to check that $\varphi \circ l = \psi$.

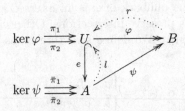

$(\ker \varphi, e \circ \pi_1, e \circ \pi_2)$ is a competitor to the pullback $\ker \psi$. This yields a map $m : \ker \varphi \longrightarrow \ker \psi$ with $\bar{\pi}_i \circ m = e \circ \pi_i$ for $i = 1, 2$. Similarly, $(\ker \psi, l \circ \bar{\pi}_1, l \circ \bar{\pi}_2)$ is a competitor to the pullback $\ker \varphi$, providing another map $\bar{m} : \ker \psi \longrightarrow \ker \varphi$ with $l \circ \bar{\pi}_i = \pi_i \circ \bar{m}$. Now let us assume that $\ker \psi$ is a bisimulation, then there exists a structure map $\rho : \ker \psi \longrightarrow F \ker \psi$ such that $\alpha_{\mathcal{A}} \circ \bar{\pi}_i = F \bar{\pi}_i \circ \rho$ for $i = 1, 2$.

$$
\begin{array}{ccc}
\ker \varphi & \underset{\pi_2}{\overset{\pi_1}{\rightrightarrows}} & U \\
m \big\downarrow \big\uparrow \bar{m} & & e \big\downarrow \big\uparrow l \\
\ker \psi & \underset{\pi_2}{\overset{\bar{\pi}_1}{\rightrightarrows}} & A
\end{array}
$$

We claim that $\rho' := F\bar{m} \circ \rho \circ m$ is a structure map witnessing that $\ker \varphi$ is a bisimulation.

$$
\begin{aligned}
F\pi_i \circ \rho' &= F\pi_i \circ F\bar{m} \circ \rho \circ m \\
&= Fl \circ F\bar{\pi}_i \circ \rho \circ m \\
&= Fl \circ \alpha_{\mathcal{A}} \circ \bar{\pi}_i \circ m \\
&= Fl \circ \alpha_{\mathcal{A}} \circ e \circ \pi_i \\
&= Fl \circ Fe \circ \alpha_{\mathcal{U}} \circ \pi_i \\
&= \alpha_{\mathcal{U}} \circ \pi_i
\end{aligned}
$$

Lemma 9. *Let* $\varphi : \mathcal{A} \twoheadrightarrow \mathcal{B}$ *be an epimorphism.* φ *is strictly regular epi iff* $[\varphi, id_B]$ *is strictly regular epi.*

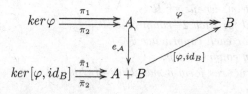

Proof. "⇒" : $\ker [\varphi, id_B] = \ker \varphi \cup \operatorname{graph} \varphi \cup (\operatorname{graph} \varphi)^- \cup \Delta_{A+B}$. Since $\ker \varphi$ is bisimulation, $\ker [\varphi, id_B]$ is bisimulation, too.

"⇐" $[\varphi, id_B]$ is regular epi by Lemma 1 and $e_{\mathcal{A}}$ is regular mono since it is injective. The rest follows from Lemma 8.

Proposition 1. *Let* $\varphi : \mathcal{A} \longrightarrow \mathcal{B}$ *and* $\psi : \mathcal{C} \longrightarrow \mathcal{D}$ *be two homomorphisms.* φ *and* ψ *are regular epi iff* $\varphi + \psi$ *is regular epi.*

Proof. "⇒" This holds in every cocomplete category [2].

"⇐" By Corollary 2 $[\ker(\varphi + \psi)]_{\mathcal{A}+\mathcal{C}} = [\ker \varphi]_{\mathcal{A}} \cup [\ker \psi]_{\mathcal{C}}$. By Lemma 2 $\ker \varphi \cup \ker \psi = \ker(\varphi + \psi) = [\ker(\varphi + \psi)]^*_{\mathcal{A}+\mathcal{C}} = [\ker \varphi]^*_{\mathcal{A}} \cup [\ker \psi]^*_{\mathcal{C}}$. Then $\ker \varphi = [\ker \varphi]^*_{\mathcal{A}}$ and $\ker \psi = [\ker \psi]^*_{\mathcal{A}}$. From Lemma 2 it then follows that φ and ψ are regular epi.

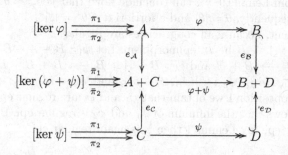

We define an order over all homomorphisms with the same domain by $\varphi \leq \psi :\Leftrightarrow \ker \varphi \subseteq \ker \psi$

Lemma 10. *For any epimorphism* $\varphi : A \twoheadrightarrow B$, *the infimum of* $id_B + [\varphi, id_B]$ *and* $[id_B, \varphi] + id_B$ *is* $id_B + \varphi + id_B$.

Proof. Mark the two isomorphic copies of B as B_1 and B_2 and abbreviate $\psi_1 := id_{B_1} + [\varphi, id_{B_2}]$, $\psi_2 := [id_{B_1}, \varphi] + id_{B_2}$ and $\psi := id_{B_1} + \varphi + id_{B_2}$. We need to check that $\ker \psi = \ker \psi_1 \cap \ker \psi_2$. Given $x \neq y$, then

$$(x, y) \in \ker \psi_1 \cap \ker \psi_2 \iff \exists (a, a') \in \ker \varphi . \, x = e_A(a) \wedge y = e_A(a')$$
$$\iff (x, y) \in \ker \psi.$$

Theorem 6. *The following are equivalent:*

1. *F preserves weak kernel pairs,*
2. *R^d is a bisimulation, whenever R is,*
3. *$[\theta]$ is difunctional for each $2-$congruence θ,*
4. *$[\theta]$ is transitive for each congruence θ,*
5. *$[\theta]^* = \theta$ for every congruence θ,*
6. *the regular congruences form a sublattice of $Cong(\mathcal{A})$.*

Proof. (1)$\Rightarrow$(2) Let R be a bisimulation between two coalgebras $\mathcal{A}_1$ and $\mathcal{A}_2$. We factor the projections $\pi_i^R : R \to \mathcal{A}_i$ into $\iota_i \circ \bar{\pi}_i^R$ as an epi followed by a regular mono and produce the pushout (p_1, p_2) of $(\bar{\pi}_1^R, \bar{\pi}_2^R)$. Then p_1 and p_2 are epi and $R^d = \ker(p_1, p_2)$. It follows from Theorem 4 that R^d is a bisimulation.

(2) $\Rightarrow$ (3) is evident

(3) $\Rightarrow$ (4) If θ is a congruence, $[\theta]$ also must be reflexive and symmetric. Considered as a $2-$congruence between $\mathcal{A}$ and itself, the hypothesis yields that $[\theta]$ is difunctional, too. Difunctionality with reflexivity and symmetry implies transitivity.

(4) $\Rightarrow$ (5) $\theta = \ker\varphi$ for some epimorphism $\varphi : \mathcal{A} \twoheadrightarrow \mathcal{B}$. By Lemma 1, $\ker[\varphi, id_\mathcal{B}]$ is a regular congruence, so $\ker[\varphi, id_\mathcal{B}] = [\ker[\varphi, id_\mathcal{B}]]^*$. By hypothesis, $[\ker[\varphi, id_\mathcal{B}]]$ is transitive, so in fact $\ker[\varphi, id_\mathcal{B}] = [\ker[\varphi, id_\mathcal{B}]]$, witnessing that $\ker[\varphi, id_\mathcal{B}]$ is a bisimulation. From Lemma 9, we can conclude now, that $\ker\varphi = \theta$ is a bisimulation, too. Consequently, $\theta = [\theta]$ and a fortiori $\theta = \theta^* = [\theta]^*$.

(5) $\Rightarrow$ (6) is evident, because all congruences are regular.

(6) $\Rightarrow$ (1) Let $\varphi : A \twoheadrightarrow B$ be an epimorphism. Let $\psi_1 : B + A + B \longrightarrow B + B$, $\psi_2 : B + A + B \longrightarrow B + B$ and $\psi : B + A + B \longrightarrow B + B + B$ be defined as $\psi_1 := id_B + [\varphi, id_B]$, $\psi_2 := [id_B, \varphi] + id_B$ and $\psi := id_B + \varphi + id_B$. From Lemma 1 and Proposition 1 we obtain that ψ_1 and ψ_2 are regular epis. From the hypothesis it follows that the infimum of ψ_1 and ψ_2 is regular epi. By Lemma 10 then ψ is regular epi. By Lemma 1 then φ is again regular epi.

4.2 Transitivity of $\sim$

The next theorem clarifies which properties assure the transitivity of bisimilarity. The equivalence (1) $\Leftrightarrow$ (3) appears in [8] under the additional assumption that F should preserve preimages. Its proof will need a simple lemma allowing us to extend bisimulations in certain situations.

Theorem 7. *The following are equivalent:*
(1) $\sim_\mathcal{A} = \nabla_\mathcal{A}$ for each $F-$coalgebra $\mathcal{A}$.
(2) $\sim_\mathcal{A}^* = \nabla_\mathcal{A}$ for each $F-$coalgebra $\mathcal{A}$.
(3) $\sim_\mathcal{A}$ is transitive for each $F-$coalgebra $\mathcal{A}$:

Proof. (1) $\Rightarrow$ (2) and (1) $\Rightarrow$ (3) are evident, since $\nabla_\mathcal{A}$ is transitive.

(2) $\Rightarrow$ (1): Given $\mathcal{A} = (A, \alpha)$, we have to show that $\nabla_\mathcal{A}$ is a bisimulation. Hence for arbitrary $(x_0, y_0) \in \nabla_\mathcal{A}$ we need to find some $p \in F\nabla_\mathcal{A}$ with $F\pi_1(p) = \alpha(x_0)$ and $F\pi_2(p) = \alpha(y_0)$ where $\pi_1, \pi_2 : \nabla_\mathcal{A} \to A$ are the projection maps.

If $x_0 = y_0$ then $(x_0, y_0) \in \Delta_A$ which is already a bisimulation contained in ∇_A. But for each bisimulation S contained in ∇_A with $(x_0, y_0) \in S$ there is already some $q \in F(S)$ with $F\pi_A^S(q) = \alpha(x_0)$ and $F\pi_A^S(q) = \alpha(x_0)$. The inclusion map $\iota_S^\nabla : S \hookrightarrow \nabla_A$ yields an element $p := F\iota_S^\nabla(q) \in F(\nabla_A)$ for which $F\pi_1(p) = F\pi_1(F\iota_S^\nabla(q) = F(\pi_1 \circ \iota_S^\nabla)(q) = F(\pi_1^S)(q) = \alpha(x_0)$, and likewise $F\pi_2(p) = \alpha(y_0)$.

If $x_0 \neq y_0$, we consider $\bar{A} := A_{x_0}^U$ (see Definition 9), where we choose $U := \{x \mid x_0 \nabla_A x \neq y_0\}$. Invoking Lemma 6 with $\pi_{\nabla_A} : A \to A/\nabla_A$ and recalling that A/∇_A is simple, we obtain $\nabla_{\bar{A}} = \nabla_A =: \nabla$. By assumption (2), there is a bisimulation S on $\bar{A}$ with $S^\star = \nabla$, so in particular $x_0 S^\star y_0$. As $x_0 \neq y_0$, there is some $z \neq y_0$ with $x_0 S^\star z S y_0$ and a fortiori $x_0 \nabla z S y_0$. Since S is a bisimulation on $\bar{A}$, there is some $q \in F(S)$ with $F\pi_1(q) = \bar{\alpha}(z) = \alpha(x_0)$ and $F\pi_2(q) = \bar{\alpha}(y_0) = \alpha(y_0)$.

(3) $\Rightarrow$ (1):

$$\nabla \underset{\pi_2}{\overset{\pi_1}{\rightrightarrows}} A \xrightarrow{\ \pi_\nabla\ } A/\nabla$$

$$e_A \downarrow \qquad \nearrow [\pi_\nabla, id_{A/\nabla}]$$

$$\ker[\pi_\nabla, id_{A/\nabla}] \underset{\bar{\pi}_2}{\overset{\bar{\pi}_1}{\rightrightarrows}} A + A/\nabla$$

By Lemma 1 $[\pi_\nabla, id_{A/\nabla}]$ is regular epi. From Theorem 2 we obtain $\sim_{A+A\nabla}^\star = \nabla_{A+A/\nabla}$. Then

$$[\ker[\pi_\nabla, id_{A\nabla}]] = [\nabla_{A+A/\nabla}] \qquad \text{lemma 3}$$
$$= \sim_{A+A\nabla}$$
$$= \sim_{A+A\nabla}^\star \qquad \text{hypothesis}$$
$$= \ker[\pi_\nabla, id_{A/\nabla}].$$

Therefore, $\ker[\pi_\nabla, id_{A/\nabla}]$ is a bisimulation, and by Lemma 9, ∇ is a bisimulation.

For image finite Kripke structures A, i.e. coalgebras for $D \times \mathbb{P}_{fin}$ where D is a fixed output set and $\mathbb{P}_{fin}$ the finite-powerset functor, it is well known that $\sim_A = \nabla_A$. However, considered as a coalgebra of $D \times \mathbb{P}_{\leq k}$ where $\mathbb{P}_{\leq k} X := \{U \subseteq X \mid |U| \leq k\}$ the same Kripke structure may fail to satisfy $\sim_A = \nabla_A$. The fact that for $k \geq 3$ the functor $\mathbb{P}_{\leq k}$, defined as a subfunctor of $\mathbb{P}$, does not preserve weak kernel pairs was noticed in [8].

4.3 Difunctionality of $\sim$

In an attempt to generalize the results of the previous subsection to relations between two coalgebras, we consider the following conditions:

1. For all F−coalgebras A, B: $\sim_{A,B} = \nabla_{A,B}$
2. For all F−coalgebras A, B: $\sim_{A,B}^d = \nabla_{A,B}$

3. For all F−coalgebras $\mathcal{A}, \mathcal{B}$: $\sim_{A,B}$ is difunctional.

We will show in the next section that the direction (2) $\Rightarrow$ (1) holds more generally, yet the implication (3) $\Rightarrow$ (1) fails in general. We provide a counterexample (Example 2) in the next section. We will use the following characterization of preimages preservation.

Theorem 8 [6]. *The following are equivalent:*

(1) F preserves preimages
(2) If $\mathcal{U}, \mathcal{V}$ are subcoalgebras of $\mathcal{A}, \mathcal{B}$ then bisimulations between $\mathcal{U}$ and $\mathcal{V}$ are just the restrictions to $U \times V$ of bisimulations between $\mathcal{A}$ and $\mathcal{B}$.

Under the assumption that F should preserves preimages we will see that difunctionality and transitivity of bisimulations are equivalent.

Theorem 9. *If F preserves preimages, then the following are equivalent:*

(1) For all F−coalgebras $\mathcal{A}, \mathcal{B}$: $\sim_{A,B} = \nabla_{A,B}$
(2) For all F−coalgebras $\mathcal{A}, \mathcal{B}$: $\sim_{A,B}$ is difunctional.
(3) For all F−coalgebras $\mathcal{A}$: $\sim_{A} = \nabla_{A}$.

Proof. (1)$\Rightarrow$ (2) is evident because $\nabla_{A,B}$ is difunctional.
(2) $\Rightarrow$ (3) From the hypothesis it follows that $\sim_{A}$ is transitive for all F−coalgebras $\mathcal{A}$, because $\sim_{A}$ is reflexive and symmetric. By Theorem 7 we obtain that for all F−coalgebras $\mathcal{A}$: $\sim_{A} = \nabla_{A}$.
(3) $\Rightarrow$ (1) Let $\mathcal{A} = (A, \alpha)$ and $\mathcal{B} = (B, \beta)$ two coalgebra. We have to show $\sim_{A,B} = \nabla_{A,B}$.

$$
\begin{aligned}
(a, b) \in \nabla_{A,B} &\iff (e_A(a), e_B(b)) \in \nabla_{A+B} && \text{by lemma 4} \\
&\iff (e_A(a), e_B(b)) \in \sim_{A+B} && \text{by hypothesis} \\
&\iff (a, b) \sim_{A,B} && \text{by theorem 8.}
\end{aligned}
$$

4.4   Relation Liftings

An alternative definition of bisimulation was given by C. Hermida and B. Jacobs in [13]. The idea is to define a bisimulation R as a relation $R \subseteq A \times B$ satisfying:

$$ x\, R\, y \implies \alpha(x)\, \bar{F}(R)\, \alpha(y) $$

where $\bar{F}$ is the "lifting" of the relation $R \subseteq A \times B$ to a relation $\bar{F}(R) \subseteq F(A) \times F(B)$, known as Barr-extension:

$$ \bar{F}(R) := \{(F\pi_A^R(u), F\pi_B^R(u)) \mid u \in F(R)\}. $$

Choosing the Barr-extension may be one particular method for extending a relation between A and B to a relation between $F(A)$ and $F(B)$, but there might likely be others, so we define:

Definition 11. *A relation lifting L is a transformation sending every relation $R \subseteq A \times A$ into a relation $L(R) \subseteq FA \times FB$. It is called monotonic, if $R \subseteq S$ implies $L(R) \subseteq L(S)$. For a given relation lifting L and coalgebras $\mathcal{A} = (A, \alpha_{\mathcal{A}})$ and $\mathcal{B} = (B, \alpha_{\mathcal{B}})$, an L-simulation is a relation $R \subseteq A \times B$ such*

$$x \, R \, y \implies \alpha(x) \, L(R) \, \alpha(y).$$

Diagrammatically, R is an L-simulation if and only if the map sending (x, y) to $(\alpha_{\mathcal{A}}(x), \alpha_{\mathcal{B}}(y))$ factors through $L(R)$, which is the same as saying that there exists a map m such that the following diagram commutes:

$$
\begin{array}{ccccc}
A & \xleftarrow{\;\pi_A\;} & R & \xrightarrow{\;\pi_B\;} & B \\
{\scriptstyle \alpha}\downarrow & \circ & {\scriptstyle m}\downarrow & \circ & \downarrow{\scriptstyle \alpha} \\
F(A) & \xleftarrow{\pi_{F(A)}} & LR & \xrightarrow{\pi_{F(B)}} & F(B)
\end{array}
$$

Thus a Hermida-Jacobs-bisimulation is the same as an L-simulation where $L = \bar{F}$. Observe, that categorically, $\bar{F}(R)$ arises by applying the functor F to the source $\pi_A^R : R \to A$ and $\pi_B^R : R \to B$, then factoring the resulting source $F\pi_A^R : F(R) \to F(A)$ and $F\pi_B^R : F(R) \to F(B)$, into an epi q followed by a mono-source:

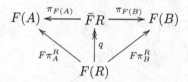

$$
F(A) \xleftarrow{\pi_{F(A)}} \bar{F}R \xrightarrow{\pi_{F(B)}} F(B)
$$

This diagram can be readily used to demonstrate the well known fact that our earlier definition of bisimulation agrees with that of Hermida and Jacobs in the presence of the axiom of choice. When $L = \bar{F}$, we can paste it to the bottom line of the previous diagram in order to see that any bisimulation $\rho : R \to F(R)$ yields an $\bar{F}$-simulation and conversely, any choice of right inverse e for q yields a bisimulation $\rho := e \circ m$.

Thijs in [22] defined relators as relation liftings with additional properties and generalized the notion of coalgebraic simulation. In [15] Marti and Venema introduced further properties, in an attempt to achieve that L-similarity should capture observational equivalence. The union of all L-simulations between given coalgebra $\mathcal{A}$ and $\mathcal{B}$ is denoted by $\approx_{\mathcal{A},\mathcal{B}}^{L}$. If L is monotonic then $\approx_{\mathcal{A},\mathcal{B}}^{L}$ is again an L-simulation. For $L = \bar{F}$, of course, $\approx_{\mathcal{A},\mathcal{B}}^{L}$ agrees with $\sim_{\mathcal{A},\mathcal{B}}$.

In [15] it was also shown that there is no relation lifting L for the neighborhood functor which captures observational equivalence in the sense that $\approx_{\mathcal{A},\mathcal{B}}^{L} = \nabla_{\mathcal{A},\mathcal{B}}$ for all coalgebras $\mathcal{A}, \mathcal{B}$. In particular, for this functor there are coalgebras $\mathcal{A}$ and $\mathcal{B}$ such that $\sim_{\mathcal{A},\mathcal{B}} \neq \nabla_{\mathcal{A},\mathcal{B}}$.

Example 2. Consider the neighborhood functor 2^{2^-}. From Theorem 6 $\sim_{\mathcal{A},\mathcal{B}}$ is difunctional for all coalgebras $\mathcal{A}, \mathcal{B}$, since 2^{2^-} weakly preserves kernel pairs [10]. But $\sim_{\mathcal{A},\mathcal{B}} = \nabla_{\mathcal{A},\mathcal{B}}$ does not hold.

In [9] we have shown, that bisimulations can be enlarged as long as the structure maps are not affected in the following sense:

Proposition 2 [9]. *Let $\mathcal{A}_1$ and $\mathcal{A}_2$ be coalgebras with corresponding structure maps α_1 and α_2. Let $R \subseteq \mathcal{A}_1 \times \mathcal{A}_2$ be a bisimulation and R' an enlargement i.e. $R \subseteq R' \subseteq \ker \alpha_1 \circ R \circ \ker \alpha_2$. Then R' is also a bisimulation.*

A relation lifting L is called *extensible*, if for all coalgebras $\mathcal{A}_1$ and $\mathcal{A}_2$ the statement of the above proposition holds with "bisimulation" replaced by "L−simulation". It turns out, that this property precisely captures monotonicity, i.e.:

Proposition 3. *A relation lifting L is monotonic iff it is extensible.*

Proof. The proof of the *if*-direction closely follows, but is not identical to, the proof of Proposition 2 from [9]: R is a L−simulation, so there exists a map $\rho : R \to L(R)$ with $\alpha_i \circ \pi_i^R = \pi_i^{LR} \circ \rho$. Let $\iota : R \to R'$ be the inclusion map, then clearly $\pi_i^R = \pi_i^{R'} \circ \iota$. By assumption, we find for every $(x', y') \in R'$ a pair $(x, y) \in R$ such that $\alpha_1(x) = \alpha_1(x')$ and $\alpha_2(y) = \alpha_2(y')$. The axiom of choice provides for a map $\mu : R' \to R$ satisfying

$$\alpha_i \circ \pi_i^{R'} \circ \iota \circ \mu = \alpha_i \circ \pi_i^{R'}.$$

We now define $\rho' : R' \to L(R')$ by $\rho' := \subseteq \circ \rho \circ \mu$.

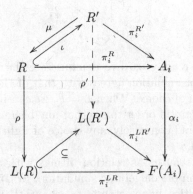

For the *only-if* direction, let $R \subseteq R' \subseteq A_1 \times A_2$ and $(\tilde{x}, \tilde{y}) \in LR$. We have to show $(\tilde{x}, \tilde{y}) \in LR'$. Since $(\tilde{x}, \tilde{y}) \in LR$ then R is a L−simulation between $(A_1, c_{\tilde{x}})$ and $(A_2, c_{\tilde{y}})$. By the hypothesis R' is also a L−simulation. Then $(\tilde{x}, \tilde{y}) \in LR'$.

With the help of this proposition, we can now somehow simplify the task of proving that $\approx^L_{\mathcal{A},\mathcal{B}}$ captures observational equivalence, provided that L is monotonic:

Theorem 10. *For a monotonic relation lifting L, the following are equivalent:*

1. For all coalgebras $\mathcal{A}, \mathcal{B}$: $\approx_{\mathcal{A},\mathcal{B}} = \nabla_{\mathcal{A},\mathcal{B}}$
2. For all coalgebras $\mathcal{A}, \mathcal{B}$: $\approx^{d}_{\mathcal{A},\mathcal{B}} = \nabla_{\mathcal{A},\mathcal{B}}$

Proof. (1) $\Rightarrow$ (2) is evident as $\nabla_{\mathcal{A},\mathcal{B}}$ is difunctional.
(2) $\Rightarrow$ (1): For any $\mathcal{A} = (A, \alpha)$ and $\mathcal{B} = (B, \beta)$, we have to show that $\nabla_{\mathcal{A},\mathcal{B}}$ is an L–simulation. Given $(x_0, y_0) \in \nabla_{\mathcal{A},\mathcal{B}}$, let $U := [x_0]\nabla_{\mathcal{A}}$ and $V := [y_0]\nabla_{\mathcal{B}}$, let $\varphi := \pi_{\nabla_{\mathcal{A}+\mathcal{B}}} \circ e_{\mathcal{A}}$ and $\psi := \pi_{\nabla_{\mathcal{A}+\mathcal{B}}} \circ e_{\mathcal{B}}$. Since $\mathcal{A} + \mathcal{B}/\nabla_{\mathcal{A}+\mathcal{B}}$ is simple, it follows from Lemma 5 that $\ker \varphi = \nabla_{\mathcal{A}}$ and $\ker \psi = \nabla_{\mathcal{B}}$. We define now two coalgebras $\bar{\mathcal{A}} := \mathcal{A}^{U}_{x_0}$ and $\bar{\mathcal{B}} := \mathcal{B}^{V}_{y_0}$ as in Definition 9. Since $U \subseteq [x_0]\ker \varphi$ and $V \subseteq [x_0]\ker \psi$, we can use Lemma 6 to see that both $\varphi : \bar{\mathcal{A}} \longrightarrow \mathcal{A} + \mathcal{B}/\nabla_{\mathcal{A}+\mathcal{B}}$ and $\psi : \bar{\mathcal{B}} \longrightarrow \mathcal{A} + \mathcal{B}/\nabla_{\mathcal{A}+\mathcal{B}}$ are homomorphisms to a simple coalgebra, so by Lemma 5, $\nabla_{\bar{\mathcal{A}},\bar{\mathcal{B}}} = \ker(\varphi, \psi) = \nabla_{\mathcal{A},\mathcal{B}}$. By hypothesis $(x_0, y_0) \in \approx^{d}_{\bar{\mathcal{A}},\bar{\mathcal{B}}}$, so there is $x_1 \in A$ with $x_1 \approx_{\bar{\mathcal{A}},\bar{\mathcal{B}}} y_0$. From $\approx^{d}_{\bar{\mathcal{A}},\bar{\mathcal{B}}} = \nabla_{\mathcal{A},\mathcal{B}}$ it follows that $\approx_{\bar{\mathcal{A}},\bar{\mathcal{B}}} \subseteq \nabla_{\mathcal{A},\mathcal{B}}$, so $(x_1, y_0) \in \nabla_{\mathcal{A},\mathcal{B}}$. Hence $\varphi(x_1) = \psi(y_0) = \varphi(x_0)$ and consequently $(x_0, x_1) \in \ker \varphi = \nabla_{\mathcal{A}}$. By construction of $\bar{\mathcal{A}}$ then $\bar{\alpha}(x_1) = \alpha(x_0) = \bar{\alpha}(x_0)$. With the help of Proposition 3 we obtain $x_0 \approx_{\bar{\mathcal{A}},\bar{\mathcal{B}}} y_0$. Hence $(\bar{\alpha}(x_0), \bar{\beta}(y_0)) \in L \approx_{\bar{\mathcal{A}},\bar{\mathcal{B}}}$ and finally $(\alpha(x_0), \beta(y_0)) \in L\nabla_{\mathcal{A},\mathcal{B}}$ since $\approx_{\bar{\mathcal{A}},\bar{\mathcal{B}}} \subseteq \nabla_{\mathcal{A},\mathcal{B}}$ and L is monotonic.

The Barr extension $\bar{F}$ is an example of a monotonic relation lifting, so we obtain:

Corollary 3. *The following are equivalent:*

1. For all coalgebras $\mathcal{A}, \mathcal{B}$: $\sim_{\mathcal{A},\mathcal{B}} = \nabla_{\mathcal{A},\mathcal{B}}$.
2. For all coalgebras $\mathcal{A}, \mathcal{B}$: $\sim^{d}_{\mathcal{A},\mathcal{B}} = \nabla_{\mathcal{A},\mathcal{B}}$.

5   Conclusion and Further Work

In this paper we exhibited conditions under which bisimulations restrict to sub-coalgebras without requiring the type functor to preserve preimages. Further, we have shown that if the transitive, resp. difunctional hull of bisimilarity covers observational equivalence then bisimilarity and observational equivalence agree. If bisimilarity is transitive for all F-coalgebras, then it agrees with observational equivalence.

A negative result is that difunctionality is not enough to cover observational equivalence between two coalgebras. Assuming preimage preservation for the type functor F, transitivity and difunctionality of bisimilarity are equivalent.

We also show that F weakly preserves kernel pairs if and only if every epi is regular epi. While it is known that this implies that every mono is regular mono, the converse remains an open question.

A further open question is: If every extensional coalgebra is simple, does this mean that $\sim_{\mathcal{A}} = \nabla_{\mathcal{A}}$ for all coalgebras $\mathcal{A}$? In order to put this question into a more general framework, we plan to investigate L–*extensionality* for arbitrary relation liftings L.

References

1. Aczel, P., Mendler, N.: A final coalgebra theorem. In: Pitt, D.H., Rydeheard, D.E., Dybjer, P., Pitts, A.M., Poigné, A. (eds.) Category Theory and Computer Science. LNCS, vol. 389, pp. 357–365. Springer, Heidelberg (1989)
2. Adámek, J., Herrlich, H., Strecker, G.E.: Abstract and Concrete Categories. Wiley, New York (1990)
3. Barr, M.: Terminal coalgebras in well-founded set theory. Theor. Comput. Sci. 144(2), 299–315 (1993)
4. Gumm, H.P.: Elements of the General Theory of Coalgebras. LUATCS 1999. Rand Afrikaans University, Johannesburg (1999)
5. Gumm, H.P.: Functors for coalgebras. Algebra Univers. 45, 135–147 (2001)
6. Gumm, H.P., Schröder, T.: Coalgebraic structure from weak limit preserving functors. Electron. Notes Theor. Comput. Sci. 33, 113–133 (2000)
7. Gumm, H.P., Schröder, T.: Products of coalgebras. Algebra Univers. 46, 163–185 (2001)
8. Gumm, H.P., Schröder, T.: Types and coalgebraic structure. Algebra Univers. 53, 229–252 (2005)
9. Gumm, H.P., Zarrad, M.: Coalgebraic simulations and congruences. In: Bonsangue, M.M. (ed.) CMCS 2014. LNCS, vol. 8446, pp. 118–134. Springer, Heidelberg (2014)
10. Hansen, H.H., Kupke, C., Pacuit, E.: Bisimulation for neighbourhood structures. In: Mossakowski, T., Montanari, U., Haveraaen, M. (eds.) CALCO 2007. LNCS, vol. 4624, pp. 279–293. Springer, Heidelberg (2007)
11. Henkel, C.: Klassifikation coalgebraischer Typfunktoren. Diplomarbeit, Universität Marburg (2010)
12. Hennessy, M., Milner, R.: Algebraic laws for nondeterminism and concurrency. J. Assoc. Comput. Mach. 32, 137–161 (1985)
13. Hermida, C., Jacobs, B.: Structural induction and coinduction in a fibrational setting. Inform. Comput 145(2), 107–152 (1998)
14. Ihringer, T., Gumm, H.P.: Allgemeine Algebra. Heldermann Verlag, Wiesbaden (2003)
15. Marti, J., Venema, Y.: Lax extensions of coalgebra functors. In: Pattinson, D., Schröder, L. (eds.) CMCS 2012. LNCS, vol. 7399, pp. 150–169. Springer, Heidelberg (2012)
16. Lawrence, S.: Moss: coalgebraic logic. Ann. Pure Appl. Logic 96, 277–317 (1999)
17. Pattinson, D.: Expressive logics for coalgebras via terminal sequence induction. Notre Dame J. Formal Logic 45, 19–33 (2004)
18. Riguet, J.: Relations binaires, fermetures, correspondances de Galois. Bulletin de la Societe Mathematique de France 76, 114–155 (1948)
19. Rutten, J.J.M.M.: Universal coalgebra: a theory of systems. Theoret. Comput. Sci. 249(1), 3–80 (2000)
20. Schröder, T.: Coalgebren und Funktoren. Doktorarbeit, Universität Marburg (2001)
21. Staton, S.: Relating coalgebraic notions of bisimulaions. Log. Methods Comput. Sci. 7(1), 1–18 (2011)
22. Thijs, A.: Simulation and fixpoint semantics. Ph.D. thesis, University of Groningen (1996)
23. Trnková, V.: Some properties of set functors. Comm. Math. Univ. Carolinae 10(2), 323–352 (1969)
24. Zarrad, M.: Verträgliche Relationen auf Coalgebren. Diplomarbeit, Universität Marburg (2012)

Affine Monads and Side-Effect-Freeness

Bart Jacobs[✉]

Institute for Computing and Information Sciences,
Radboud Universiteit, Nijmegen, The Netherlands
bart@cs.ru.nl

Abstract. The notions of side-effect-freeness and commutativity are typical for probabilistic models, as subclass of quantum models. This paper connects these notions to properties in the theory of monads. A new property of a monad ('strongly affine') is introduced. It is shown that for such strongly affine monads predicates are in bijective correspondence with side-effect-free instruments. Also it is shown that these instruments are commutative, in a suitable sense, for monads which are commutative (monoidal).

1 Introduction

In a recent line of work in categorical quantum foundations [3–6,13,16] the notion of effectus has been proposed. Within that context one associates an instrument with each predicate, which performs measurement. These instruments are coalgebras, of a particular form, which may change the state. Indeed, it is one of the key features of the quantum world that measurement can change the object under observation. Thus, observation may have a side-effect.

In [6] a subclass of *commutative* effectuses is defined where there is a one-to-one correspondence between predicates and side-effect-free instruments. These commutative effectuses capture the probabilistic models, as special case of quantum models. Examples of commutative effectuses are the Kleisli categories $\mathcal{K}\ell(\mathcal{D})$ and $\mathcal{K}\ell(\mathcal{G})$ of the distribution monad $\mathcal{D}$ and the Giry monad $\mathcal{G}$, and the category of commutative von Neumann algebras.

The starting point for the work presented here is: can we translate these notions of side-effect-freeness and commutativity from effectus theory to the theory of monads — and coalgebras of monads — since they are instrumental in the semantics of programming languages? Especially, is there a connection between:

1. side-effect-freeness · of measurment-instruments and the property that a monad is affine (that is, preserves the final object);
2. commutativity as in effectus theory and commutativity of a monad?

B. Jacobs—The research leading to these results has received funding from the European Research Council under the European Union's Seventh Framework Programme (FP7/2007-2013)/ERC grant agreement nr. 320571.

© IFIP International Federation for Information Processing 2016
Published by Springer International Publishing Switzerland 2016. All Rights Reserved
I. Hasuo (Ed.): CMCS 2016, LNCS 9608, pp. 53–72, 2016.
DOI: 10.1007/978-3-319-40370-0_5

The main point of the paper is that these questions can be answered positively.

The first question makes sense because both the distribution and the Giry monad are affine, and it seems that this property is typical for monads that are relevant in probability theory. We shall see below that we actually need a slightly stronger property than 'affine', namely what we call 'strongly affine'.

Given the terminological coincidence, the second question may seem natural, but the settings are quite different and *a priori* unrelated. Here we do establish a connection, via a non-trivial calculation.

The relation between predicates and associated actions (instruments / coalgebras) comes from quantum theory in general, and effectus theory in particular. This relationship is complicated in the quantum case, but quite simple in the probabilistic case (see Theorem 1 below). It is the basis for a novel logic and type theory for probabilism in [4] (see also [17]).

The background of this work is effectus theory [6] in which logic (in terms of effect modules) and instruments play an important role. Here we concentrate on these instruments, and show that they can be studied in the theory of monads, independent of the logic of effect modules. Including these effect modules in the theory (for special monads) is left to future work.

2   Preliminaries

We assume that the reader is familiar with the notion of monad. We recall that a monad $T = (T, \eta, \mu)$ on a category with finite products $(\times, 1)$ is called *strong* if there is a 'strength' natural transformation st_1 with components $(\mathrm{st}_1)_{X,Y} \colon T(X) \times Y \to T(X \times Y)$ making the following diagrams commute — in which we omit indices, for convenience.

$$
\begin{array}{ccc}
T(X) \times Y \xrightarrow{\mathrm{st}_1} T(X \times Y) \\
_{\pi_1}\searrow \quad \downarrow T(\pi_1) \\
T(X)
\end{array}
\qquad
\begin{array}{ccc}
(T(X) \times Y) \times Z \xrightarrow{\cong} T(X) \times (Y \times Z) \\
\mathrm{st}_1 \times \mathrm{id} \downarrow \qquad\qquad\qquad \downarrow \mathrm{st}_1 \\
T(X \times Y) \times Z \qquad\qquad \\
\mathrm{st}_1 \downarrow \qquad\qquad\qquad\qquad \\
T((X \times Y) \times Z) \xrightarrow{\cong} T(X \times (Y \times Z))
\end{array}
\qquad (1)
$$

$$
\begin{array}{ccc}
X \times Y = X \times Y \\
\eta \times \mathrm{id} \downarrow \qquad \downarrow \eta \\
T(X) \times Y \xrightarrow[\mathrm{st}_1]{} T(X \times Y)
\end{array}
\qquad
\begin{array}{ccc}
T^2(X) \times Y \xrightarrow{\mathrm{st}_1} T(T(X) \times Y) \xrightarrow{T(\mathrm{st}_1)} T^2(X \times Y) \\
\mu \times \mathrm{id} \downarrow \qquad\qquad\qquad\qquad\qquad \downarrow \mu \\
T(X) \times Y \xrightarrow[\hspace{4cm}\mathrm{st}_1\hspace{4cm}]{} T(X \times Y)
\end{array}
\qquad (2)
$$

Each monad on the category **Sets** of sets and functions is automatically strong, via the definition $\mathrm{st}_1(u, y) = T(\lambda x.\, \langle x, y \rangle)(u)$.

Given a strength map $\mathrm{st}_1 \colon T(X) \times Y \to T(X \times Y)$ we define an associated version st_2 via swapping:

$$\mathrm{st}_2 = \left(X \times T(Y) \xrightarrow[\cong]{\gamma} T(Y) \times X \xrightarrow{\mathrm{st}_1} T(Y \times X) \xrightarrow[\cong]{T(\gamma)} T(X \times Y) \right)$$

where $\gamma = \langle \pi_2, \pi_1 \rangle$ is the swap map.

The monad T is called *commutative* (following [19]) when the order of applying strength in two coordinates does not matter, as expressed by commutation of the following diagram.

$$
\begin{array}{ccccccc}
 & \xrightarrow{\mathrm{st}_1} & T(X \times T(Y)) & \xrightarrow{T(\mathrm{st}_2)} & T^2(X \times Y) & \searrow^{\mu} & \\
T(X) \times T(Y) & & & & & & T(X \times Y) \quad (3) \\
 & \xrightarrow{\mathrm{st}_2} & T(T(X) \times Y) & \xrightarrow[T(\mathrm{st}_1)]{} & T^2(X \times Y) & \nearrow_{\mu} & \\
\end{array}
$$

We then write $\mathrm{dst} \colon T(X) \times T(Y) \to T(X \times Y)$ for 'double strength', to indicate the resulting single map, from left to right. Notice that $\mathrm{dst} \circ \gamma = T(\gamma) \circ \mathrm{dst}$.

Below we shall use distributive categories. They have finite products $(\times, 1)$ and coproducts $(+, 0)$, where products distribute over coproducts, in the sense that the following maps are isomorphisms.

$$0 \xrightarrow{\;!\;} 0 \times X \qquad (A \times X) + (B \times X) \xrightarrow{\mathrm{dis}_1 = [\kappa_1 \times \mathrm{id}, \, \kappa_2 \times \mathrm{id}]} (A + B) \times X$$
$$(4)$$

Swapping yields an associated distributivity map:

$$(X \times A) + (X \times B) \xrightarrow[= \gamma \circ \mathrm{dis}_1 \circ (\gamma + \gamma)]{\mathrm{dis}_2 = [\mathrm{id} \times \kappa_1, \, \mathrm{id} \times \kappa_2]} X \times (A + B)$$

It is an easy exercise to show that dis_1 and dis_2 interact in the following way.

$$
\begin{array}{ccc}
((A \times X) + (B \times X)) + ((A \times Y) + (B \times Y)) & \xrightarrow{\mathrm{dis}_1 + \mathrm{dis}_1} & ((A + B) \times X) + ((A + B) \times Y) \\
\Big\uparrow{\scriptstyle [\kappa_1 + \kappa_1, \, \kappa_2 + \kappa_2]} \; {\scriptstyle\cong} & & \Big\downarrow{\scriptstyle \mathrm{dis}_2} \\
 & & (A + B) \times (X + Y) \quad (5) \\
 & & \Big\uparrow{\scriptstyle \mathrm{dis}_1} \\
((A \times X) + (A \times Y)) + ((B \times X) + (B \times Y)) & \xrightarrow[\mathrm{dis}_2 + \mathrm{dis}_2]{} & (A \times (X + Y)) + (B \times (X + Y))
\end{array}
$$

The strength and distributivity maps also interact in the obvious way. There are two equivalent versions, with st_1 and dis_2 and with st_2 and dis_1. We describe the version that we actually need later on — and leave the verification to the meticulous reader.

$$A \times T(X) + B \times T(X) \xrightarrow{\text{dis}_1} (A+B) \times T(X)$$

with $\text{st}_2 + \text{st}_2$ downward on left, st_2 downward on right:

$$\begin{array}{ccc}
A \times T(X) + B \times T(X) & \xrightarrow{\text{dis}_1} & (A+B) \times T(X) \\
\downarrow{\scriptstyle \text{st}_2 + \text{st}_2} & & \\
T(A \times X) + T(B \times X) & & \downarrow{\scriptstyle \text{st}_2} \\
\downarrow{\scriptstyle [T(\kappa_1), T(\kappa_2)]} & & \\
T((A \times X) + (B \times X)) & \xrightarrow{T(\text{dis}_1)} & T((A+B) \times X)
\end{array} \tag{6}$$

The object $2 = 1 + 1$ will play a special role below. In a distributive category we have two 'separation' isomorphisms written as:

$$2 \times X \xrightarrow[\cong]{\text{sep}_1} X + X \xleftarrow[\cong]{\text{sep}_2} X \times 2 \tag{7}$$

Explicitly, they are defined as:

$$2 \times X \xrightarrow{\text{sep}_1} X + X \xleftarrow{\text{sep}_2} X \times 2$$

$$\text{dis}_1^{-1} \searrow 1 \times X + 1 \times X \xrightarrow{\pi_2 + \pi_2} \quad \pi_1 + \pi_1 \searrow X \times 1 + X \times 1 \xleftarrow{\text{dis}_2^{-1}}$$

These separation maps are natural in X and satisfy for instance:

$$\begin{array}{ll}
\nabla \circ \text{sep}_1 = \pi_2 & \nabla \circ \text{sep}_2 = \pi_1 \\
(! + !) \circ \text{sep}_1 = \pi_1 & (! + !) \circ \text{sep}_2 = \pi_2 \\
[\kappa_2, \kappa_1] \circ \text{sep}_1 = \text{sep}_1 \circ ([\kappa_2, \kappa_1] \times \text{id}) & [\kappa_2, \kappa_1] \circ \text{sep}_1 = \text{sep}_2 \circ (\text{id} \times [\kappa_2, \kappa_1])
\end{array} \tag{8}$$

These two maps are related via: $\text{sep}_1 \circ \gamma = \text{sep}_2$, for $\gamma = \langle \pi_2, \pi_1 \rangle$.

In the special case where $X = 2$ we have inverses

$$2 \times 2 \underset{[[\langle \kappa_1, \kappa_1 \rangle, \langle \kappa_1, \kappa_2 \rangle], [\langle \kappa_2, \kappa_1 \rangle, \langle \kappa_2, \kappa_2 \rangle]]}{\overset{\text{sep}_1}{\underset{\cong}{\rightleftarrows}}} 2 + 2 \underset{[[\langle \kappa_1, \kappa_1 \rangle, \langle \kappa_2, \kappa_1 \rangle], [\langle \kappa_1, \kappa_2 \rangle, \langle \kappa_2, \kappa_2 \rangle]]}{\overset{\text{sep}_2}{\underset{\cong}{\rightleftarrows}}} 2 \times 2$$

It is not hard to see that:

$$\text{sep}_1 \circ \gamma = \text{sep}_2 = [\kappa_1 + \kappa_1, \kappa_2 + \kappa_2] \circ \text{sep}_1 \tag{9}$$

The isomorphism $[\kappa_1 + \kappa_1, \kappa_2 + \kappa_2] : 2 + 2 \xrightarrow{\cong} 2 + 2$ can be illustrated as:

$$\begin{array}{ccccc}
2 + 2 & = & (1 + 1) & + & (1 + 1) \\
& & \downarrow & \times & \downarrow \\
2 + 2 & = & (1 + 1) & + & (1 + 1)
\end{array}$$

3   Affine and Strongly Affine Monads

In this section we recall what it means for a monad to be affine (see [11, 20, 21]), and introduce a slightly stronger notion. We describe basic properties and examples.

Definition 1. *Let $\mathbf{C}$ be a category with a monad $T \colon \mathbf{C} \to \mathbf{C}$.*

1. *Assuming that $\mathbf{C}$ has a final object 1, one calls T affine if the map $T(1) \to 1$ is an isomorphism, or simply, if $T(1) \cong 1$.*
2. *Assuming that $\mathbf{C}$ has binary products $\times$ and T is a strong monad, we call T strongly affine if the squares below are pullbacks.*

$$
\begin{array}{ccc}
T(X) \times Y & \xrightarrow{\ \pi_2\ } & Y \\
{\scriptstyle \mathrm{st}_1} \downarrow & & \downarrow {\scriptstyle \eta_Y} \\
T(X \times Y) & \xrightarrow[T(\pi_2)]{} & T(Y)
\end{array}
\tag{10}
$$

The notion of an 'affine monad' is well-known. What we call 'strongly affine' is new. The relationship with ordinary affine monads is quite subtle. Example 2 below show that 'strongly affine' is really stronger than 'affine'. But first we describe some properties and examples.

Lemma 1. *Let T be a strong monad on a category $\mathbf{C}$ with finite products $(\times, 1)$.*

1. *The monad T is affine iff the diagrams (10) commute. As a result, a strongly affine monad is affine.*
2. *There is at most one mediating (pullback) map for the diagram (10).*

The first point gives an alternative formulation of affineness. An older alternative formulation is: $\langle T(\pi_1), T(\pi_2)\rangle \circ \mathrm{dst} = \mathrm{id}$, see [20, Theorem 2.1], where dst is the double strength map from (3), for a commutative monad T.

The second point is useful when we wish to prove that a particular monad is strongly affine: we only need to prove existence of a mediating map, since uniqueness holds in general, see Example 1.

Proof. For the first point, let T be affine. We stretch Diagram (10) as follows.

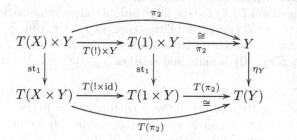

The square on the left commutes by naturality of strength. For the one on the right we use that $T(1)$ is final, so that $\pi_2 \colon T(1) \times Y \to Y$ is an isomorphism, with inverse $\langle \eta_1 \circ !_Y, \mathrm{id} \rangle$. Hence:

$$
\begin{aligned}
T(\pi_2) \circ \mathrm{st}_1 &= T(\pi_2) \circ \mathrm{st}_1 \circ \langle \eta_1 \circ !_Y, \mathrm{id} \rangle \circ \pi_2 \\
&= T(\pi_2) \circ \mathrm{st}_1 \circ (\eta_1 \times \mathrm{id}) \circ \langle !_Y, \mathrm{id} \rangle \circ \pi_2 \\
&\overset{(2)}{=} T(\pi_2) \circ \eta_{1 \times Y} \circ \langle !_Y, \mathrm{id} \rangle \circ \pi_2 \\
&= \eta_Y \circ \pi_2 \circ \langle !_Y, \mathrm{id} \rangle \circ \pi_2 \\
&= \eta_Y \circ \pi_2.
\end{aligned}
$$

In the other direction, assume that diagrams (10) commute. We consider the special case $X = Y = 1$.

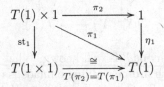

The lower triangle commutes by (1). We need to prove that $T(1)$ is final. It suffices to prove that the composite $\eta_1 \circ ! \colon T(1) \to 1 \to T(1)$ is the identity. This is obtained from the upper triangle:

$$
\eta_1 \circ ! = \eta_1 \circ \pi_2 \circ \langle \mathrm{id}, ! \rangle = \pi_1 \circ \langle \mathrm{id}, ! \rangle = \mathrm{id}.
$$

For the second point in the lemma we prove uniqueness of mediating maps. Assume we have two maps $f, g \colon Z \to T(X) \times Y$ with $\pi_2 \circ f = \pi_2 \circ g$ and $\mathrm{st}_1 \circ f = \mathrm{st}_1 \circ g$. We then obtain $\pi_1 \circ f = \pi_1 \circ g$ from:

$$
\pi_1 \circ f \overset{(1)}{=} T(\pi_1) \circ \mathrm{st}_1 \circ f = T(\pi_1) \circ \mathrm{st}_1 \circ g \overset{(1)}{=} \pi_1 \circ g. \qquad \square
$$

Example 1. Three examples of affine monads are the distribution monad $\mathcal{D}$ on **Sets** for discrete probability, the Giry monad $\mathcal{G}$ on the category **Meas** of measurable spaces, for continuous probability, and the expectation monad $\mathcal{E}$ on **Sets**. We show that all of them are strongly affine.

(1) The elements of $\mathcal{D}(X)$ are the finite formal convex combinations $\sum_i r_i | x_i \rangle$ with elements $x_i \in X$ and probabilities $r_i \in [0, 1]$ satisfying $\sum_i r_i = 1$. We can identify such a convex sum with a function $\varphi \colon X \to [0, 1]$ whose support $\mathrm{supp}(\varphi) = \{x \mid \varphi(x) \neq 0\}$ is finite and satisfies $\sum_x \varphi(x) = 1$. We can thus write $\varphi = \sum_x \varphi(x) | x \rangle$.

We have $\mathcal{D}(1) \cong 1$, since the sole element of $\mathcal{D}(1)$ is the distribution $1| * \rangle$, where we write $*$ for the element of the singleton set $1 = \{*\}$.

We show that this monad is also strongly affine. So let in Diagram (10) $\varphi \in \mathcal{D}(X \times Y)$ be a given distribution with $\mathcal{D}(\pi_2)(\varphi) = 1|z\rangle$ for a given element

$z \in Y$. Let's write $\varphi = \sum_{x,y} \varphi(x,y)|x,y\rangle$, so that $\mathcal{D}(\pi_2)(\varphi)$ is the marginal distribution:

$$\mathcal{D}(\pi_2)(\varphi) = \sum_y \left(\sum_x \varphi(x,y) \right) |y\rangle.$$

If this is the trivial distribution $1|z\rangle$, then $\varphi(x,y) = 0$ for all x and $y \neq z$. We obtain a new distribution $\psi = \mathcal{D}(\pi_1)(\varphi) \in \mathcal{D}(X)$, which takes the simple form $\psi(x) = \varphi(x,z)$. The pair $(\psi, z) \in \mathcal{D}(X) \times Y$ is the unique element giving us the pullback (10), since:

$$\mathrm{st}_1(\psi, z) = \sum_x \psi(x)|x,z\rangle = \sum_x \varphi(x,z)|x,z\rangle = \sum_{x,y} \varphi(x,y)|x,y\rangle = \psi.$$

(2) Next we consider the Giry monad $\mathcal{G}$ on the category **Meas** of measurable spaces. The elements of $\mathcal{G}(X)$ are probability measures $\omega \colon \Sigma_X \to [0,1]$. The unit $\eta \colon X \to \mathcal{G}(X)$ is given by $\eta(x)(M) = 1$ if $x \in M$ and $\eta(x)(M) = 0$ if $x \notin M$, for each $M \in \Sigma_X$. The strength map $\mathrm{st}_1 \colon \mathcal{G}(X) \times Y \to \mathcal{G}(X \times Y)$ is defined as the probability measure $\mathrm{st}_1(\omega, y) \colon \Sigma_{X \times Y} \to [0,1]$ determined by $M \times N \mapsto \omega(M) \cdot \eta(y)(N)$, see also [10,12,22].

So let's consider the situation (10) for $T = \mathcal{G}$, with a joint probability measure $\omega \in \mathcal{G}(X \times Y)$ and an element $z \in Y$ with

$$\mathcal{G}(\pi_2)(\omega)(N) = \omega(X \times N) = \eta(z)(N), \tag{11}$$

for all $N \in \Sigma_Y$. We prove 'non-entwinedness' of ω, that is, ω is the product of its marginals. Abstractly this means $\omega = \mathrm{dst}\big(\mathcal{G}(\pi_1)(\omega), \mathcal{G}(\pi_2)(\omega)\big)$, and concretely:

$$\omega(M \times N) = \omega(M \times Y) \cdot \omega(X \times N), \tag{12}$$

for all $M \in \Sigma_X$ and $Y \in \Sigma_Y$. We distinguish two cases.

- If $z \notin N$, then, by monotonicity the probability measure ω,

$$\omega(M \times N) \leq \omega(X \times N) \stackrel{(11)}{=} \eta(z)(N) = 0.$$

Hence $\omega(M \times N) = 0$. But also:

$$\omega(M \times Y) \cdot \omega(X \times N) \stackrel{(11)}{=} \omega(M \times Y) \cdot \eta(z)(N) = \omega(M \times Y) \cdot 0 = 0.$$

- If $z \in N$, then $z \notin \neg N$, so that:

$$
\begin{aligned}
\omega(M \times N) &= \omega(M \times N) + 0 \\
&= \omega(M \times N) + \omega(M \times \neg N) && \text{as just shown} \\
&= \omega\big((M \times N) \cup (M \times \neg N)\big) && \text{by additivity} \\
&= \omega(M \times Y) \\
&= \omega(M \times Y) \cdot \eta(z)(N) \\
&\stackrel{(11)}{=} \omega(M \times Y) \cdot \omega(X \times N).
\end{aligned}
$$

We now take $\phi \in \mathcal{G}(X)$ defined by $\phi(M) = \mathcal{G}(\pi_1)(\omega)(M) = \omega(M \times Y)$. The pair $(\phi, z) \in \mathcal{G}(X) \times Y$ is then mediating in (10):

$$
\begin{aligned}
\mathrm{st}_1(\phi, z)(M \times N) = \phi(M) \cdot \eta(z)(N) &= \omega(M \times Y) \cdot \eta(z)(N) \\
&\overset{(11)}{=} \omega(M \times Y) \cdot \omega(X \times N) \\
&\overset{(12)}{=} \omega(M \times N).
\end{aligned}
$$

Hence the Giry monad $\mathcal{G}$ is strongly affine.

(3) We turn to the expectation monad $\mathcal{E}(X) = \mathbf{EMod}([0, 1]^X, [0, 1])$ on $\mathbf{Sets}$, where $\mathbf{EMod}$ is the category of effect modules, see [15] for details. Let $h \in \mathcal{E}(X \times Y)$ satisfy $\mathcal{E}(\pi_2)(h) = \eta(z)$, for some $z \in Y$. This means that for each predicate $q \in [0, 1]^Y$ we have $h(q \circ \pi_2) = q(z)$.

Our first aim is to prove the analogue of the non-entwinedness equation (12) for $\mathcal{E}$, namely:

$$
h(1_{U \times V}) = h(1_{U \times Y}) \cdot h(1_{X \times V}), \tag{13}
$$

for arbitrary subsets $U \subseteq X$ and $V \subseteq Y$. Here we write $1_{U \times V} : X \times Y \to [0, 1]$ for the obvious indicator function. We distinguish:

- if $z \notin V$, then $h(1_{U \times V}) \leq h(1_{X \times V}) = h(1_V \circ \pi_2) = 1_V(z) = 0$. Hence (13) holds since both sides are 0.
- if $z \in V$, then $h(1_{U \times V}) = h(1_{U \times V}) + h(1_{U \times \neg V}) = h(1_{U \times Y}) = h(1_{U \times Y}) \cdot h(1_{X \times V})$.

By [15, Lemma 12] each predicate can be written as limit of step functions. It suffices to prove the result for such step functions, since by [15, Lemma 10] the map of effect modules h is automatically continuous.

Hence we concentrate on an arbitrary step function $p \in [0, 1]^{X \times Y}$ of the form $p = \sum_{i,j} r_{i,j} 1_{U_i \times V_j}$, where the $U_i \subseteq X$ and $V_j \subseteq Y$ form disjoint covers, and $r_{i,j} \in [0, 1]$. We prove that $h(p) = \mathrm{st}_1(\mathcal{E}(\pi_1)(h), z)(p)$, so that we can take $\mathcal{E}(\pi_1)(h) \in \mathcal{E}(X)$ to obtain a pullback in (10).

Let j_0 be the (unique) index with $z \in V_{j_0}$, so that $p(x, z) = \sum_i r_{i,j_0} 1_{U_i}(x)$. Then:

$$
\begin{aligned}
h(p) = h\big(\sum_{i,j} r_{i,j} 1_{U_i \times V_j}\big) &= \sum_{i,j} r_{i,j} h\big(1_{U_i \times V_j}\big) \\
&\overset{(13)}{=} \sum_{i,j} r_{i,j} h\big(1_{U_i \times Y}\big) \cdot h\big(1_{X \times V_j}\big) \\
&= \sum_{i,j} r_{i,j} h\big(1_{U_i \times Y}\big) \cdot 1_{V_j}(z) \\
&= \sum_i r_{i,j_0} h\big(1_{U_i \times Y}\big) \\
&= h\big(\sum_i r_{i,j_0} 1_{U_i \times Y}\big) \\
&= h\big(\lambda(x, y). \, p(x, z)\big) \\
&= \mathrm{st}_1\big(\mathcal{E}(\pi_1)(h), z\big)(p).
\end{aligned}
$$

The following (counter) example is due to Kenta Cho.

Example 2. An example of an affine but not strongly affine monad is the 'generalised distribution' monad $\mathcal{D}_\pm$ on $\mathbf{Sets}$. Elements of $\mathcal{D}_\pm(X)$ are finite formal sums $\sum_i r_i |x_i\rangle$ with $r_i \in \mathbb{R}$ and $x_i \in X$ satisfying $\sum_i r_i = 1$. The other data of a (strong) monad are similar to the ordinary distribution monad $\mathcal{D}$. Clearly $\mathcal{D}_{\pm}(1) \cong 1$, *i.e.* $\mathcal{D}_{\pm}$ is affine.

Now consider the square (10) with $X = \{x_1, x_2\}$ and $Y = \{y_1, y_2\}$. Define:

$$\varphi = 1|x_1, y_1\rangle + 1|x_1, y_2\rangle + (-1)|x_2, y_2\rangle \in \mathcal{D}_{\pm}(X \times Y).$$

We have $\mathcal{D}_{\pm}(\pi_2)(\varphi) = 1|y_1\rangle = \eta(y_1)$, since the terms with y_2 cancel each other out. But there is no element $\psi \in \mathcal{D}_{\pm}(X)$ such that $\mathrm{st}_1(\psi, y_1) = \varphi$. Hence the square (10) is not a pullback.

The fact that the terms in this example cancel each other out is known as 'interference' in the quantum world. It already happens with negative coefficients. This same monad $\mathcal{D}_{\pm}$ is used in [1]. How the notions of non-locality and contextuality that are studied there relate to strong affineness requires further investigation.

For the record we recall from [2,23] that each endofunctor on **Sets** can be written as a coproduct of affine functors.

The following result gives a 'graph' construction that is useful in conditional constructions in probability, see the subsequent discussion.

Proposition 1. *For a strongly affine monad T there is a canonical bijective correspondence:*

$$\frac{Y \xrightarrow{\ f\ } T(X)}{Y \xrightarrow{\ g\ } T(X \times Y) \ \text{with} \ T(\pi_2) \circ g = \eta}$$

What we mean by 'canonical' is that the mapping downwards is given by $f \mapsto \mathrm{st}_1 \circ \langle f, \mathrm{id} \rangle$.

Proof. The if-part of the statement is obvious, since the correspondence is a reformulation of the pullback property of the diagram (10). In the other direction, let T be strongly affine. As stated, the mapping downwards is given by $\overline{f} = \mathrm{st}_1 \circ \langle f, \mathrm{id} \rangle$. Then:

$$T(\pi_2) \circ \overline{f} = T(\pi_2) \circ \mathrm{st}_1 \circ \langle f, \mathrm{id} \rangle \stackrel{(10)}{=} \eta \circ \pi_2 \circ \langle f, \mathrm{id} \rangle = \eta.$$

In the other direction we map $g \colon Y \to T(X \times Y)$ to $\overline{g} = T(\pi_1) \circ g$. Then:

$$\overline{\overline{f}} = T(\pi_1) \circ \mathrm{st}_1 \circ \langle f, \mathrm{id} \rangle \stackrel{(1)}{=} \pi_1 \circ \langle f, \mathrm{id} \rangle = f.$$

In order to prove $\overline{\overline{g}} = g$ we notice that by the pullback property of diagram (10) we know that there is a unique $h \colon Y \to T(X)$ with $g = \mathrm{st}_1 \circ \langle h, \mathrm{id} \rangle = \overline{h}$. But then $\overline{\overline{h}} = h$, by what we have just shown, so that:

$$\overline{\overline{g}} = \overline{\overline{\overline{h}}} = \overline{h} = g. \qquad \square$$

The correspondence in this proposition is used (for the distribution monad $\mathcal{D}$) as Lemma 1 in [9]. There, the map $\mathrm{st}_1 \circ \langle f, \mathrm{id} \rangle$ is written as $\mathrm{gr}(f)$, and called the graph of f. It is used in the description of conditional probability. It is also used (implicitly) in [8, §3.1], where a measure/state $\omega \in \mathcal{G}(X)$ and a Kleisli map $f \colon X \to \mathcal{G}(Y)$ give rise to a joint probability measure $\mathrm{gr}(f) \bullet \omega$ in $\mathcal{G}(X \times Y)$.

4   Affine Parts of Monads, and Causal Maps

It is known for a long time that the 'affine part' of a monad can be extracted via pullbacks, see [21] (or also [11]). Here we shall relate this affine part to 'causal' maps in Kleisli categories of monads.

Proposition 2. *Let T be a monad on a category $\mathbf{C}$ with a final object 1. Assume that the following pullbacks exist in $\mathbf{C}$, for each object X.*

$$
\begin{array}{ccc}
T_a(X) & \xrightarrow{\;!\;} & 1 \\
{\scriptstyle \iota_X}\Big\downarrow & & \Big\downarrow{\scriptstyle \eta} \\
T(X) & \xrightarrow[T(!)]{} & T(1)
\end{array}
\tag{14}
$$

Then:

1. *the mapping $X \mapsto T_a(X)$ is a monad on $\mathbf{C}$;*
2. *the mappings $\iota_X \colon T_a(X) \to T(X)$ are monic, and form a map of monads $T_a \Rightarrow T$;*
3. *T_a is an affine monad, and in fact the universal (greatest) affine submonad of T;*
4. *if T is a strong resp. commutative monad, then so is T_a.*

Proof. These results are standard. We shall illustrate point (3). If we take $X = 1$ in Diagram (14), then the bottom arrow $T(!_X) \colon T(X) \to T(1)$ is the identity. Hence top arrow $T_a(1) \to 1$ is an isomorphism, since isomorphisms are preserved under pullback.

To see that $T_a \Rightarrow T$ is universal, let $\sigma \colon S \Rightarrow T$ be a map of monads, where S is affine, then we obtain a map $\overline{\sigma}_X$ in:

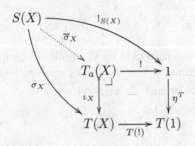

The outer diagram commutes since S is affine, so that $\eta_1^S \circ !_{S(1)} = \mathrm{id}_{S(1)}$; then:

$$T(!_X) \circ \sigma_X = \sigma_1 \circ S(!) = \sigma_1 \circ \eta_1^S \circ !_{S(1)} \circ S(!_X) = \eta_1^T \circ !_{S(X)}. \qquad \square$$

Example 3. We list several examples of affine parts of monads.

1. Let $\mathcal{M} = \mathcal{M}_{\mathbb{R}_{\geq 0}}$ be the multiset monad on **Sets** with the non-negative real numbers $\mathbb{R}_{\geq 0}$ as scalars. Elements of $\mathcal{M}(X)$ are thus finite formal sums $\sum_i r_i | x_i \rangle$ with $r_i \in \mathbb{R}_{\geq 0}$ and $x_i \in X$. The affine part $\mathcal{M}_a$ of this monad is the distribution monad $\mathcal{D}$ since $1| * \rangle = \mathcal{M}(!)(\sum_i r_i | x_i \rangle) = (\sum_i r_i)| * \rangle$ iff $\sum_i r_i = 1$. Thus $\mathcal{D}(X) = \mathcal{M}_a(X)$ yields a pullback in Diagram (14).
   The monad $\mathcal{D}_{\pm}$ used in Example 2 can be obtained in a similar manner as an affine part, not of the multiset monad $\mathcal{M}_{\mathbb{R}_{\geq 0}}$ with non-negative coefficients, but from the multiset monad $\mathcal{M}_{\mathbb{R}}$ with arbitrary coefficients: its multisets are formal sums $\sum_i r_i | x_i \rangle$ where the r_i are arbitrary real numbers.
2. For the powerset monad $\mathcal{P}$ on **Sets** the affine submonad $\mathcal{P}_a \rightarrowtail \mathcal{P}$ is given by the *non-empty* powerset monad. Indeed, for a subset $U \subseteq X$ we have:

$$\mathcal{P}(!)(U) = \begin{cases} \{*\} & \text{if } U \neq \emptyset \\ \emptyset & \text{if } U = \emptyset \end{cases}$$

   Hence $\mathcal{P}(!)(U) = \{*\} = \eta(*)$ iff U is non-empty. It is not hard to see that the non-empty powerset monad $\mathcal{P}_a$ is strongly affine.
3. Let $T(X) = (S \times X)^S$ be the state monad on **Sets**, for a fixed set of states S. The unit $\eta \colon X \to T(X)$ is defined as $\eta(x) = \lambda s \in S.(s, x)$ so that the pullback (14) is given by:

$$\begin{aligned} T_a(X) &= \{f \in (S \times X)^S \mid T(!)(f) = \eta(*)\} \\ &= \{f \in (S \times X)^S \mid \forall s. (\mathrm{id} \times !)(f(s)) = (s, *)\} \\ &= \{f \in (S \times X)^S \mid \forall s. \pi_1 f(s) = s\} \\ &\cong X^S. \end{aligned}$$

   Thus, Kleisli maps $Y \to T_a(X) = X^S$ may use states $s \in S$ to compute the output in X, but they cannot change states: they are side-effect-free.
   In a similar way one shows that the list monad $X \mapsto X^\star$ and the lift monad $X \mapsto X + 1$ have the identity monad as affine submonad.
4. Fix a set C and consider the continuation, (or double-dual) monad $\mathcal{C}$ on **Sets** given by $\mathcal{C}(X) = C^{(C^X)}$, with unit $\eta \colon X \to \mathcal{C}(X)$ given by $\eta(x)(f) = f(x)$. The pullback (14) is then:

$$\begin{aligned} \mathcal{C}_a(X) &= \{h \in C^{(C^X)} \mid \mathcal{C}(!)(h) = \eta(*)\} \\ &= \{h \in C^{(C^X)} \mid \forall f \in C^1. h(f \circ !) = f(*)\} \\ &= \{h \in C^{(C^X)} \mid \forall c \in C. h(\lambda x. c) = c\}. \end{aligned}$$

   This is the submonad of functions $h \colon C^X \to C$ which have output $c \in C$ on the constant function $\lambda x. c \colon X \to C$.

We write $\mathcal{K}\ell(T)$ for the Kleisli category of a monad T, and we write a fat bullet $\bullet$ for Kleisli composition $g \bullet f = \mu \circ T(g) \circ f$. For each object X there is a special 'ground' map:

$$\bar{\top}_X = \left(X \xrightarrow{!_X} 1 \xrightarrow{\eta_1} T(1) \right) \tag{15}$$

This is the result of applying the standard functor $\mathbf{C} \to \mathcal{K}\ell(T)$ to the map $! \colon X \to 1$ in the underlying category $\mathbf{C}$.

Causal maps have been introduced in the context of CP*-categories, see [7], where they express that measurements in the future, given by $\bar{\top}$, cannot influence the past.

Definition 2. *A Kleisli map $f \colon X \to T(Y)$ will be called* causal *or* unital *if it preserves ground, in the sense that:*

$$\bar{\top}_Y \bullet f = \bar{\top}_X \qquad \text{that is} \qquad T(!_Y) \circ f = \bar{\top}_X.$$

Causal maps are used in [6] to construct effectuses. Here we define them quite generally, for an arbitrary monad. Notice that each map $f \colon X \to T(Y)$ is causal when T is an affine monad. The following elementary observation gives a more precise description.

Lemma 2. *A Kleisli map $f \colon X \to T(Y)$ is causal if and only if it restricts to a (necessarily unique) map $f' \colon X \to T_a(Y)$ for the affine submonad $\iota \colon T_a \rightarrowtail T$, where $\iota_Y \circ f' = f$.*

Proof. Obviously, the causality requirement $\bar{\top}_Y \bullet f = T(!) \circ f = \eta_1 \circ \, ! = \bar{\top}_X$ means that the outer diagram commutes in:

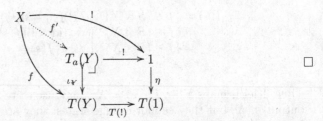

As a result, a Kleisli map $X \to \mathcal{D}(X)$ for the distribution monad $\mathcal{D}$ can equivalently be described as a causal map $X \to \mathcal{M}(X)$ for the multiset monad $\mathcal{M}$, see Example 3 (1). This gives a more systematic approach than the "constrained" description from [18], which restricts multisets to a certain subset.

5   Predicates and Instruments

In a very general sense we can define a *predicate* on an object X in the Kleisli category $\mathcal{K}\ell(T)$ of a monad T as a map $p \colon X \to 2$, where $2 = 1 + 1$, that is as a map $p \colon X \to T(2)$ in the underlying category. There is always a 'truth' predicate $\mathbf{1} = T(\kappa_1) \circ \bar{\top} = \eta \circ \kappa_1 \circ \, ! \colon X \to 1 \to 1 + 1 \to T(1 + 1)$. Similarly there is falsity predicate $\mathbf{0} = \eta \circ \kappa_2 \circ \,!$, and a negation operation $p^\perp = T([\kappa_2, \kappa_1]) \circ p$ obtained by swapping. Clearly, $p^{\perp\perp} = p$ and $\mathbf{1}^\perp = \mathbf{0}$. In certain cases there is more algebraic structure, see [6], where predicates form effect modules.

At this stage we informally describe an *instrument* associated with a predicate $p\colon X \to T(1+1)$ as a map $\mathrm{instr}_p\colon X \to T(X+X)$ with $T(!+!) \circ \mathrm{instr}_p = p$. Such an instrument is called *side-effect-free* if the following diagram commutes in $\mathcal{K}\ell(T)$.

$$
\begin{array}{ccc}
X & \xrightarrow{\ \mathrm{instr}_p\ } & X + X \\
 & \searrow & \downarrow{\scriptstyle \nabla = [\mathrm{id}, \mathrm{id}]} \\
 & & X
\end{array}
$$

Equivalently, if $T(\nabla) \circ \mathrm{instr}_p = \eta$ in the underlying category.

This instrument terminology comes from [13] (see also [6]), where it is used in a setting for quantum computation. Here we adapt the terminology to a monad setting. The instrument is used to interpret, for instance, a conditional statement as composite:

$$
\text{if } p \text{ then } f \text{ else } g = \Big(X \xrightarrow{\ \mathrm{instr}_p\ } X + X \xrightarrow{\ [f,g]\ } Y \Big).
$$

For example, for the distribution monad $\mathcal{D}$ a predicate on a set X is a function $p\colon X \to \mathcal{D}(1+1) \cong [0,1]$. For such a 'fuzzy' predicate there is an instrument map $\mathrm{instr}_p\colon X \to \mathcal{D}(X+X)$ given by the convex sum:

$$
\mathrm{instr}_p(x) = p(x)|\kappa_1 x\rangle + (1 - p(x))|\kappa_2 x\rangle.
$$

The associated if-then-else statement gives a weighted combination of the two options, where the weights are determined by the probability $p(x) \in [0,1]$.

Next we describe how such instruments can be obtained via a general construction in distributive categories.

Definition 3. *Let T be a strong monad on a distributive category* **C**. *For a predicate $p\colon X \to T(1+1)$ we define an instrument $\mathrm{instr}_p\colon X \to T(X+X)$ as composite:*

$$
\mathrm{instr}_p = \Big(X \xrightarrow{\langle p, \mathrm{id}\rangle} T(2)\times X \xrightarrow{\ \mathrm{st}_1\ } T(2\times X) \xrightarrow[\cong]{T(\mathrm{sep}_1)} T(X+X) \Big)
$$

where sep_1 is the separation isomorphism from (7).

We collect some basic results about instrument maps.

Lemma 3. *In the context of the previous definition we have:*

1. *$T(!+!) \circ \mathrm{instr}_p = p$; in particular, $\mathrm{instr}_p = p$ for each $p\colon 1 \to T(2)$;*
2. *if p is causal, then instr_p is side-effect-free and causal;*
3. *$\mathrm{instr}_1 = \eta \circ \kappa_1$ and $\mathrm{instr}_0 = \eta \circ \kappa_2$, and $\mathrm{instr}_{p^\perp} = T([\kappa_2, \kappa_1]) \circ \mathrm{instr}_p$;*

4. *for a map* $f\colon Y \to X$ *in the underlying category,*

$$T(f + f) \circ instr_{p \circ f} = instr_p \circ f.$$

5. *for predicates* $p\colon X \to T(2)$ *and* $q\colon Y \to T(2)$,

$$instr_{[p,q]} = [T(\kappa_1 + \kappa_1), T(\kappa_2 + \kappa_2)] \circ (instr_p + instr_q).$$

Proof. We handle these points one by one.

1. We have:

$$
\begin{aligned}
T(! + !) \circ instr_p &= T(! + !) \circ T(\mathrm{sep}_1) \circ \mathrm{st}_1 \circ \langle p, \mathrm{id}\rangle \\
&\overset{(8)}{=} T(\pi_1) \circ \mathrm{st}_1 \circ \langle p, \mathrm{id}\rangle \\
&\overset{(1)}{=} \pi_1 \circ \langle p, \mathrm{id}\rangle \\
&= p.
\end{aligned}
$$

2. Assume that the predicate p is causal, that is $T(!) \circ p = \bar{\top}$. We first show that the instrument $instr_p$ is side-effect-free:

$$
\begin{aligned}
T(\nabla) \circ instr_p &= T(\nabla) \circ T(\mathrm{sep}_1) \circ \mathrm{st}_1 \circ \langle p, \mathrm{id}\rangle \\
&= T(\nabla) \circ T(\pi_2 + \pi_2) \circ T(\mathrm{dis}_1^{-1}) \circ \mathrm{st}_1 \circ \langle p, \mathrm{id}\rangle \\
&= T(\pi_2) \circ T(\nabla) \circ T(\mathrm{dis}_1^{-1}) \circ \mathrm{st}_1 \circ \langle p, \mathrm{id}\rangle \\
&= T(\pi_2) \circ T(\nabla \times \mathrm{id}) \circ \mathrm{st}_1 \circ \langle p, \mathrm{id}\rangle \\
&= T(\pi_2) \circ \mathrm{st}_1 \circ (T(\nabla) \times \mathrm{id}) \circ \langle p, \mathrm{id}\rangle \\
&= T(\pi_2) \circ \mathrm{st}_1 \circ \langle T(!) \circ p, \mathrm{id}\rangle \\
&= T(\pi_2) \circ \mathrm{st}_1 \circ \langle \eta \circ !, \mathrm{id}\rangle \quad\quad \text{since } p \text{ is causal} \\
&\overset{(2)}{=} T(\pi_2) \circ \eta \circ \langle !, \mathrm{id}\rangle \\
&= \eta \circ \pi_2 \circ \langle !, \mathrm{id}\rangle \\
&= \eta.
\end{aligned}
$$

The instrument $instr_p$ is causal too:

$$
\begin{aligned}
\bar{\top} \bullet instr_p &= T(!) \circ instr_p \\
&= T(!) \circ T(\mathrm{sep}_1) \circ \mathrm{st}_1 \circ \langle p, \mathrm{id}\rangle \\
&= T(!) \circ T(\pi_1) \circ \mathrm{st}_1 \circ \langle p, \mathrm{id}\rangle \\
&\overset{(1)}{=} T(!) \circ \pi_1 \circ \langle p, \mathrm{id}\rangle \\
&= T(!) \circ p \\
&= \bar{\top}.
\end{aligned}
$$

3. For the truth predicate $1 = \eta \circ \kappa_1 \circ !$ we have:

$$
\begin{aligned}
instr_1 &= T(\mathrm{sep}_1) \circ \mathrm{st}_1 \circ \langle \eta \circ \kappa_1 \circ !, \mathrm{id}\rangle \\
&\overset{(2)}{=} T(\mathrm{sep}_1) \circ \eta \circ \langle \kappa_1 \circ !, \mathrm{id}\rangle \\
&= \eta \circ \mathrm{sep}_1 \circ \langle \kappa_1 \circ !, \mathrm{id}\rangle \\
&= \eta \circ \kappa_1.
\end{aligned}
$$

Similarly one obtains $instr_0 = \eta \circ \kappa_2$. Next,

$$T([\kappa_2, \kappa_1]) \circ instr_p$$
$$= T([\kappa_2, \kappa_1]) \circ T(sep_1) \circ st_1 \circ \langle p, id \rangle$$
$$\overset{(8)}{=} T(sep_1) \circ T([\kappa_2, \kappa_1] \times id) \circ st_1 \circ \langle p, id \rangle$$
$$= T(sep_1) \circ st_1 \circ (T([\kappa_2, \kappa_1]) \times id) \circ \langle p, id \rangle$$
$$= T(sep_1) \circ st_1 \circ \langle p^{\perp}, id \rangle$$
$$= instr_{p^{\perp}}.$$

4. In a straightforward manner we obtain for a map f in the underlying category:

$$T(f + f) \circ instr_{p \circ f}$$
$$= T(f + f) \circ T(sep_1) \circ st_1 \circ \langle p \circ f, id \rangle$$
$$= T(sep_1) \circ T(id \times f) \circ st_1 \circ \langle p \circ f, id \rangle \qquad \text{by naturality of } sep_1$$
$$= T(sep_1) \circ st_1 \circ (id \times f) \circ \langle p \circ f, id \rangle$$
$$= T(sep_1) \circ st_1 \circ \langle p, id \rangle \circ f$$
$$= instr_p \circ f.$$

5. Via point (4) we get:

$$[T(\kappa_1 + \kappa_1), T(\kappa_2 + \kappa_2)] \circ (instr_p + instr_q)$$
$$= [T(\kappa_1 + \kappa_1) \circ instr_{[p,q] \circ \kappa_1}, T(\kappa_2 + \kappa_2) \circ instr_{[p,q] \circ \kappa_2}]$$
$$= [instr_{[p,q]} \circ \kappa_1, instr_{[p,q]} \circ \kappa_2]$$
$$= instr_{[p,q]}. \qquad \square$$

The main result of this section gives, for strongly affine monads, a bijective correspondence between predicates and side-effect-free instruments.

Theorem 1. *Let T be a strongly affine monad on a distributive category. Then there is a bijective correspondence between:*

$$\frac{\text{predicates } X \xrightarrow{\ p\ } T(1 + 1)}{X \xrightarrow[f]{} T(X + X) \text{ with } T(\nabla) \circ f = \eta}$$

Proof. The mapping downwards is $p \mapsto instr_p$, and upwards is $f \mapsto T(! + !) \circ f$. Point (2) in Lemma 3 says that $T(\nabla) \circ instr_p = \eta$, since p is causal (because T is affine); point (1) tells that going down-up is the identity. For the up-down part we need to show that $f = instr_p$, for $p = T(! + !) \circ f$. We use the 'strongly affine' pullback (10) to get a predicate q in:

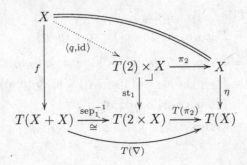

The outer diagram commutes by (8). By construction we have $f = \text{instr}_q$, see Definition 3. We thus need to prove that $q = p$. But this follows from Lemma 3 (1):

$$p \overset{\text{def}}{=} T(!+!) \circ f = T(!+!) \circ \text{instr}_q = q. \qquad \square$$

Example 4. We shall illustrate the situation for the powerset monad $\mathcal{P}$ on the (distributive) category **Sets**. We write $1 + 1 = 2 = \{0, 1\}$, where we identify the element $0 \in 2$ with $\kappa_2 *$ and $1 \in 2$ with $\kappa_1 *$. Hence $\mathcal{P}(2) = \{\emptyset, \{0\}, \{1\}, \{0, 1\}\}$ and $\mathcal{P}_a(2) = \{\{0\}, \{1\}, \{0, 1\}\}$, where $\mathcal{P}_a \rightarrowtail \mathcal{P}$ is the affine submonad of non-empty subsets, see Example 3 (2).

For a predicate $p \colon X \to \mathcal{P}(2)$ the associated instrument $\text{instr}_p \colon X \to \mathcal{P}(X + X)$ is, according to Definition 3, given by:

$$\text{instr}_p(x) = \{\kappa_1 x \mid 1 \in p(x)\} \cup \{\kappa_2 x \mid 0 \in p(x)\}$$
$$= \begin{cases} \emptyset & \text{if } p(x) = \emptyset \\ \{\kappa_1 x\} & \text{if } p(x) = \{1\} \\ \{\kappa_2 x\} & \text{if } p(x) = \{0\} \\ \{\kappa_1 x, \kappa_2 x\} & \text{if } p(x) = \{0, 1\}. \end{cases}$$

We thus see:

$$\big(\mathcal{P}(\nabla) \circ \text{instr}_p\big)(x) = \{x \mid 0 \in p(x) \text{ or } 1 \in p(x)\} = \begin{cases} \{x\} & \text{if } p(x) \neq \emptyset \\ \emptyset & \text{if } p(x) = \emptyset. \end{cases}$$

Hence these instruments are not side-effect-free, in general. But if we restrict ourselves to the (strongly affine) submonad $\mathcal{P}_a$ of non-emptyset subsets, then we do have side-effect-freeness — as shown in general in Lemma 3 (2).

In that case we have a bijective correspondence between maps $f \colon X \to \mathcal{P}_a(X + X)$ with $\mathcal{P}_a(\nabla) \circ f = \{-\}$ and predicates $p \colon X \to \mathcal{P}_a(2)$ — as shown in general in Theorem 1.

6   Commutativity

In this section we assume that T is a strong monad on a distributive category $\mathbf{C}$, so that we can associate an instrument $\mathrm{instr}_p\colon X \to T(X+X)$ with a predicate $p\colon X \to T(2)$, like in Definition 3.

Given such a predicate p we define the *assert* map $\mathrm{asrt}_p\colon X \to T(X+\bar{1})$ as:

$$\mathrm{asrt}_p = T(\mathrm{id} +\,!) \circ \mathrm{instr}_p = T(\pi_2 + \pi_1) \circ T(\mathrm{dis}_1^{-1}) \circ \mathrm{st}_1 \circ \langle p, \mathrm{id}\rangle.$$

These assert maps play an important role to define conditional probabilities (after normalisation), see [4]. Here we illustrate how one can define, via these assert maps, a sequential composition operation — called 'andthen' — on predicates $p, q\colon X \to T(2)$ as:

$$\begin{aligned} p\ \&\ q &= [q, \kappa_2] \bullet \mathrm{asrt}_p && \text{in } \mathcal{K}\ell(T) \\ &= \mu \circ T([q, T(\kappa_2) \circ \eta]) \circ \mathrm{asrt}_p && \text{in } \mathbf{C}. \end{aligned}$$

This operation incorporates the side-effect of p, if any. Hence, in principle, this is not a commutative operation.

Example 5. We elaborate the situation described above for the state monad $T(X) = (S \times X)^S$ from Example 3 (3). A predicate on X can be identified with a map $p\colon X \to (S+S)^S$, since:

$$T(2) = (S \times 2)^S \cong (S+S)^S.$$

For $x \in X$ and $s \in S$ the value $p(x)(s) \in S + S$ describes the 'true' case via the left component, and the 'false' case via the right component. Clearly, the predicate can also change the state, and thus have a side-effect.

The associated instrument $\mathrm{instr}_p\colon X \to (S \times (X+X))^S \cong (S \times X + S \times X)^S$ is described by:

$$\mathrm{instr}_p(x)(s) = \begin{cases} \kappa_1(s', x) & \text{if } p(x)(s) = \kappa_1 s' \\ \kappa_2(s', x) & \text{if } p(x)(s) = \kappa_2 s' \end{cases}$$

Similarly, $\mathrm{asrt}_p\colon X \to (S \times (X+1))^S \cong (S \times X + S)^S$ is:

$$\mathrm{asrt}_p(x)(s) = \begin{cases} \kappa_1(s', x) & \text{if } p(x)(s) = \kappa_1 s' \\ \kappa_2 s' & \text{if } p(x)(s) = \kappa_2 s' \end{cases}$$

Hence for predicates $p, q\colon X \to (S+S)^S$ we have $p\ \&\ q\colon X \to (S+S)^S$ described by:

$$(p\ \&\ q)(x)(s) = \begin{cases} q(x)(s') & \text{if } p(x)(s) = \kappa_1 s' \\ \kappa_2 s' & \text{if } p(x)(s) = \kappa_2 s' \end{cases}$$

The side-effect s' of p is passed on to q, if p holds. Clearly, $\&$ is not commutative for the state monad.

The theorem below plays a central role for commutativity of the andthen operation &. It establishes a connection between commutativity of sequential composition and commutativity of the monad, as described in Diagram (3).

Theorem 2. *If T is a commutative monad, then instruments commute: for predicates $p, q \colon X \to T(2)$, the following diagram commutes in $\mathcal{K}\ell(T)$.*

$$
\begin{array}{ccccc}
X & \xrightarrow{\ instr_p\ } & X + X & \xrightarrow{\ q+q\ } & 2 + 2 \\
\| & & & & \cong \downarrow {\scriptstyle [\kappa_1 + \kappa_1, \kappa_2 + \kappa_2]} \\
X & \xrightarrow[\ instr_q\ ]{} & X + X & \xrightarrow[\ p+p\ ]{} & 2 + 2
\end{array}
\tag{16}
$$

Proof. The structure of the proof is given by the following diagram in the underlying category.

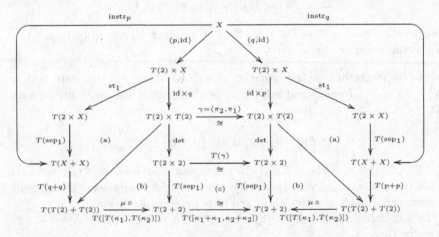

The sub-diagrams (a) commute by naturality, and sub-diagrams (b) by (6) commutation of (c) is equation (9), and the square in the middle is commutativity of the monad T, see (3). Details are left to the interested reader.   □

Corollary 1. *For a commutative monad (on a distributive category), sequential composition & is commutative on causal predicates.*

Proof. We first note that in $\mathcal{K}\ell(T)$ we can write $\mathrm{asrt}_p = (\mathrm{id} + \bar{\mp}) \bullet \mathrm{instr}_p$. Hence if p, q are both causal, then:

$$
\begin{aligned}
p \mathbin{\&} q &= [q, \kappa_2] \bullet (\mathrm{id} + \bar{\mp}) \bullet \mathrm{instr}_p \\
&= [\mathrm{id}, \kappa_2 \bullet \bar{\mp}] \bullet (q + q) \bullet \mathrm{instr}_p && \text{since } q \text{ is causal} \\
&= [\mathrm{id}, \kappa_2 \bullet \bar{\mp}] \bullet (p + p) \bullet \mathrm{instr}_q && \text{by Theorem 2} \\
&= [p, \kappa_2] \bullet (\mathrm{id} + \bar{\mp}) \bullet \mathrm{instr}_q && \text{since } p \text{ is causal} \\
&= q \mathbin{\&} p.
\end{aligned}
$$

7   Conclusions

We have translated the notions of side-effect-freeness and commutativity from quantum foundations (in the form of effectus theory) to monad theory, and proven some elementary results. This is only a starting point. Expecially, connections between (strong) affineness and non-locality need to be clarified.

Further, the current work forms the basis for a categorical description (that is in the making) of probability theory using strongly affine monads.

We should point out that the setting of the current work is given by distributive categories, with finite cartesian products, and not tensor products. They form in themselves already a classical setting.

Acknowledgements. Thanks are due to Kenta Cho and Fabio Zanasi for helpful discussions on the topic of the paper, and to the anonymous referees for suggesting several improvements.

References

1. Abramsky, S., Brandenburger, A.: The sheaf-theoretic structure of non-locality and contextuality. New J. Phys. **13**, 113036 (2011)
2. Adámek, J., Velebil, J.: Analytic functors and weak pullbacks. Theor. Appl. Categ. **21**(11), 191–209 (2008)
3. Adams, R.: QPEL: quantum program and effect language. In: Coecke, B., Hasuo, I., Panangaden, P. (eds.) Electrical Proceedings in Theoretical Computer Science on Quantum Physics and Logic (QPL) 2014, no. 172, pp. 133–153 (2014)
4. Adams, R., Jacobs, B.: A type theory for probabilistic and Bayesian reasoning (2015). arXiv.org/abs/1511.09230
5. Cho, K.: Total and partial computation in categorical quantum foundations. In: Heunen, C., Selinger, P., Vicary, J. (eds.) Electrical Proceedings in Theoretical Computer Science on Quantum Physics and Logic (QPL) 2015, no. 195, pp. 116–135 (2015)
6. Cho, K., Jacobs, B., Westerbaan, A., Westerbaan, B.: An introduction to effectus theory (2015). arXiv.org/abs/1512.05813
7. Coecke, B., Heunen, C., Kissinger, A.: Categories of quantum and classical channels. Quantum Inf. Process. 1–31 (2014)
8. Fong, B.: Causal theories: a categorical perspective on Bayesian networks. Master's thesis, Univ. of Oxford (2012). arXiv.org/abs/1301.6201
9. Furber, R., Jacobs, B.: Towards a categorical account of conditional probability. In: Heunen, C., Selinger, P., Vicary, J. (eds.) Electronic Proceedings in Theoretical Computer Science of Quantum Physics and Logic (QPL) 2015, no. 195, pp. 179–195 (2015)
10. Giry, M.: A categorical approach to probability theory. In: Banaschewski, B. (ed.) Categorical Aspects of Topology and Analysis. Lecture Notes in Mathematics, vol. 915, pp. 68–85. Springer, Berlin (1982)
11. Jacobs, B.: Semantics of weakening and contraction. Ann. Pure Appl. Logic **69**(1), 73–106 (1994)
12. Jacobs, B.: Measurable spaces and their effect logic. In: Logic in Computer Science. IEEE, Computer Science Press (2013)

13. Jacobs, B.: New directions in categorical logic, for classical, probabilistic and quantum logic. Logical Methods in Comp. Sci. **11**(3), 1–76 (2015)
14. Jacobs, B.: Effectuses from monads. In: MFPS 2016 (2016, to appear)
15. Jacobs, B., Mandemaker, J.: The expectation monad in quantum foundations. In: Jacobs, B., Selinger, P., Spitters, B. (eds.) Elecronic Proccedings in Theoretical Computer Science of Quantum Physics and Logic (QPL) 2011, no. 95, pp. 143–182 (2012)
16. Jacobs, B., Westerbaan, B., Westerbaan, B.: States of convex sets. In: Pitts, A. (ed.) FOSSACS 2015. LNCS, vol. 9034, pp. 87–101. Springer, Heidelberg (2015)
17. Jacobs, B., Zanasi, F.: A predicate/state transformer semantics for Bayesian learning. In: MFPS 2016 (2016, to appear)
18. Klin, Bartek: Structural operational semantics for weighted transition systems. In: Palsberg, Jens (ed.) Semantics and Algebraic Specification. LNCS, vol. 5700, pp. 121–139. Springer, Heidelberg (2009)
19. Kock, A.: Monads on symmetric monoidal closed categories. Arch. Math. **XXI**, 1–10 (1970)
20. Kock, A.: Bilinearity and cartesian closed monads. Math. Scand. **29**, 161–174 (1971)
21. Lindner, H.: Affine parts of monads. Arch. Math. **XXXIII**, 437–443 (1979)
22. Panangaden, P.: Labelled Markov Processes. Imperial College Press, London (2009)
23. Trnková, V.: Some properties of set functors. Comment. Math. Univ. Carolinae **10**, 323–352 (1969)

Duality of Equations and Coequations
via Contravariant Adjunctions

Julian Salamanca[1]([⊠]), Marcello Bonsangue[1,2], and Jurriaan Rot[3]

[1] CWI, Amsterdam, The Netherlands
salamanc@cwi.nl
[2] LIACS - Leiden University, Leiden, The Netherlands
[3] LIP, Université de Lyon, CNRS, Ecole Normale Supérieure de Lyon, INRIA,
Université Claude-Bernard Lyon 1, Lyon, France

Abstract. In this paper we show duality results between categories of equations and categories of coequations. These dualities are obtained as restrictions of dualities between categories of algebras and coalgebras, which arise by lifting contravariant adjunctions on the base categories. By extending this approach to (co)algebras for (co)monads, we retrieve the duality between equations and coequations for automata proved by Ballester-Bolinches, Cosme-Llópez and Rutten, and generalize it to dynamical systems.

1 Introduction

Equations play a fundamental role in (universal) algebra. Their categorical dual in universal coalgebra is the notion of *coequations*. Coequations were studied extensively in the search for a dual of Birkhoff's theorem and the specification of classes of coalgebras (see, e.g., [1, 2, 5, 9, 11, 12, 18, 19, 21–23]).

The aim of the current paper is a different one: to relate equations to coequations and vice versa. Our starting point is the abstract definition of (co)equations on (co)algebras for an endofunctor. These definitions give rise to categories of equations and coequations; we seek sufficient conditions to obtain dual equivalences between such categories.

We start with a more general concept than a duality, namely, a contravariant adjunction. Our approach is to lift adjunctions to categories of algebras and coalgebras [13]. In the setting of a contravariant adjunction, and by using preservation of limits by adjoints, we have that sets of equations are sent to sets of coequations. To guarantee the converse, i.e., that coequations are also mapped

---

J. Salamanca—The research of this author is funded by the Dutch NWO project 612.001.210.

J. Rot—Supported by the LABEX MILYON (ANR-10-LABX-0070) of Université de Lyon, within the program "Investissements d'Avenir" (ANR-11-IDEX-0007) operated by the French National Research Agency (ANR).

I. Hasuo (Ed.): CMCS 2016, LNCS 9608, pp. 73–93, 2016.
DOI: 10.1007/978-3-319-40370-0_6

to equations, we assume that the contravariant adjunction is a duality. This gives us a duality result between equations and coequations. We derive known dualities between equations and coequations for automata [7, 24, 25] as a special case of this abstract approach, and we generalize the duality shown in [7] to include (general) dynamical systems.

As a natural next step in this study we include monads and comonads into the picture and prove a lifting theorem to lift contravariant adjunctions to Eilenberg-Moore categories. From this lifting theorem we show the following results:

- Dualities between equations and coequations for Eilenberg-Moore categories.
- Lifting of contravariant adjunctions to Eilenberg-Moore categories where, given a contravariant adjunction and a comonad, we define a canonical monad.
- Lifting of dualities to Eilenberg-Moore categories where, given a duality and a monad, we define a canonical comonad.

The paper is organized as follows. Section 2 is a preliminary section in which we introduce some notation we use in the paper. In Sect. 3 we introduce the abstract definitions of equations and coequations, satisfaction of equations for algebras and satisfaction of coequations for coalgebras. Section 4 introduces the notion of a contravariant adjunction. We state a theorem for lifting contravariant adjunctions (Theorem 3), which is essentially a special case of [13, 2.14. Theorem], and then illustrate this lifting theorem through several examples. In Sect. 5 we focus on the particular case that the contravariant adjunction is a duality to show a general duality result between equations and coequations. Further, we show how to get a canonical notion of satisfaction of equations for coalgebras. In Sect. 6 we include monads and comonads in our setting to prove a lifting theorem (Theorem 11) that allows us to lift contravariant adjunctions to a contravariant adjunction between Eilenberg-Moore categories. We show how to construct a comonad from a given monad and vice versa to get respective lifting theorems. Finally, in Sect. 7 we apply the lifting theorems (to Eilenberg-Moore) to the study of equations and coequations for dynamical systems and deterministic automata.

2   Preliminaries

In this section we introduce the notation for categories of algebras and coalgebras that we will use in the paper. We assume that the reader is familiar with basic concepts from category theory and coalgebra, see, e.g., [6, 22].

Given a category $\mathcal{D}$ and an endofunctor $L \colon \mathcal{D} \to \mathcal{D}$, we denote by $\mathrm{alg}(L)$ the category of L-algebras and their homomorphisms, i.e., objects in $\mathrm{alg}(L)$ are pairs (X, α) where X is an object in $\mathcal{D}$ and $\alpha \in \mathcal{D}(LX, X)$, and a homomorphism from an L-algebra (X_1, α_1) to an L-algebra (X_2, α_2) is a morphism $h \in \mathcal{D}(X_1, X_2)$ such that $h \circ \alpha_1 = \alpha_2 \circ Lh$.

Dually, for a given endofunctor $B \colon \mathcal{C} \to \mathcal{C}$ on a category $\mathcal{C}$, $\mathrm{coalg}(B)$ denotes the category of B-coalgebras, i.e., objects in $\mathrm{coalg}(B)$ are pairs (Y, β) where Y

is an object in $\mathcal{C}$ and $\beta \in \mathcal{C}(Y, BY)$, and a homomorphism from a B-coalgebra (Y_1, β_1) to a B-coalgebra (Y_2, β_2) is a morphism $h \in \mathcal{C}(Y_1, Y_2)$ such that $\beta_2 \circ h = Bh \circ \beta_1$.

In case that we have a monad $\mathsf{L} = (L, \eta, \mu)$, we let $\mathrm{Alg}(\mathsf{L})$ denote the category of (Eilenberg-Moore) L-algebras, i.e., algebras for the monad L. Similarly, for a comonad $\mathsf{B} = (B, \epsilon, \delta)$, the category $\mathrm{Coalg}(\mathsf{B})$ consists of Eilenberg-Moore coalgebras for the comonad B. Notice that we use the notation L, B to refer to (co)monads, and L, B to refer to the underlying functors.

Each of the categories $\mathrm{alg}(L)$, $\mathrm{Alg}(\mathsf{L})$, $\mathrm{coalg}(B)$, and $\mathrm{Coalg}(\mathsf{B})$ has a canonical forgetful functor into the underlying category. For instance, the forgetful functor for $\mathrm{Alg}(\mathsf{L})$ is the functor $U \colon \mathrm{Alg}(\mathsf{L}) \to \mathcal{D}$ defined as $U(X, \alpha) = X$ and $Uf = f$ for any L-algebra morphism f. We will refer to those forgetful functors without giving them a specific name.

3   Equations and Coequations

We introduce the abstract definitions of equations and coequations. Let L be an endofunctor on $\mathcal{D}$ and S be an object in $\mathcal{D}$. The *free L-algebra on S generators* is an algebra $\mathfrak{F}(S) = (\mathfrak{F}(S), \tau) \in \mathrm{alg}(L)$ together with a morphism $\eta \in \mathcal{D}(S, \mathfrak{F}(S))$, called *unit*, satisfying the following universal property: for any L-algebra $X = (X, \alpha)$ and any morphism $f \in \mathcal{D}(S, X)$ there is a unique morphism $f^\sharp \in \mathrm{alg}(L)(\mathfrak{F}(S), X)$ such that $f^\sharp \circ \eta = f$, i.e., the following diagram commutes:

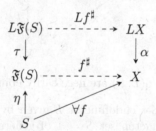

We define *equations* for L on S generators as epimorphisms with domain $\mathfrak{F}(S)$, i.e., elements $e_P \in \mathrm{alg}(L)(\mathfrak{F}(S), P)$ that are epimorphisms for some $P = (P, \zeta) \in \mathrm{alg}(L)$. Observe that if L is a polynomial functor on Set (see, e.g., [22, Section 10]) then equations can be identified with L-congruences C of $\mathfrak{F}(S)$, since $\mathfrak{F}(S)/C \cong P$ for $C = \ker(e_P)$, and elements in C are pairs of terms with variables on the set S. This corresponds to the classical definition of equations in universal algebra. Finally, we say that an L-algebra $X = (X, \alpha)$ satisfies the equation e_P, denoted as $(X, \alpha) \models e_P$, if for any morphism $f \in \mathcal{D}(S, X)$ the morphism $f^\sharp$ factors through e_P, i.e., there exists $g_f \in \mathrm{alg}(L)(P, X)$ such that the following diagram commutes:

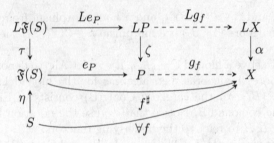

Now, assuming that the free L-algebra on S generators $\mathfrak{F}(S) = (\mathfrak{F}(S), \tau)$ exists, we can define the category $\mathrm{eq}(L, S)$ of equations for L on S generators as follows:

Objects of $\mathrm{eq}(L, S)$: epimorphisms $e_X \in \mathrm{alg}(L)(\mathfrak{F}(S), X)$ for some
$$X = (X, \alpha) \in \mathrm{alg}(L).$$
Arrows of $\mathrm{eq}(L, S)$: for $e_{X_i} \in \mathrm{eq}(L, S), i = 1, 2$, a morphism
$$f \in \mathrm{eq}(L, S)(e_{X_1}, e_{X_2}) \text{ is a morphism } f \in \mathrm{alg}(L)(X_1, X_2)$$
such that the following diagram commutes:

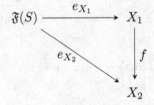

Notice that morphisms in $\mathrm{eq}(L, S)$ are necessarily epimorphisms.

Example 1. Consider the Set endofunctor L given by $LX = A \times X$, where A is a fixed set, and the singleton set $S = 1$ of generators. Then an L-algebra together with an assignment of the single generator is a *pointed deterministic automaton*, i.e., a triple (X, α, x) consisting of a set of states X, a transition function $\alpha \colon A \times X \to X$ and an element $x \in X$.

The free L-algebra on 1 is given by $A^* = (A^*, \tau)$ where $\tau \colon A \times A^* \to A^*$ is defined by $\tau(a, w) = wa$ and the unit $\eta \colon 1 \to A^*$ maps the single generator to the empty word $\varepsilon \in A^*$, i.e., $\eta = \varepsilon$. Given a pointed automaton (X, α, x) we obtain a unique homomorphism $r_x \colon A^* \to X$, given by $r_x(\varepsilon) = x$ and $r_x(wa) = \alpha(a, r_x(w))$. In the sequel we sometimes denote $r_x(w)$ by $w(x)$, the state we reach from the state x by processing the word w.

A *right congruence on A^** is an equivalence relation $C \subseteq A^* \times A^*$ such that for any $a \in A$ and $(u, v) \in C$ we have that $(ua, va) \in C$. Right congruences C correspond to equations as defined above, by letting $A^*/C = (A^*/C, [\tau]) \in \mathrm{alg}(L)$ where $[\tau]$ is given by $[\tau](a, [w]) = [wa]$ and the epimorphism (equation) $e_C \in \mathrm{alg}(L)(A^*, A^*/C)$ maps every word to its equivalence class.

An L-algebra (X, α) satisfies the equation e_C, i.e., $(X, \alpha) \models C$, if and only if for every $(u, v) \in C$ and any $x \in X$, we have $r_x(u) = r_x(v)$. This coincides with satisfaction of equations as defined in [7].

Notice that the function $\tau' : A \times A^* \to A^*$ defined as $\tau'(a, w) = aw$ is such that the algebra (A^*, τ') is also a free L-algebra, which gives us the notion of left congruence as a corresponding notion of equation.   □

We dualize the definition of equations to obtain the definition of coequations, e.g., [18,19,22]. Let B be an endofunctor on $\mathcal{C}$ and R be an object in $\mathcal{C}$. The *cofree B-coalgebra on R colours* is a coalgebra $\mathfrak{C}(R) = (\mathfrak{C}(R), \upsilon) \in \mathrm{coalg}(B)$ together with a morphism $\epsilon \in \mathcal{C}(\mathfrak{C}(R), R)$, called *counit*, satisfying the following universal property: for any B-coalgebra $Y = (Y, \beta)$ and any morphism (colouring) $f \in \mathcal{C}(Y, R)$ there is a unique morphism $f^\flat \in \mathrm{coalg}(L)(Y, \mathfrak{C}(R))$ such that $\epsilon \circ f^\flat = f$, i.e., the following diagram commutes:

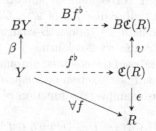

We define *coequations* for B on R colours as monomorphisms with codomain $\mathfrak{C}(R)$, i.e., elements $m_Q \in \mathrm{coalg}(B)(Q, \mathfrak{C}(R))$ that are monomorphisms for some $Q = (Q, \delta) \in \mathrm{coalg}(B)$. We say that a B-coalgebra $Y = (Y, \beta)$ satisfies the coequation m_Q, denoted as $(Y, \beta) \Vvdash m_Q$ (notice the difference between the symbols: $\models$ for equations and $\Vvdash$ for coequations), if for any morphism (colouring) $f \in \mathcal{C}(Y, R)$ the morphism $f^\flat$ factors through m_Q, i.e., there exists $g_f \in \mathrm{coalg}(B)(Y, Q)$ such that the following diagram commutes:

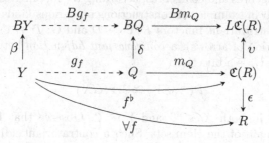

Assuming that the cofree B-coalgebra on R colours $\mathfrak{C}(R) = (\mathfrak{C}(R), \upsilon)$ exists, define the category $\mathrm{coeq}(B, R)$ of coequations for B on R colours whose objects are monomorphisms $m_Y \in \mathrm{coalg}(B)(Y, \mathfrak{C}(R))$ for some $Y = (Y, \beta) \in \mathrm{coalg}(B)$, and, a morphism between two objects m_{Y_1} and m_{Y_2} in $\mathrm{coeq}(B, R)$ is a morphism $g \in \mathrm{coalg}(B)(Y_1, Y_2)$ such that $m_{Y_2} \circ g = m_{Y_1}$. Notice that morphisms in $\mathrm{coeq}(B, R)$ are necessarily monomorphisms.

Example 2. For a given set A, consider the Set endofunctor B defined by $BX = X^A$, and consider the two-element set $R = 2$ of colours. Then a B-coalgebra together with an assignment of colours to states is a *coloured deterministic automaton*: a triple (Y, β, f) consisting of a set of states Y, a transition function $\beta \colon Y \to Y^A$ and an assignment of final states $f \colon Y \to 2$.

The cofree B-coalgebra on 2 colours is given by $2^{A^*} = (2^{A^*}, \upsilon)$ where $\upsilon \colon 2^{A^*} \to (2^{A^*})^A$ is given by right derivative

$$\upsilon(L)(a) = L_a = \{w \mid aw \in L\}$$

and the counit $\epsilon \colon 2^{A^*} \to 2$ is given by $\epsilon(L) = L(\varepsilon)$. Given a coloured deterministic automaton (Y, β, f), we obtain a unique B-coalgebra morphism $l \colon Y \to 2^{A^*}$ that maps every state to the language it accepts, i.e., $l(x)(\varepsilon) = f(x)$ and $l(x)(aw) = l(\beta(x)(a))(w)$.

Coequations for B on R correspond to subsets of 2^{A^*} that are closed under right derivatives, i.e., subcoalgebras of 2^{A^*}. Given any monomorphism (coequation) m_Q with codomain 2^{A^*} and a B-coalgebra (Y, β), we have that $(Y, \beta) \models m_Q$ if and only if for every 2-colouring $f \in \mathrm{Set}(Y, 2)$ the set of those languages accepted by the states of the coloured automaton (Y, β, f) is contained in $\mathrm{Im}(m_Q)$. This coincides with satisfaction of coequations as defined in [7].

Similarly to the previous example, the function $\upsilon' \colon 2^{A^*} \to (2^{A^*})^A$ given by left derivative

$$\upsilon'(L)(a) = {}_aL = \{w \mid wa \in L\}$$

is such that $(2^{A^*}, \upsilon')$ is also a cofree B-coalgebra for which the corresponding notion of coequations are subsets of 2^{A^*} closed under left derivatives. $\square$

4   Lifting Contravariant Adjunctions

In this section we recall the notion of a contravariant adjunction and how to lift it to categories of algebras and coalgebras, according to [13,14]. We instantiate this abstract approach in examples of constructions on various kinds of automata.

Given two contravariant functors $F \colon \mathcal{C} \to \mathcal{D}$ and $G \colon \mathcal{D} \to \mathcal{C}$ (i.e., F and G reverse the direction of arrows), a *contravariant adjunction* between F and G, denoted by $F \dashv\vdash G$, is a bijection

$$\mathcal{D}(X, FY) \cong \mathcal{C}(Y, GX)$$

which is natural in both $X \in \mathcal{D}$ and $Y \in \mathcal{C}$. Observe that both F and G are on the codomain of the Hom-sets. Such a contravariant adjunction can be equivalently defined by two units $\eta^{FG} \colon Id_{\mathcal{D}} \Rightarrow FG$ and $\eta^{GF} \colon Id_{\mathcal{C}} \Rightarrow GF$ that satisfy the triangle identities $G\eta^{FG} \circ \eta^{GF}_G = Id_G$ and $F\eta^{GF} \circ \eta^{FG}_F = Id_F$. By standard preservation properties, both F and G map colimits to limits, in particular initial objects to final objects and epimorphisms to monomorphisms.

Given a contravariant adjunction as above, if η^{GF} and η^{FG} are isomorphisms then we say F, G form a *duality*, and denote it by $F \cong G$. In this case, limits are mapped to colimits and vice versa.

Our basic setting consists of a contravariant adjunction between F and G, an endofunctor B on $\mathcal{C}$ and an endofunctor L on $\mathcal{D}$, depicted in the diagram below. Throughout this paper we depict contravariant functors in diagrams with an '×' at the beginning of the arrow.

$$B \circlearrowright \mathcal{C} \underset{G}{\overset{F}{\rightleftarrows}} \mathcal{D} \circlearrowleft L$$

In this setting, we are interested in lifting the adjunction to a contravariant adjunction between lifted functors $\widehat{F}\colon \mathrm{coalg}(B) \to \mathrm{alg}(L)$ and $\widehat{G}\colon \mathrm{alg}(L) \to \mathrm{coalg}(B)$ of F and G, respectively, as in the following picture:

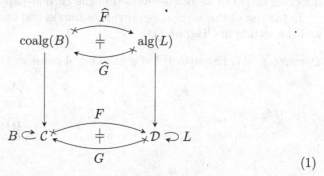

$$(1)$$

where the vertical arrows are forgetful functors. An important consequence of such a lifting is that, if L has an initial algebra, then it is mapped by $\widehat{G}$ to a final B-coalgebra.

In [13, 2.14.Theorem] it is shown that a sufficient condition for such a lifting is the existence of a natural isomorphism $\gamma\colon GL \Rightarrow BG$. This is summarized by the theorem below.

Theorem 3. *Let $F\colon \mathcal{C} \to \mathcal{D}$ and $G\colon \mathcal{D} \to \mathcal{C}$ be contravariant functors that form a contravariant adjunction. Let B be an endofunctor on $\mathcal{C}$ and L be an endofunctor on $\mathcal{D}$. If there is a natural isomorphism $\gamma\colon GL \Rightarrow BG$, then*

1. *The adjunction $F \dashv G$ lifts to an adjunction as in Diagram (1), i.e., to a contravariant adjunction between functors $\widehat{F}\colon \mathrm{coalg}(B) \to \mathrm{alg}(L)$ and $\widehat{G}\colon \mathrm{alg}(L) \to \mathrm{coalg}(B)$.*
2. *If F, G form a duality then $\widehat{F}, \widehat{G}$ form a duality as well.*

The functors $\widehat{F}\colon \mathrm{coalg}(B) \to \mathrm{alg}(L)$ and $\widehat{G}\colon \mathrm{alg}(L) \to \mathrm{coalg}(B)$ are defined on objects as:

$$(Y \xrightarrow{\beta} BY) \xmapsto{\widehat{F}} (LFY \xrightarrow{\rho_Y} FBY \xrightarrow{F\beta} FY)$$
$$(LX \xrightarrow{\alpha} X) \xmapsto{\widehat{G}} (GX \xrightarrow{G\alpha} GLX \xrightarrow{\gamma_X} BGX)$$

and on morphisms as $\widehat{F} = F$ and $\widehat{G} = G$. The natural transformation $\rho\colon LF \Rightarrow FB$ in the definition of $\widehat{F}$ is defined as the *mate* of the inverse $\gamma^{-1}\colon BG \Rightarrow GL$:

$$\rho \overset{\text{def.}}{=} FB\eta^{GF} \circ F\gamma^{-1}F \circ \eta^{FG}LF,$$

using the units η^{GF} and η^{FG} of the adjunction. Natural transformations of the form $\rho\colon LF \Rightarrow FB$ and the definition of $\widehat{F}$ form the heart of the approach to *coalgebraic modal logic* based on contravariant adjunctions/dualities (see, e.g., [8,15,16,20]). There is a one-to-one correspondence between such natural transformations and those of the form $BG \Rightarrow GL$, using the above construction. We are only interested in the case where the natural transformation $BG \Rightarrow GL$ is an isomorphism, to lift adjunctions, as in [17]. For notational convenience, the direction in $\gamma\colon GL \Rightarrow BG$ is reversed in the current paper.

In the rest of this section we provide examples and applications of Theorem 3 and the setting in Diagram (1).

Example 4. [17, Example 4] For a fixed set A consider the following situation:

$$FX = GX = 2^X$$
$$LX = (A \times X) + 1$$
$$BY = 2 \times Y^A$$
$$\gamma_X \colon 2^{A \times X + 1} \to 2 \times (2^X)^A$$

Here L-algebras are *pointed deterministic automata* on A (Example 1) and B-coalgebras are *two-coloured deterministic automata* on A (Example 2). The contravariant functors F and G form a contravariant adjunction which, by Theorem 3, can be lifted to an adjunction between $\widehat{F}$ and $\widehat{G}$. The isomorphism $\gamma\colon GL \Rightarrow BG$ is defined for any X as the function $\gamma_X\colon 2^{A \times X + 1} \to 2 \times (2^X)^A$ such that $\gamma_X(f) = (f(\cdot), \lambda a.\ \lambda x.f(a,x))$.

Given a B-coalgebra $(Y, \langle c, \beta \rangle)$ we have that $\widehat{F}(Y, \langle c, \beta \rangle) = (2^Y, [\alpha, i])$ where $\alpha\colon A \times 2^Y \to 2^Y$ and $i\colon 1 \to 2^Y$ are functions defined as follows:

$$i(\cdot) = c^{-1}(\{1\}) = \text{ accepting states of } (Y, \langle c, \beta \rangle),$$
$$\alpha(a, Z) = \{y \in Y \mid \beta(y)(a) \in Z\}.$$

Given an L-algebra $(X, [\alpha, i])$ we have that $\widehat{G}(X, [\alpha, i]) = (2^X, \langle c, \beta \rangle)$ where the functions $c\colon 2^X \to 2$ and $\beta\colon 2^X \to (2^X)^A$ are defined as:

$$c(Z) = 1 \text{ iff } i(\cdot) \in Z$$
$$\beta(Z)(a) = \{x \in X \mid \alpha(a, x) \in Z\}$$

Recall from Example 1 that the initial L-algebra is given by $(A^*, [\eta, \tau'])$, where A^* is the free monoid with generators A and identity element ε, $\eta\colon 1 \to A^*$ is the empty word ε and $\tau'\colon A \times A^* \to A^*$ is the concatenation function given by $\tau'(a, w) = aw$. Because of the contravariant adjunction, the initial L-algebra is sent by $\widehat{G}$ to the final B-coalgebra, given by $\widehat{G}(A^*, [\eta, \tau']) = (2^{A^*}, \langle \hat{\epsilon}, \hat{\tau} \rangle)$ where $\hat{\epsilon}(L) = L(\epsilon)$ and $\hat{\tau}(L)(a)(w) = L(aw)$. Note that the final B-coalgebra is not sent by $\widehat{F}$ to the initial L-algebra. □

Example 5. Let CABA be the category of complete atomic Boolean algebras whose morphisms are complete Boolean algebra homomorphisms. For a fixed set A consider the following situation:

$$B \supset \text{CABA} \underset{G}{\overset{F}{\underset{\cong}{\rightleftarrows}}} \text{Set} \subset L \qquad \begin{aligned} FY &= \text{At}(Y) \\ GX &= 2^X \\ LX &= (A \times X) + 1 \\ BY &= 2 \times Y^A \\ \gamma_X &: 2^{A \times X + 1} \to 2 \times (2^X)^A \end{aligned}$$

Here $\text{At}(Y)$ denotes the set of atoms of the object Y in CABA. The contravariant functors F and G form a contravariant adjunction, in fact a duality, which, by Theorem 3, can be lifted to a duality between $\widehat{F}$ and $\widehat{G}$ if we consider the canonical natural isomorphism $\gamma: GL \Rightarrow BG$ defined for every X as the morphism $\gamma_X: 2^{A \times X + 1} \to 2 \times (2^X)^A$ such that $\gamma_X(f) = (f(\cdot), \lambda a.\ \lambda x. f(a, x))$.

Given a B-coalgebra $(Y, \langle c, \beta \rangle)$, we have that $\widehat{F}(Y, \langle c, \beta \rangle) = (\text{At}(Y), [\alpha, i])$ where the functions $\alpha: A \times \text{At}(Y) \to \text{At}(Y)$ and $i: 1 \to \text{At}(Y)$ are defined as follows:

$$i(\cdot) = \text{ the unique element } y_0 \in \text{At}(Y) \text{ s.t. } c(y_0) = 1,$$
$$\alpha(a, y) = \text{ the unique element } y' \in \text{At}(Y) \text{ s.t. } \beta(y')(a) \geq y.$$

In particular, if $P \subseteq 2^{A^*}$ is a *preformation of languages* [7, Definition 11], i.e., $P \in \text{CABA}$ and it is closed under left and right derivatives[1], then $(P, \langle \hat{\epsilon}, \hat{\tau}' \rangle) \in \text{coalg}(B)$ where $\hat{\epsilon}(L) = L(\epsilon)$ and $\hat{\tau}'(L)(a) = {}_aL$. In this case, $\widehat{F}(P, \langle \hat{\epsilon}, \hat{\tau}' \rangle) = \mathbf{free}(P)$ which is the quotient A^*/C where C is the set, in fact congruence, of all equations satisfied by the automaton $(P, \hat{\tau})$, where $\hat{\tau}(L)(a) = L_a$ (see [7]).

Given an L-algebra $(X, [\alpha, i])$, we have that $\widehat{G}(X, [\alpha, i]) = (2^X, \langle c, \beta \rangle)$ where the CABA morphisms $c: 2^X \to 2$ and $\beta: 2^X \to (2^X)^A$ are defined as

$$c(Z) = 1 \text{ iff } i(\cdot) \in Z$$
$$\beta(Z)(a) = \{x \in X \mid \alpha(a, x) \in Z\}.$$

In particular, if C is a congruence of the monoid A^* then $(A^*/C, [[\tau'], [\epsilon]]) \in \text{alg}(L)$ where $[\tau'](a, [w]) = [aw]$. In this case, $\widehat{G}(A^*/C, [[\tau'], [\epsilon]]) \cong \mathbf{cofree}(A^*/C)$ which is the minimum set of coequations that the automaton $(A^*/C, [\tau])$ satisfies, where $[\tau](a, [w]) = [wa]$ (see [7]).

---

[1] $P \subseteq 2^{A^*}$ is closed under right (left) derivatives if for every $L \in P$ and $a \in A$, $L_a \in P$ (${}_aL \in P$). Here $L_a(w) = L(aw)$, and ${}_aL(w) = L(wa)$, $w \in A^*$.

Similarly to the previous example, the initial L-algebra $A^* = (A^*, [\eta, \tau'])$, where $\eta = \varepsilon$ and $\tau'(a, w) = aw$, is sent by $\widehat{G}$ to the final B-coalgebra $2^{A^*} = (2^{A^*}, \langle \hat{\varepsilon}, \hat{\tau} \rangle)$, where $\hat{\varepsilon}(L) = L(\varepsilon)$ and $\hat{\tau}(L)(a)(w) = L(aw)$. Also, because the contravariant adjunction is a duality, the final B-coalgebra 2^{A^*} is sent by $\widehat{F}$ to the initial L-algebra A^*. We will explore this case further in Sect. 5 to get dualities between sets of equations and sets of coequations.   □

Example 6. For a fixed field $\mathbb{K}$, let $\mathrm{Vec}_{\mathbb{K}}$ be the category of vector spaces over $\mathbb{K}$ with linear maps. Let A be a fixed set and consider the following situation:

$$B \subset \mathrm{Vec}_{\mathbb{K}} \overset{F}{\underset{G}{\rightleftarrows}} \mathrm{Vec}_{\mathbb{K}} \supset L \qquad \begin{aligned} FX = GX &= X^\partial \\ LX &= \mathbb{K} + (A \times X) \\ BY &= \mathbb{K} \times Y^A \\ \gamma_X \colon (\mathbb{K} + A \times X)^\partial &\to \mathbb{K} \times (X^\partial)^A \end{aligned}$$

Here $X^\partial = \mathrm{Vec}_{\mathbb{K}}(X, \mathbb{K})$, the *dual space* of X, and $A \times X := \coprod_{a \in A} X$. We have that the contravariant functors F and G form a contravariant adjunction which, by Theorem 3, can be lifted to a contravariant adjunction between $\widehat{F}$ and $\widehat{G}$ if we consider the canonical natural isomorphism $\gamma \colon GL \Rightarrow BG$ defined for every X as the map $\gamma_X \colon (\mathbb{K} + A \times X)^\partial \to \mathbb{K} \times (X^\partial)^A$ such that $\gamma_X(\varphi) = (\varphi(1), \lambda a. \lambda x. \varphi(a, x))$.

Given a B-coalgebra $(Y, \langle c, \beta \rangle)$, we have that $\widehat{F}(Y, \langle c, \beta \rangle) = (Y^\partial, [i, \alpha])$ where $i \colon \mathbb{K} \to Y^\partial$ and $\alpha \colon A \times Y^\partial \to Y^\partial$ are linear maps which are defined on the canonical basis as:

$$i(1)(y) = c(y)$$
$$\alpha(a, \varphi)(y) = (\varphi \circ \beta(y))(a) = \varphi(\beta(y)(a))$$

In particular, if $S \subseteq \mathbb{K}^{A^*}$ is a subsystem such that for every $f \in S$ and $a \in A$, $f_a, {}_af \in S$, where $f_a(w) = f(aw)$ and ${}_af(w) = f(wa)$, $w \in A^*$, then we have that $(S, \langle \hat{\varepsilon}, \hat{\tau}' \rangle) \in \mathrm{coalg}(B)$ where $\hat{\varepsilon}(f) = f(\varepsilon)$ and $\hat{\tau}'(f)(a) = {}_af$. In this case, $\widehat{F}(S, \langle \hat{\varepsilon}, \hat{\tau}' \rangle) \cong \mathbf{free}(S)$ which is the quotient $V(A^*)/C$ where C is the set, in fact linear congruence, of all linear equations satisfied by the automaton $(S, \hat{\tau})$. Here $V(A^*) = \{\phi \colon A^* \to \mathbb{K} \mid \mathrm{supp}(\phi) \text{ is finite}\}$, where $\mathrm{supp}(\phi) = \{w \in A^* \mid \phi(w) \neq 0\}$ is the support of ϕ, and the function $\hat{\tau}$ is defined as $\hat{\tau}(f)(a) = f_a$ (see [25]).

Given an L-algebra $(X, [i, \alpha])$, we have that $\widehat{G}(X, [i, \alpha]) = (X^\partial, \langle c, \beta \rangle)$ where the linear maps $c \colon X^\partial \to \mathbb{K}$ and $\beta \colon X^\partial \to (X^\partial)^A$ are defined as

$$c(\varphi) = \varphi(i(1))$$
$$\beta(\varphi)(a)(x) = \varphi(\alpha(a, x))$$

In particular, if $C \subseteq V(A^*) \times V(A^*)$ is a linear congruence on $V(A^*)$, then we have that $(V(A^*)/C, [[\tau'], [\varepsilon]]) \in \mathrm{alg}(L)$, where $[\tau'](a, [\phi]) = [a\phi]$, and we have that $\widehat{G}(V(A^*)/C, [[\tau'], [\varepsilon]]) \cong \mathbf{cofree}(V(A^*)/C)$ which is the minimum set of coequations (power series) satisfied by the automaton $(V(A^*)/C, [\tau])$, where $[\tau](a, [\phi]) = [\phi a]$ (see [25]).

Notice that the contravariant adjunction is not a duality, but if we restrict to vector spaces of finite dimension then we get a duality. In the latter case there is no initial L-algebra or, equivalently, there is no final B-coalgebra.   □

5   Duality Between Equations and Coequations

In Sect. 3, we defined equations as epimorphisms from an initial algebra and coequations as monomorphisms into a final coalgebra. In the previous section, we have seen how to relate initial algebras and final coalgebras by lifting contravariant adjunctions and dualities. Next, we describe how to apply these liftings to obtain a correspondence between equations and coequations.

If we lift the contravariant adjunction on the base categories to a contravariant adjunction $\widehat{F}\colon \mathrm{coalg}(B) \to \mathrm{alg}(L)$ and $\widehat{G}\colon \mathrm{alg}(L) \to \mathrm{coalg}(B)$ as in the previous section, then $\widehat{G}$ sends the initial L-algebra to the final B-coalgebra, and $\widehat{G}$ sends epimorphisms to monomorphisms. As a consequence, equations are sent by $\widehat{G}$ to coequations. However, $\widehat{F}$ does not map coequations to equations, in general.

In order to obtain a full correspondence between equations and coequations, suppose that the contravariant adjunction between F and G is a duality (and that there is a natural isomorphism $\gamma\colon GL \Rightarrow BG$). Then, by Theorem 3, the duality between F and G lifts to a duality between $\widehat{F}$ and $\widehat{G}$. In this case, we can add another level to the picture in (1), yielding a duality between equations and coequations:

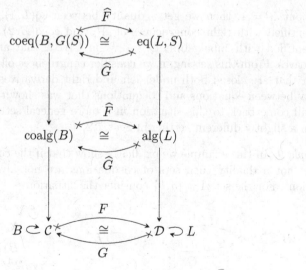

$$\tag{2}$$

where $\mathrm{eq}(L, S)$ and $\mathrm{coeq}(B, G(S))$ are the categories of equations for L on S generators and coequations for B on $G(S)$ colours respectively, as defined in

Sect. 3, lower vertical arrows are forgetful functors, and upper vertical arrows are the canonical functors $U\colon \mathrm{coeq}(B, G(S)) \to \mathrm{coalg}(B)$ and $V\colon \mathrm{eq}(L, S) \to \mathrm{alg}(S)$ which are defined as $U(m_Y) = Y$ and $V(e_X) = X$ on objects and $Uf = f$ and $Vg = g$ on morphisms.

Theorem 7. *Let $F\colon \mathcal{C} \to \mathcal{D}$ and $G\colon \mathcal{D} \to \mathcal{C}$ be contravariant functors that form a duality. Let B be an endofunctor on $\mathcal{C}$, L be an endofunctor on $\mathcal{D}$ with an object S in $\mathcal{D}$ such that the free L-algebra $\mathfrak{F}(S)$ on S generators exists. If there is a natural isomorphism $\gamma\colon GL \Rightarrow BG$ then:*

1. *The duality between F and G lifts to a duality $\widehat{F}\colon \mathrm{coeq}(B, G(S)) \to \mathrm{eq}(L, S)$ and $\widehat{G}\colon \mathrm{eq}(L, S) \to \mathrm{coeq}(B, G(S))$, as in Diagram (2).*
2. *Given $e_P \in \mathrm{eq}(L, S)$, $m_Q \in \mathrm{coeq}(B, G(S))$, $(X, \alpha) \in \mathrm{alg}(L)$, and $(Y, \beta) \in \mathrm{coalg}(B)$ we have:*
   (i) *$(X, \alpha) \models e_P$ if and only if $\widehat{G}(X, \alpha) \models \widehat{G}(e_P)$.*
   (ii) *$\widehat{F}(Y, \beta) \models \widehat{F}(m_Q)$ if and only if $(Y, \beta) \models m_Q$.*

As an application of the previous theorem we have the following.

Example 8. (cf. Example 5) For a fixed set A consider the following situation:

$$FY = \mathrm{At}(Y)$$
$$GX = 2^X$$
$$B \, \circlearrowright \, \mathbf{CABA} \overset{F}{\underset{G}{\rightleftarrows}} \cong \, \mathbf{Set} \, \circlearrowleft L \qquad LX = A \times X$$
$$BY = Y^A$$
$$\gamma_X \colon 2^{A \times X} \to (2^X)^A$$

If we put $S = 1$, then we get a duality between $\mathrm{eq}(L, 1)$, whose objects can be identified with right congruences of A^*, and $\mathrm{coeq}(B, 2)$, whose objects can be identified with subalgebras $Q \subseteq 2^{A^*}$ in $\mathbf{CABA}$ that are closed under left derivatives. From this setting, if we consider congruences of A^* and subalgebras of 2^{A^*} that are closed both under left and right derivatives, we can derive the duality between equations and coequations that was shown in [7, Theorem 22]. We will come back to this situation in a more general setting in Sect. 7.1 and also in a slightly different setting in Sect. 7.2. □

Example 9. In this example we explicitly show that if the contravariant adjunction is not a duality then sets of coequations are not always sent to sets of equations. For the set $A = \{a, b\}$ consider the situation:

$$FX = GX = 2^X$$
$$B \, \circlearrowright \, \mathbf{Set} \overset{F}{\underset{G}{\rightleftarrows}} \dashv \, \mathbf{Set} \, \circlearrowleft L \qquad LX = A \times X$$
$$BY = Y^A$$
$$\gamma_X \colon 2^{A \times X} \to (2^X)^A$$

In this case, consider the set $S = 1$ of generators. The free L-algebra on S generators is given by $A^* = (A^*, \tau')$, where $\tau'(a, w) = aw$, and unit $\eta = \varepsilon$. The cofree B-coalgebra on $G(S) = 2$ colours is the coalgebra $2^{A^*} = (2^{A^*}, v) = \widehat{G}(A^*)$, where $v(L)(a)(w) = L(aw)$, and counit $\epsilon(L) = L(\varepsilon)$. Now, consider the element $m_Q \in \mathrm{coeq}(B, 2)$ where $Q = \{\emptyset, A^*\}$ and m_Q is the inclusion map $m_Q \colon Q \to 2^{A^*}$. Then the codomain of $\widehat{F}(m_Q)$ is $(2^Q, \alpha)$ where $\alpha(a, f) = \alpha(b, f) = f$ for all $f \in 2^Q$ (this definition of α follows from Example 4).

We have that $2^Q = (2^Q, \alpha)$ cannot be a homomorphic image of A^*. In fact, if there exists an epimorphism $e \in \mathrm{alg}(L)(A^*, 2^Q)$ then there is a right congruence C of A^* such that $(A^*/C, [\tau']) \cong (2^Q, \dot\alpha)$ which means that A^*/C has four equivalence classes and for each equivalence class $[w] \in A^*/C$ we have that $[w] = [aw] = [bw]$, which is a contradiction since the last equality implies that there is only one equivalence class. □

5.1   Equations for Coalgebras

In this section we show how to define equations for coalgebras by using liftings of contravariant adjunctions. The concepts presented here can be dualized to define coequations for algebras.

Assume that we have lifted a contravariant adjunction between functors $F \colon \mathcal{C} \to \mathcal{D}$ and $G \colon \mathcal{D} \to \mathcal{C}$ to a contravariant adjunction between $\widehat{F} \colon \mathrm{coalg}(B) \to \mathrm{alg}(L)$ and $\widehat{G} \colon \mathrm{alg}(L) \to \mathrm{coalg}(B)$ for an endofunctor B on $\mathcal{C}$ and an endofunctor L on $\mathcal{D}$. Given an equation $e_P \in \mathrm{eq}(L, S)$ for some S in $\mathcal{D}$, we define, for a given coalgebra (Y, β) in $\mathrm{coalg}(B)$, $(Y, \beta) \models e_P$, and say that the coalgebra (Y, β) satisfies the equation e_P, as:

$$(Y, \beta) \models e_P \overset{def}{\Leftrightarrow} \widehat{F}(Y, \beta) \models e_P.$$

Notice that if $\widehat{F}$ and $\widehat{G}$ form a duality then $\widehat{F}(Y, \beta) \models e_P$ is equivalent to $(Y, \beta) \models \widehat{G}(e_P)$. One could be tempted to use $(Y, \beta) \models \widehat{G}(e_P)$ as a definition for $(Y, \beta) \models e_P$ since $\widehat{G}(e_P) \in \mathrm{coeq}(B, G(S))$ but we prefer to avoid this since the dual argument is not true in general, i.e., given $m_Q \in \mathrm{coeq}(B, G(S))$, $\widehat{F}(m_Q)$ is not always in $\mathrm{eq}(L, S)$, as it was shown in Example 9.

Example 10. Consider the situation given in Example 4 and let $S = \emptyset$. Then we have that for a B-coalgebra (deterministic automaton) $Y = (Y, \langle c, \beta \rangle)$ and a right congruence C on A^*:

$$(Y, \langle c, \beta \rangle) \models C \iff \forall (u, v) \in C \ \ u(i(\cdot)) = v(i(\cdot)) \text{ in } \widehat{F}(Y, \langle c, \beta \rangle)$$
$$\iff \forall (u, v) \in C \ \ \{x \in X \mid c(u(x)) = 1\} = \{x \in X \mid c(v(x)) = 1\}.$$

In words, a right congruence C on A^* is satisfied by $Y = (Y, \langle c, \beta \rangle)$ if for every pair $(u, v) \in C$ the set of states that accept u coincides with the set of states that accept v.

In Example 1 we also defined satisfaction of right congruences for deterministic automata, as the canonical notion that arises by viewing (the transition structure of) automata as algebras. According to this, if we consider (Y, β) as an $A \times Id_{\mathrm{Set}}$-algebra, we have a direct definition for $(Y, \beta) \models C$. We conclude this example by showing the relation between $(Y, \langle c, \beta \rangle) \models C$ and $(Y, \beta) \models C$.

Consider the coloured automaton $(Y, \langle c, \beta \rangle)$ on $A = \{a\}$ given by:

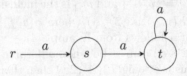

If we denote by $\langle u = v \rangle$ the least right congruence containing the pair $(u, v) \in A^* \times A^*$, then we have that $(Y, \langle c, \beta \rangle) \models \langle a = aa \rangle$ since

$$\{y \in Y \mid c(a(y)) = 1\} = \{r, s, t\} = \{y \in Y \mid c(aa(y)) = 1\}$$

but $(Y, \beta) \not\models \langle a = aa \rangle$ since $a(r) = s \neq t = aa(r)$. One can prove that $(Y, \beta) \models \langle u = v \rangle$ implies $(Y, \langle c, \beta \rangle) \models \langle u = v \rangle$ and that the converse holds if $(Y, \langle c, \beta \rangle)$ is minimal.   □

6 Lifting Contravariant Adjunctions to Eilenberg-Moore Categories

In this section we extend the results from the previous sections, on lifting adjunctions and dualities, to the case that the endofunctor L is a monad and the functor B is a comonad. We state the main theorem for lifting contravariant adjunctions to Eilenberg-Moore categories (Theorem 11), and obtain a theorem for dualities between equations and coequations as a consequence. Further, given either a monad or a comonad, we show how to derive a corresponding canonical comonad or monad, respectively.

Assume a contravariant adjunction between $F \colon \mathcal{C} \to \mathcal{D}$ and $G \colon \mathcal{D} \to \mathcal{C}$, a monad $\mathsf{L} = (L, \eta, \mu)$ on $\mathcal{D}$, and a comonad $\mathsf{B} = (B, \epsilon, \delta)$ on $\mathcal{C}$, as summarized in the following picture:

$$\mathsf{B} = (B, \epsilon, \delta) \circlearrowright \mathcal{C} \underset{G}{\overset{F}{\rightleftarrows}} \mathcal{D} \circlearrowleft \mathsf{L} = (L, \eta, \mu)$$

Then we can ask under what conditions the contravariant adjunction can be lifted to functors $\widehat{F} \colon \mathrm{Coalg}(\mathsf{B}) \to \mathrm{Alg}(\mathsf{L})$ and $\widehat{G} \colon \mathrm{Alg}(\mathsf{L}) \to \mathrm{Coalg}(\mathsf{B})$ on the Eilenberg-Moore categories. Similar to the approach in Sect. 4, we require a natural isomorphism $\gamma \colon GL \Rightarrow BG$, but for the current case we also require γ to satisfy certain conditions that relate the monad L and the comonad B.

Theorem 11. *Let $F \colon \mathcal{C} \to \mathcal{D}$ and $G \colon \mathcal{D} \to \mathcal{C}$ be contravariant functors that form a contravariant adjunction. Let $\mathsf{L} = (L, \eta, \mu)$ be a monad on $\mathcal{D}$, and $\mathsf{B} = (B, \epsilon, \delta)$*

a comonad on $\mathcal{C}$. *If there is a natural isomorphism* $\gamma \colon GL \Rightarrow BG$ *such that the following two diagrams commute:*

$$
\begin{array}{ccc}
G & \xleftarrow{\;\epsilon_G\;} & BG \\
\;\;\nwarrow{\scriptstyle G\eta} & & \uparrow{\scriptstyle \gamma} \\
& GL &
\end{array}
\qquad
\begin{array}{ccc}
BBG & \xleftarrow{\quad\quad \delta_G \quad\quad} & BG \\
\uparrow{\scriptstyle B\gamma} & & \uparrow{\scriptstyle \gamma} \\
BGL & \xleftarrow{\gamma_L} \; GLL \xleftarrow{G\mu} & GL
\end{array}
\qquad (3)
$$

then F *lifts to a functor* $\widehat{F} \colon \mathrm{Coalg(B)} \to \mathrm{Alg(L)}$ *and* G *lifts to a functor* $\widehat{G} \colon \mathrm{Alg(L)} \to \mathrm{Coalg(B)}$, *such that* $\widehat{F}$ *and* $\widehat{G}$ *form a contravariant adjunction. Additionally, if* F *and* G *form a duality then* $\widehat{F}$ *and* $\widehat{G}$ *form a duality.*

As an application of the previous theorem we can derive dualities between equations and coequations in Eilenberg-Moore categories, whose general result is obtained in a similar way as in Sect. 3. Notice that we do not need to explicitly assume the existence of free algebras since for any object S in $\mathcal{D}$ the algebra $(LS, \mu_S) \in \mathrm{Alg(L)}$ has the universal property that characterizes free objects, with the unit given by $\eta_S \colon S \to LS$.

Theorem 12. *Let* $F \colon \mathcal{C} \to \mathcal{D}$ *and* $G \colon \mathcal{D} \to \mathcal{C}$ *be contravariant functors that form a duality. Let* $\mathsf{L} = (L, \eta, \mu)$ *be a monad on* $\mathcal{D}$, *and* $\mathsf{B} = (B, \epsilon, \delta)$ *a comonad on* $\mathcal{C}$. *If there is a natural isomorphism* $\gamma \colon GL \Rightarrow BG$ *making the diagrams (3) commute, then:*

1. *The duality between* F *and* G *lifts to a duality* $\widehat{F} \colon \mathrm{Coeq}(\mathsf{B}, G(S)) \to \mathrm{Eq}(\mathsf{L}, S)$ *and* $\widehat{G} \colon \mathrm{Eq}(\mathsf{L}, S) \to \mathrm{Coeq}(\mathsf{B}, G(S))$.
2. *Given* $e_P \in \mathrm{Eq}(\mathsf{L}, S)$, $m_Q \in \mathrm{Coeq}(\mathsf{B}, G(S))$, $(X, \alpha) \in \mathrm{Alg(L)}$, *and* $(Y, \beta) \in \mathrm{Coalg(B)}$ *we have that:*
   (i) $(X, \alpha) \models e_P$ *if and only if* $\widehat{G}(X, \alpha) \models \widehat{G}(e_P)$.
   (ii) $\widehat{F}(Y, \beta) \models \widehat{F}(m_Q)$ *if and only if* $(Y, \beta) \models m_Q$.

We proceed with special cases of our setting where, given the contravariant adjunction and a comonad B on $\mathcal{C}$, we can canonically define a monad L on $\mathcal{D}$ such that the contravariant adjunction lifts (Sect. 6.1). We can also do it in the opposite way, i.e., define a comonad from a given monad, but in this case additional assumptions are required (Sect. 6.2).

6.1 Defining a Monad from a Comonad

In this part we start with a contravariant adjunction between contravariant functors $F \colon \mathcal{C} \to \mathcal{D}$ and $G \colon \mathcal{D} \to \mathcal{C}$ and a comonad $\mathsf{B} = (B, \epsilon, \delta)$ on $\mathcal{C}$. That is, we have the following setting:

$$
\mathsf{B} = (B, \epsilon, \delta) \circlearrowleft \mathcal{C} \underset{G}{\overset{F}{\rightleftharpoons}} \mathcal{D}
$$

The purpose is to find a canonical monad $\mathsf{L} = (L, \eta, \mu)$ on $\mathcal{D}$ together with a lifting $\widehat{F}\colon \mathrm{Coalg}(\mathsf{B}) \to \mathrm{Alg}(\mathsf{L})$ and $\widehat{G}\colon \mathrm{Alg}(\mathsf{L}) \to \mathrm{Coalg}(\mathsf{B})$ of the contravariant adjunction. We choose $L = FBG$, and define $\eta\colon Id_{\mathcal{D}} \Rightarrow L$ and $\mu\colon LL \Rightarrow L$ by:

$$\eta = (Id_{\mathcal{D}} \xoverset{\eta^{FG}}{\Longrightarrow} FG \xoverset{F\epsilon_G}{\longrightarrow} FBG)$$

$$\mu = (FBGFBG \xoverset{FB\eta^{GF}_{BG}}{\Longrightarrow} FBBG \xoverset{F\delta_G}{\Longrightarrow} FBG) \tag{4}$$

where η^{FG} and η^{GF} are the units of the contravariant adjunction. With this choice of (L, η, μ) we have the following result.

Proposition 13. *Let* $F\colon \mathcal{C} \to \mathcal{D}$ *and* $G\colon \mathcal{D} \to \mathcal{C}$ *be contravariant functors that form a contravariant adjunction. Let* $\mathsf{B} = (B, \epsilon, \delta)$ *be a comonad on* $\mathcal{C}$. *Then* (L, η, μ) *with* $L = FBG$ *and* η, μ *defined as in* (4) *is a monad on* $\mathcal{D}$.

Additionally, if η^{GF} *is a natural isomorphism, then the contravariant adjunction between* F *and* G *lifts to a contravariant adjunction between* $\widehat{F}\colon \mathrm{Coalg}(\mathsf{B}) \to \mathrm{Alg}(\mathsf{L})$ *and* $\widehat{G}\colon \mathrm{Alg}(\mathsf{L}) \to \mathrm{Coalg}(\mathsf{B})$. *In this case, if* F *and* G *form a duality then the lifting* $\widehat{F}$ *and* $\widehat{G}$ *is also a duality.*

6.2 Defining a Comonad from a Monad

We can dualize the previous proposition in order to define a comonad on $\mathcal{C}$ if we have a monad on $\mathcal{D}$. In order to do this we will assume that the contravariant adjunction is a duality so we can use the fact that the units of the contravariant adjunction are isomorphisms.

Assume that we have a contravariant adjunction between two contravariant functors $F\colon \mathcal{C} \to \mathcal{D}$ and $G\colon \mathcal{D} \to \mathcal{C}$, and $\mathsf{L} = (L, \eta, \mu)$ a monad on $\mathcal{D}$. Define the endofunctor B on $\mathcal{C}$ as $B = GLF$. Now, if we assume that the contravariant adjunction is a duality with units $\eta^{FG}\colon Id_{\mathcal{D}} \Rightarrow FG$ and $\eta^{GF}\colon Id_{\mathcal{C}} \Rightarrow GF$ that are natural isomorphisms. Then we can define natural transformations $\epsilon\colon B \Rightarrow Id_{\mathcal{C}}$ and $\delta\colon B \Rightarrow BB$ as:

$$\epsilon = (GLF \xoverset{G\eta_F}{\Longrightarrow} GF \xoverset{(\eta^{GF})^{-1}}{\longrightarrow} Id_{\mathcal{C}})$$

$$\delta = (GLF \xoverset{G\mu_F}{\Longrightarrow} GLLF \xoverset{GL(\eta^{FG})^{-1}_{LF}}{\longrightarrow} GLFGLF) \tag{5}$$

Under the previous assumptions and choice of (B, ϵ, δ) we get:

Proposition 14. *Let* $F\colon \mathcal{C} \to \mathcal{D}$ *and* $G\colon \mathcal{D} \to \mathcal{C}$ *be contravariant functors that form a duality. Let* $\mathsf{L} = (L, \eta, \mu)$ *be a monad on* $\mathcal{D}$. *Then* (B, ϵ, δ), *where* $B = GLF$ *and* ϵ, δ *are defined as in* (5), *is a comonad on* $\mathcal{C}$. *Further, the duality between* F *and* G *lifts to a duality between* $\widehat{F}\colon \mathrm{Coalg}(\mathsf{B}) \to \mathrm{Alg}(\mathsf{L})$ *and* $\widehat{G}\colon \mathrm{Alg}(\mathsf{L}) \to \mathrm{Coalg}(\mathsf{B})$.

7 Applications

In this section we will apply results from the previous section to study equations and coequations for dynamical systems and deterministic automata.

7.1 Equations and Coequations for Dynamical Systems

Let $M = (M, \cdot, e)$ be a monoid, let $\mathsf{L} = (L, \eta, \mu)$ be the monad on Set defined as:

$$LX = X \times M \qquad \eta_X \colon X \to X \times M \qquad \mu_X \colon (X \times M) \times M \to X \times M$$
$$x \mapsto (x, e) \qquad\qquad (x, m, n) \mapsto (x, m \cdot n)$$

and let $\mathsf{B} = (B, \epsilon, \delta)$ be the comonad on CABA defined as:

$$BY = Y^M \qquad \epsilon_Y \colon Y^M \to Y \qquad \delta_Y \colon Y^M \to (Y^M)^M$$
$$f \mapsto f(e) \qquad\qquad f \mapsto \lambda_m.\lambda_n f(n \cdot m)$$

Consider the duality between CABA and Set given by the contravariant functors $F \colon \text{CABA} \to \text{Set}$ and $G \colon \text{Set} \to \text{CABA}$ defined as $FY = \text{At}(Y)$ and $GX = 2^X$, if we consider the natural isomorphism $\gamma \colon GL \Rightarrow BG$ given by the canonical isomorphism $\gamma_X \colon 2^{X \times M} \to (2^X)^M$ then we can easily verify the hypothesis of Theorem 11 to lift the duality between F and G from the following setting:

$$FY = \text{At}(Y)$$
$$GX = 2^X$$
$$\mathsf{B} = (B, \epsilon, \delta) \ \overset{\curvearrowleft}{\ } \ \text{CABA} \ \overset{F}{\underset{G}{\overset{\cong}{\rightleftarrows}}} \ \text{Set} \ \overset{\curvearrowright}{\ } \ \mathsf{L} = (L, \eta, \mu) \quad LX = X \times M$$
$$BY = Y^M$$
$$\gamma_X \colon 2^{X \times M} \to (2^X)^M$$

Observe that elements $(X, \alpha) \in \text{Alg}(\mathsf{L})$ are dynamical systems (monoid actions) on Set, that is, an L-algebra is a set X together with a map $\alpha \colon X \times M \to X$ that satisfies the properties $\alpha(x, e) = x$ and $\alpha(\alpha(x, m), n) = \alpha(x, m \cdot n)$. Further, a B-coalgebra is a set Y with a map $\beta \colon Y \to Y^M$ such that $\beta(x)(e) = x$ and $\beta(\beta(x)(m))(n) = \beta(x)(n \cdot m)$.

We are going to consider equations and coequations for dynamical systems for the particular case that the set of generators is $S = 1$. We have that the free algebra $\mathfrak{F}(1)$ in $\text{Alg}(\mathsf{L})$ on $S = 1$ generators is $\mathfrak{F}(1) = (M, \tau)$ where $\tau \colon M \times M \to M$ is given by $\tau(m, n) = m \cdot n$ and the unit $\eta \colon 1 \to M$ is given by $\eta = e$, the identity element in M. On the other hand, the cofree coalgebra $\mathfrak{C}(G(1)) = \mathfrak{C}(2)$ in $\text{Coalg}(\mathsf{B})$ on 2 colours is $\mathfrak{C}(2) = (2^M, \hat{\tau})$, where $\hat{\tau} \colon 2^M \to (2^M)^M$ is given by

$$\hat{\tau}(f)(n) = \{m \in M \mid f(m \cdot n) = 1\}$$

and the counit $\epsilon \colon 2^M \to 2$ is given by $\epsilon(f) = f(e)$.

According to this, equations in $\text{Eq}(\mathsf{L}, 1)$ correspond to quotients $M/C = (M/C, [\tau])$ where $C \subseteq M \times M$ is a *right congruence* on M, i.e. an equivalence relation such that for any $p \in M$, $(m, n) \in C$ implies $(m \cdot p, n \cdot p) \in M$, and the function $[\tau] \colon M/C \times M \to M/C$ is given by $[\tau]([m], n) = [m \cdot n]$. On the other hand coequations in $\text{Coeq}(\mathsf{L}, G(S))$ correspond to *left-closed-subsystems* $Q = (Q, \hat{\tau})$, i.e. subalgebras Q of the complete atomic Boolean algebra 2^M such that for any $f \in Q$ and $m \in M$, $\hat{\tau}(f)(m) \in Q$.

Now, by using Theorem 12, we have as a consequence a correspondence between right congruences and left-closed-subsystems for dynamical systems.

Proposition 15. *There is a duality between* $\mathrm{Eq}(\mathsf{L}, 1)$ *and* $\mathrm{Coeq}(\mathsf{B}, 2)$ *given by* $\widehat{F}$ *and* $\widehat{G}$ *that induces a duality between right congruences on* M *and left-closed-subsystems of* 2^M.

Using this duality one can prove that right congruences on M and left-closed subsystems of 2^M characterize the same classes of dynamical systems.

Proposition 16. *For any dynamical system* (X, α) *on* M *and any right congruence* C *on* M *let* $e_C \in \mathrm{Alg}(\mathsf{L})(M, M/C)$ *be the canonical epimorphism (equation) defined as* $e_C(m) = [m]$. *The following are equivalent:*

(i) $(X, \alpha) \models e_C$.
(ii) *For every colouring* $c \colon X \to 2$ *and any* $x \in X$ *we have that*

$$\{m \in M \mid c(\beta(x)(m)) = 1\} \in \mathrm{Im}(\widehat{G}(e_C)).$$

If M is the free monoid on A generators then we get [24, Corollary 14]. In this case, property ii) in the previous proposition is the definition for satisfaction of coequations given in [7] where the set of coequations considered is $\mathrm{Im}(\widehat{G}(e_C))$.

7.2   Equations and Coequations for Automata

Consider the following setting:

$$\mathrm{CABA} \underset{G}{\overset{F}{\underset{\cong}{\rightleftarrows}}} \mathrm{Set} \circlearrowleft \mathsf{L} = (L, \eta, \mu)$$

where $FY = \mathrm{At}(Y)$, $GX = 2^X$, and L is the monad given by:

$$LX = X^* = \coprod_{i \in \mathbb{N}} X^i \qquad \eta_X \colon X \to X^* \qquad \mu_X \colon (X^*)^* \to X^* \\ \qquad\qquad\qquad\qquad x \mapsto x \qquad\quad w_1 \cdots w_n \mapsto w_1 \cdots w_n$$

According to Proposition 14, as F and G form a duality, we get a comonad $\mathsf{B} = (B, \epsilon, \delta)$ on CABA and a duality between $\mathrm{Coalg}(\mathsf{B})$ and $\mathrm{Alg}(\mathsf{L})$. Observe that $\mathrm{Alg}(\mathsf{L})$ is isomorphic to the category of monoids.

For any set A, $LA = A^*$ is the free monoid on A generators, with unit morphism η_A and multiplication μ_A. Now we will fix the set A and show how the notion of satisfaction of equations given in [7] for a deterministic automaton on A can be equivalently defined in this setting. In fact, given a deterministic automaton $(X, \alpha \colon X \times A \to A)$ on A we can use the correspondence:

$$\frac{\alpha \colon X \times A \to X}{\alpha \colon A \to X^X}$$

to work with the monoid $X^X = (X^X, \beta) \in \mathrm{Alg}(\mathsf{L})$ with composition of functions as multiplication β. We have that homomorphic images of A^*, i.e., elements in $\mathrm{Eq}(\mathsf{L}, A)$, correspond to congruences of the monoid A^*. Given any congruence C

of A^* we have that $(X, \alpha) \models A^*/C$, if the unique extension $\alpha^\# \in \mathrm{Alg}(\mathsf{L})(A^*, X^X)$ of α factors through the canonical morphism $e \colon A^* \to A^*/C$. That is, we have that $(X, \alpha) \models A^*/C$ if there exists $g_\alpha \in \mathrm{Alg}(\mathsf{L})(A^*/C, X^X)$ such that the following diagram commutes:

$$
\begin{array}{ccccc}
(A^*)^* & \xrightarrow{\ e^*\ } & (A^*/C)^* & \dashrightarrow{\ g_\alpha^*\ } & (X^X)^* \\
\downarrow{\mu_A} & & \downarrow{[\mu_A]} & & \downarrow{\beta} \\
A^* & \xrightarrow{\ e\ } & A^*/C & \dashrightarrow{\ g_\alpha\ } & X^X \\
\uparrow{\eta_A} & & \alpha^\# & & \\
A & & & \alpha &
\end{array}
$$

this means that for any $(u, v) \in C$ the transition functions $f_u, f_v \in X^X$, where $f_w(x) = w(x)$, $w \in A^*$, are the same. This is the notion of satisfaction of equations we previously defined in Example 1, and which appears in [7].

We apply G to the previous diagram to get the following diagram:

$$
\begin{array}{ccccc}
2^{((A^*)^*)} & \xleftarrow{\ 2^{e^*}\ } & 2^{((A^*/C)^*)} & \dashleftarrow{\ 2^{g_\alpha^*}\ } & 2^{((X^X)^*)} \\
\uparrow{2^{\mu_A}} & & \uparrow{2^{[\mu_A]}} & & \uparrow{2^\beta} \\
2^{(A^*)} & \xleftarrow{\ 2^e\ } & 2^{(A^*/C)} & \dashleftarrow{\ 2^{g_\alpha}\ } & 2^{(X^X)} \\
\downarrow{2^{\eta_A}} & & 2^{\alpha^\#} & & \\
2^A & & & 2^\alpha &
\end{array}
$$

This means that $\mathrm{Im}(2^{\alpha^\#}) \subseteq \mathrm{Im}(2^e) = \{L \in 2^{(A^*)} \mid \forall\, (u, v) \in C, L(u) = L(v)\}$, which is an object in CABA and it is closed under left and right derivatives because C is a congruence.

By Theorem 12, we get a duality between $\mathrm{Eq}(\mathsf{L}, A)$ and $\mathrm{Coeq}(\mathsf{B}, G(A))$ which is the duality between equations and coequations given in [7, Theorem 22]. Additionally, using the previous commutative diagrams, one can prove the equivalence between (i) and (ii) given in Proposition 16 for the case that $M = A^*$, the congruences C are congruences of A^*, and the coequations $\mathrm{Im}(\widehat{G}(e_C))$ are subalgebras of 2^M that are closed under left and right derivatives, cf. [24, Theorem 17].

8 Conclusions

We presented duality results between categories of equations and categories of coequations, Theorem 7. We started our approach by using a more general concept than a duality, namely, contravariant adjunctions. By using this setting we can employ algebraic techniques to study coalgebras as we showed by defining equations for coalgebras and we can also do it the opposite way, define coequations for algebras. Then we showed similar results if we add monads and comonads into our setting, to this end, we proved a lifting theorem to lift contravariant adjunctions to Eilenberg-Moore categories, Theorem 12.

The work here is aimed to understand the interaction between the algebraic and coalgebraic world, including the interpretation of coequations, and the study of comonads. In the future we would like to explore more the coalgebraic aspect, either with the aid of the algebraic side or not, in order to find applications to e.g., tree automata (cf. [17]). Because of limited space we have left out from our examples fundamental dualities such as Stone or Priestley dualities, which will possibly lead to a connection with the recent Eilenberg-type correspondences studied in [3,4,10]. We also leave open for future work the question if a converse of Theorem 11 holds.

Acknowledgements. We would like to thank Alexander Kurz for his valuable comments, and Jan Rutten for his support and suggestions.

References

1. Adámek, J.: A logic of coequations. In: Ong, L. (ed.) CSL 2005. LNCS, vol. 3634, pp. 70–86. Springer, Heidelberg (2005)
2. Adámek, J.: Birkhoff's covariety theorem without limitation. Comment. Math. Univ. Carolinae **46**(2), 197–215 (2005)
3. Adámek, J., Milius, S., Myers, R.S.R., Urbat, H.: Generalized eilenberg theorem I: local varieties of languages. In: FoSSaCS 2014, pp. 366–380 (2014)
4. Adámek, J., Milius, S., Myers, R.S.R., Urbat, H.: Varieties of languages in a category. In: Proceedings of LICS 2015. IEEE (2015)
5. Awodey, S., Hughes, J.: Modal operators and the formal dual of Birkhoff's completeness theorem. Math. Struct. CS **13**(2), 233–258 (2003)
6. Awodey, S.: Category Theory. Oxford University Press, Oxford (2006)
7. Ballester-Bolinches, A., Cosme-Llópez, E., Rutten, J.J.M.M.: The dual equivalence of equations and coequations for automata. Inf. Comput. **244**, 49–75 (2015)
8. Bonsangue, M.M., Kurz, A.: Duality for logics of transition systems. In: Sassone, V. (ed.) FOSSACS 2005 and ETAPS 2005. LNCS, vol. 3441, pp. 455–469. Springer, Heidelberg (2005)
9. Clouston, R., Goldblatt, R.: Covarieties of coalgebras: comonads and coequations. In: Van Hung, D., Wirsing, M. (eds.) ICTAC 2005. LNCS, vol. 3722, pp. 288–302. Springer, Heidelberg (2005)
10. Gehrke, M., Grigorieff, S., Pin, J.É.: Duality and equational theory of regular languages. In: Aceto, L., Damgård, I., Goldberg, L.A., Halldórsson, M.M., Ingólfsdóttir, A., Walukiewicz, I. (eds.) ICALP 2008, Part II. LNCS, vol. 5126, pp. 246–257. Springer, Heidelberg (2008)
11. Gumm, H.P.: Birkhoff's variety theorem for coalgebras. Contrib. Gen. Algebra **13**, 159–173 (2000)
12. Gumm, H.P., Schröder, T.: Covarieties and complete covarieties. ENTCS **11**, 42–55 (1998)
13. Hermida, C., Jacobs, B.: Structural induction and coinduction in a fibrational setting. Inf. Comput. **145**(2), 107–152 (1998)

14. Jacobs, B., Silva, A., Sokolova, A.: Trace semantics via determinization. J. Comput. Syst. Sci. **81**(5), 859–879 (2015)
15. Jacobs, B., Sokolova, A.: Exemplaric expressivity of modal logics. J. Log. Comput. **20**(5), 1041–1068 (2010)
16. Klin, B.: ENTCS **173**, 177–201 (2007)
17. Klin, B., Rot, J.: Coalgebraic trace semantics via forgetful logics. In: Pitts, A. (ed.) FOSSACS 2015 and ETAPS 2015. LNCS, vol. 9034, pp. 151–166. Springer, Heidelberg (2015)
18. Kurz, A.: Logics for coalgebras and applications to computer science. Doctoral Thesis, Ludwigs-Maximilians-Universität München (2000)
19. Kurz, A., Rosický, J.: Operations and equations for coalgebras. Math. Struct. Comput. Sci. **15**(1), 149–166 (2005)
20. Pavlovic, D., Mislove, M.W., Worrell, J.B.: Testing semantics: connecting processes and process logics. In: Johnson, M., Vene, V. (eds.) AMAST 2006. LNCS, vol. 4019, pp. 308–322. Springer, Heidelberg (2006)
21. Roşu, G.: Equational axiomatizability for coalgebra. Theoret. Comput. Sci. **260** (1–2), 229–247 (2001)
22. Rutten, J.: Universal coalgebra: a theory of systems. Theoret. Comput. Sci. **249**(1), 3–80 (2000)
23. Schwencke, D.: Coequational logic for accessible functors. Inf. Comput. **208**(12), 1469–1489 (2010)
24. Salamanca, J., Ballester-Bolinches, A., Bonsangue, M.M., Cosme-Llópez, E., Rutten, J.J.M.M.: Regular varieties of automata and coequations. In: Hinze, R., Voigtländer, J. (eds.) MPC 2015. LNCS, vol. 9129, pp. 224–237. Springer, Heidelberg (2015)
25. Salamanca, J., Bonsangue, M., Rutten, J.: Equations and coequations for weighted automata. In: Italiano, G.F., Pighizzini, G., Sannella, D.T. (eds.) MFCS 2015. LNCS, vol. 9234, pp. 444–456. Springer, Heidelberg (2015)

Category Theoretic Semantics for Theorem Proving in Logic Programming: Embracing the Laxness

Ekaterina Komendantskaya[1]([⊠]) and John Power[2]

[1] Department of Computer Science, Heriot-Watt University, Edinburgh, UK
komendantskaya@gmail.com
[2] Department of Computer Science, University of Bath, Bath, UK

Abstract. A propositional logic program P may be identified with a $P_f P_f$-coalgebra on the set of atomic propositions in the program. The corresponding $C(P_f P_f)$-coalgebra, where $C(P_f P_f)$ is the cofree comonad on $P_f P_f$, describes derivations by resolution. Using lax semantics, that correspondence may be extended to a class of first-order logic programs without existential variables. The resulting extension captures the proofs by term-matching resolution in logic programming. Refining the lax approach, we further extend it to arbitrary logic programs. We also exhibit a refinement of Bonchi and Zanasi's saturation semantics for logic programming that complements lax semantics.

Keywords: Logic programming · Coalgebra · Term-matching resolution · Coinductive derivation tree · Lawvere theories · Lax transformations · Kan extensions

1 Introduction

Consider the following two logic programs.

Example 1. ListNat (for lists of natural numbers) denotes the logic program
1. $\texttt{nat}(0) \leftarrow$
2. $\texttt{nat}(\texttt{s}(\texttt{x})) \leftarrow \texttt{nat}(\texttt{x})$
3. $\texttt{list}(\texttt{nil}) \leftarrow$
4. $\texttt{list}(\texttt{cons}(\texttt{x}, \texttt{y})) \leftarrow \texttt{nat}(\texttt{x}), \texttt{list}(\texttt{y})$

Example 2. GC (for graph connectivity) denotes the logic program
0. $\texttt{connected}(\texttt{x}, \texttt{x}) \leftarrow$
1. $\texttt{connected}(\texttt{x}, \texttt{y}) \leftarrow \texttt{edge}(\texttt{x}, \texttt{z}), \texttt{connected}(\texttt{z}, \texttt{y})$

Ekaterina Komendantskaya would like to acknowledge the support of EPSRC Grant EP/K031864/1.
John Power would like to acknowledge the support of EPSRC grant EP/K028243/1.
No data was generated in the course of this research.

I. Hasuo (Ed.): CMCS 2016, LNCS 9608, pp. 94–113, 2016.
DOI: 10.1007/978-3-319-40370-0_7

A critical difference between ListNat and GC is that in the latter, which is a leading example in Sterling and Shapiro's book [31], there is a variable z in the tail of the second clause that does not appear in its head. The category theoretic consequences of that fact are the central concern of this paper.

It has long been observed, e.g., in [4,8], that logic programs induce coalgebras, allowing coalgebraic modelling of their operational semantics. In [20], we developed the idea for variable-free logic programs as follows. Using the definition of a logic program [25], given a set of atoms At, one can identify a variable-free logic program P built over At with a $P_f P_f$-coalgebra structure on At, where P_f is the finite powerset functor on Set: each atom is the head of finitely many clauses in P, and the body of each clause contains finitely many atoms. Our main result was that if $C(P_f P_f)$ is the cofree comonad on $P_f P_f$, then, given a logic program P qua $P_f P_f$-coalgebra, the corresponding $C(P_f P_f)$-coalgebra structure characterises the and-or derivation trees generated by P, cf [12].

This result has proved to be stable, not only having been further developed by us [9,10,15,23,24], but also forming the basis for Bonchi and Zanasi's saturation semantics for logic programming (LP) [6,7]. In Sects. 2 and 3, we give an updated account of the work, with updated definitions, proofs and detailed examples, to start our semantic analysis of derivations and proofs in LP.

In [21], we extended our analysis from variable-free logic programs to arbitrary logic programs. Following [1,4,5,19], given a signature Σ of function symbols, we let $\mathcal{L}_\Sigma$ denote the Lawvere theory generated by Σ, and, given a logic program P with function symbols in Σ, we considered the functor category $[\mathcal{L}_\Sigma^{op}, Set]$, extending the set At of atoms in a variable-free logic program to the functor from $\mathcal{L}_\Sigma^{op}$ to Set sending a natural number n to the set $At(n)$ of atomic formulae with at most n variables generated by the function symbols in Σ and the predicate symbols in P. We sought to model P by a $[\mathcal{L}_\Sigma^{op}, P_f P_f]$-coalgebra $p : At \longrightarrow P_f P_f At$ that, at n, takes an atomic formula $A(x_1, \ldots, x_n)$ with at most n variables, considers all substitutions of clauses in P into clauses with variables among $x_1, \ldots, x_n$ whose head agrees with $A(x_1, \ldots, x_n)$, and gives the set of sets of atomic formulae in antecedents, mimicking the construction for variable-free logic programs. Unfortunately, that idea was too simple.

Consider the logic program ListNat, i.e., Example 1. There is a map in $\mathcal{L}_\Sigma$ of the form $0 \to 1$ that models the nullary function symbol 0. So, naturality of the map $p : At \longrightarrow P_f P_f At$ in $[\mathcal{L}_\Sigma^{op}, Set]$ would yield commutativity of the diagram

$$
\begin{array}{ccc}
At(1) & \xrightarrow{\ p_1\ } & P_f P_f At(1) \\
{\scriptstyle At(0)}\Big\downarrow & & \Big\downarrow{\scriptstyle P_f P_f At(0)} \\
At(0) & \xrightarrow[\ p_0\ ]{} & P_f P_f At(0)
\end{array}
$$

But consider $\mathbf{nat}(x) \in At(1)$: there is no clause of the form $\mathbf{nat}(x) \leftarrow$ in ListNat, so commutativity of the diagram would imply that there cannot be a clause in ListNat of the form $\mathbf{nat}(0) \leftarrow$ either, but in fact there is one.

At that point, proposed resolutions diverged: at CALCO in 2011, we proposed one approach using lax transformations [21], then at CALCO 2013, Bonchi and Zanasi proposed another, using saturation semantics [6], an example of the positive interaction generated by CALCO! In fact, as we explain in Sect. 6, the two approaches may be seen as complementary rather than as alternatives. First we shall describe our approach.

We followed the standard category theoretic technique of relaxing the naturality condition on p to a subset condition, e.g., as in [2, 3, 13, 16, 18], so that, in general, given a map in $\mathcal{L}_\Sigma$ of the form $f : n \to m$, the diagram

$$
\begin{array}{ccc}
At(m) & \xrightarrow{\ p_m\ } & P_f P_f At(m) \\
{\scriptstyle At(f)} \big\downarrow & & \big\downarrow {\scriptstyle P_f P_f At(f)} \\
At(n) & \xrightarrow[\ p_n\ ]{} & P_f P_f At(n)
\end{array}
$$

need not commute, but rather the composite via $P_f P_f At(m)$ need only yield a subset of that via $At(n)$. So, for example, $p_1(\mathtt{nat(x)})$ could be the empty set while $p_0(\mathtt{nat(0)})$ could be non-empty in the semantics for ListNat as required. We extended *Set* to *Poset* in order to express the laxness, and we adopted established category theoretic research on laxness, notably that of [16], in order to prove that a cofree comonad exists and behaves as we wished.

For a representative class of logic programs, the above semantics describes derivations arising from restricting the usual SLD-resolution used in LP to *term-matching resolution*, cf. [22, 23]. As transpired in further studies [9, 15], this particular restriction to resolution rule captures the theorem-proving aspect of LP as opposed to the problem-solving aspect captured by SLD-resolution with unification. We explain this idea in Sect. 2. Derivation trees arising from proofs by term-matching resolution were called *coinductive trees* in [22, 23] to mark their connection to the coalgebraic semantics.

Categorical semantics introduced in [21] worked well for ListNat, allowing us to model its coinductive trees, as we show in Sect. 4 (It was not shown explicitly in [21]). However, it does not work well for GC, the key difference being that, in ListNat, no variable appears in a tail of a clause that does not also appear in its head, i.e., clauses in ListNat contain no *existential* variables. In contrast, although not expressed in these terms in [21], we were unable to model the coinductive trees generated by GC because it is an *existential* program, i.e. program containing clauses with existential variables. We worked around the problems in [21], but only inelegantly.

We give an updated account of [21] in Sect. 4, going beyond [21] to explain how coinductive trees for logic programs without existential variables are modelled, and explaining the difficulty in modelling coinductive semantics for arbitrary logic programs. We then devote Sect. 5 of the paper to resolution of the difficulty, providing lax semantics for coinductive trees generated by arbitrary logic programs.

In contrast to this, Bonchi and Zanasi, concerned by the complexity involved with laxness, proposed the use of saturation, following [4], to provide an alternative category theoretic semantics [6,7]. Saturation is indeed an established and useful construct, as Bonchi and Zanasi emphasised [6,7], with a venerable tradition, and, as they say, laxness requires careful calculation, albeit much less so in the setting of posets than that of categories. On the other hand, laxness is a standard part of category theory, one that has been accepted by computer scientists as the need has arisen, e.g., by He Jifeng and Tony Hoare to model data refinement [13,14,18,27]. More fundamentally, saturation can be seen as complementary to the use of laxness rather than as an alternative to it, as we shall explain in Sect. 6. This reflects the important connection between the theorem proving and problem solving aspects of proof search in LP, as Sect. 2 further explains. So we would suggest that both approaches are of value, with the interaction between them meriting serious consideration.

Saturation inherently yields a particular kind of compositionality, but one loses the tightness of the relationship between semantic model and operational behaviour. The latter is illustrated by the finiteness of branching in a coinductive tree, in contrast to the infinity of possible substitutions, which are inherent in saturation. To the extent that it is possible, we would like to recover operational semantics from the semantic model, along the lines of [28], requiring maintenance of intensionality where possible. We regard the distinction between ListNat and GC as a positive feature of lax semantics, as a goal of semantics is to shed light on the critical issues of programming, relation of existential programs to theorem-proving in LP being one such [9]. So we regard Sect. 6 as supporting both lax and saturation semantics, the interaction between them shedding light on logic programming.

2 Background: Theorem Proving in LP

A *signature* Σ consists of a set $\mathcal{F}$ of function symbols $f, g, \ldots$ each equipped with an arity. Nullary (0-ary) function symbols are constants. For any set *Var* of variables, the set $Ter(\Sigma)$ of terms over Σ is defined inductively as usual:

- $x \in Ter(\Sigma)$ for every $x \in Var$.
- If f is an n-ary function symbol ($n \geq 0$) and $t_1, \ldots, t_n \in Ter(\Sigma)$, then $f(t_1, \ldots, t_n) \in Ter(\Sigma)$.

A *substitution* over Σ is a total function $\sigma : Var \rightarrow \mathbf{Term}(\Sigma)$. Substitutions are extended from variables to terms as usual: if $t \in \mathbf{Term}(\Sigma)$ and σ is a substitution, then the *application* $\sigma(t)$ is a result of applying σ to all variables in t. A substitution σ is a *unifier* for t, u if $\sigma(t) = \sigma(u)$, and is a *matcher* for t against u if $\sigma(t) = u$. A substitution σ is a *most general unifier* (*mgu*) for t and u if it is a unifier for t and u and is more general than any other such unifier. A *most general matcher* (*mgm*) σ for t against u is defined analogously.

In line with LP tradition [25], we also take a set $\mathcal{P}$ of predicate symbols each equipped with an arity. It is possible to define logic programs over terms only, in line with term-rewriting (TRS) tradition [33], as we do in [15], but we will follow the usual LP tradition here. This gives us the following inductive definitions of the sets of atomic formulae, Horn clauses and logic programs (we also include the definition of terms here for convenience):

Definition 1.

> $Terms\ Ter ::= Var \mid \mathcal{F}(Ter, ..., Ter)$
> $Atomic\ formulae\ (or\ atoms)\ At ::= \mathcal{P}(Ter, ..., Ter)$
> $(Horn)\ clauses\ HC ::= At \leftarrow At, ..., At$
> $Logic\ programs\ Prog ::= HC, ..., HC$

In what follows, we will use letters A, B, C, D, possibly with subscripts, to refer to elements of At.

Given a logic program P, we may ask whether a certain atom is logically entailed by P. E.g., given the program ListNat we may ask whether list(cons(0,nil)) is entailed by ListNat. The following rule, which is a restricted form of the famous SLD-resolution, provides a semi-decision procedure to derive the entailment:

Definition 2 (Term-Matching (TM) Resolution).

$$\frac{}{P \vdash []} \qquad \frac{P \vdash \sigma A_1 \quad \cdots \quad P \vdash \sigma A_n}{P \vdash \sigma A} \ if\ (A \leftarrow A_1, ..., A_n) \in P$$

In contrast, the SLD-resolution rule could be presented in the following form:

$$B_1, ..., B_j, ..., B_n \leadsto_P \sigma B_1, ..., \sigma A_1, ..., \sigma A_n, ..., \sigma B_n,$$

if $(A \leftarrow A_1, ..., A_n) \in P$, and σ is the mgu of A and B_j. The derivation for A succeeds when $A \leadsto_P []$; we use $\leadsto_P^*$ to denote several steps of SLD-resolution.

At first sight, the difference between TM-resolution and SLD-resolution seems to be that of notation. Indeed, both $ListNat \vdash$ list(cons(0,nil)) and list(cons(0,nil)) $\leadsto_{ListNat}^* []$ by the above rules (see also Fig. 1). However, $ListNat \nvdash$ list(cons(x,y)) whereas list(cons(x,y)) $\leadsto_{ListNat}^* []$. And, even more mysteriously, $GC \nvdash$ connected(x,y) while connected(x,y) $\leadsto_{GC} []$.

As it turns out, TM-resolution reflects the *theorem proving* side of LP: rules of Definition 2 can be used to semi-decide whether a given term t is entailed by P. In contrast, SLD-resolution reflects the *problem solving* aspect of LP: using the SLD-resolution rule, one asks whether, for a given t, a substitution σ can be found such that $P \vdash \sigma(t)$. There is a subtle but important difference between these two aspects of proof search.

For example, when considering the successful derivation list(cons(x,y)) $\leadsto_{ListNat}^* []$, we assume that list(cons(x,y)) holds only relative to a computed substitution, e.g. x $\mapsto$ 0, y $\mapsto$ nil. Of course this distinction is natural from the point of view of theorem proving: list(cons(x,y)) is not a "theorem" in this generality, but its special case, list(cons(0,nil)), is. Thus, $ListNat \vdash$ list(cons(0,nil)) but $ListNat \nvdash$ list(cons(x,y)) (see also Fig. 1). Similarly, connected$(x, y) \leadsto_{GC}$ [ ] should be read as: connected(x, y) holds relative to the computed substitution $y \mapsto x$.

According to the soundness and completeness theorems for SLD-resolution [25], the derivation $\leadsto$ has *existential* meaning, i.e. when list$(cons(x, y)) \leadsto^*_{ListNat}$ [ ], the succeeded goal list$(cons(x, y))$ is not meant to be read as universally quantified over x an y. On the contrary, TM-resolution proves a universal statement. That is, $GC \vdash$ connected(x, x) reads as: connected(x, x) is entailed by GC for any x.

Much of our recent work has been devoted to formal understanding of the relation between the theorem proving and problem solving aspects of LP [9,15]. The type-theoretic semantics of TM-resolution, given by "Horn clauses as types, λ-terms as proofs" is given in [9,10].

Definition 2 gives rise to derivation trees. E.g. the derivation (or, equivalently, the proof) for $ListNat \vdash$ list$(cons(0, nil))$ can be represented by the following derivation tree:

$$\text{list(cons(0,nil))}$$
$$\diagup \quad \diagdown$$
$$\text{nat(0)} \quad \text{list(nil)}$$
$$| \qquad\quad |$$
$$[\,] \qquad\quad [\,]$$

In general, given a term t and a program P, more than one derivation for $P \vdash t$ is possible. For example, if we add a fifth clause to program $ListNat$:

5. list$(cons(0, x)) \leftarrow$ list(x)

then yet another, alternative, proof is possible for the extended program: $ListNat^+ \vdash$ list$(cons(0, nil))$ via the clause 5:

$$\text{list(cons(0,nil))}$$
$$|$$
$$\text{list(nil)}$$
$$|$$
$$[\,]$$

To reflect the choice of derivation strategies at every stage of the derivation, we can introduce a new kind of nodes, *or-nodes*. For our example, this would give us the tree shown in Fig. 1, note the •-nodes.

This intuition is made precise in the following definition of a *coinductive tree*, which first appeared in [21,23] and was later refined in [15] under the name of a rewriting tree. Note the use of mgms (rather than mgus) in the last item.

Definition 3 (Coinductive Tree). *Let P be a logic program and A be an atomic formula. The* coinductive tree *for A is the possibly infinite tree T satisfying the following properties.*

- *A is the root of T*
- *Each node in T is either an and-node or an or-node*
- *Each or-node is given by •*
- *Each and-node is an atom*

– *For every and-node A' occurring in T, if there exist exactly $m > 0$ distinct clauses $C_1, \ldots, C_m$ in P (a clause C_i has the form $B_i \leftarrow B_1^i, \ldots, B_{n_i}^i$ for some n_i), such that $A' = B_1\theta_1 = \ldots = B_m\theta_m$, for mgms $\theta_1, \ldots, \theta_m$, then A' has exactly m children given by or-nodes, such that, for every $i \in m$, the i-th or-node has n_i children given by and-nodes $B_1^i\theta_i, \ldots, B_{n_i}^i\theta_i$.*

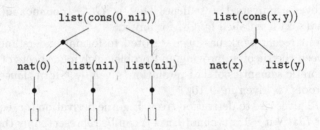

Fig. 1. Left: a coinductive tree for `list(cons(0,nil))` and the extended program $ListNat^+$. **Right:** a coinductive tree for `list(cons(x, y))` and $ListNat^+$. The •-nodes mark different clauses applicable to every atom in the tree.

Coinductive trees provide a convenient model for proofs by TM-resolution.

Let us make one final observation on TM-resolution. Generally, given a program P and an atom t, one can prove that

$$t \leadsto_P^* [\,] \text{ with computed substitution } \sigma \text{ iff } P \vdash \sigma t.$$

This simple fact may give an impression that proofs (and corresponding coinductive trees) for TM-resolution are in some sense fragments of reductions by SLD-resolution. Compare e.g. the right-hand tree of Fig. 1 before substitution and a grown left-hand tree obtained after the substitution. In this case, we could emulate the problem solving aspect of SLD-resolution by using coinductive trees and allowing to apply substitutions within coinductive trees, as was proposed in [9,15,22]. Such intuition would hold perfectly for e.g. ListNat, but would not hold for existential programs: although there is a one step SLD-derivation for `connected(x, y)` $\leadsto_{GC} [\,]$ (with $y \mapsto x$), TM-resolution proof for `connected(x, y)` diverges and gives rise to the following infinite coinductive tree:

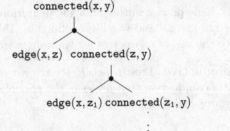

Not only the proof for $GC \vdash$ `connected(x, y)` is not in any sense a fragment of the derivation `connected(x, y)` $\leadsto_{GC} [\,]$, but it also takes larger (i.e. infinite) signature. Thus, operational semantics of TM-resolution and SLD-resolution can be very different for existential programs: both in aspects of termination and signature size.

This problem is orthogonal to non-termination. Consider the non-terminating (but not existential) program Bad:

bad(x) ← bad(x)

For Bad, operational behavior of TM-resolution and SLD-resolution are similar: derivations with both do not terminate and require finite signature. Once again, such programs can be analysed using similar coinductive methods in TM- and SLD-resolution [10, 30].

The problems caused by existential variables are known in the literature on theorem proving and term-rewriting [33]. In TRS [33], existential variables are not allowed to appear in rewriting rules, and in type inference based on term rewriting or TM-resolution, the restriction to non-existential programs is common [11].

So theorem-proving, in contrast to problem-solving, is modelled by term-matching; term-matching gives rise to coinductive trees; and as explained in the introduction and, in more detail, later, coinductive trees give rise to laxness. So in this paper, we use laxness to model coinductive trees, and thereby theorem-proving in LP, and relate our semantics with Bonchi and Zanasi's work, which we believe models primarily problem-solving aspect of logic programming.

Categorical semantics for existential programs, which are known to be challenging for theorem proving, is the main contribution of Sect. 5 and this paper.

3   Modelling Coinductive Trees for Variable-Free Logic Programs

In this section, we recall and develop the work of [20] and in particular we restrict our semantics to variable-free logic programs, i.e. we take $Var = \emptyset$ in Definition 1. Variable-free logic programs are operationally equivalent to propositional logic programs, as substitutions play no role in derivations. In this (propositional) setting, coinductive trees coincide with the and-or derivation trees known in the LP literature [12].

Proposition 1. *For any set At, there is a bijection between the set of variable-free logic programs over the set of atoms At and the set of $P_f P_f$-coalgebra structures on At, where P_f is the finite powerset functor on Set.*

Theorem 1. *Let $C(P_f P_f)$ denote the cofree comonad on $P_f P_f$. Then, for $p : At \longrightarrow P_f P_f(At)$, the corresponding $C(P_f P_f)$-coalgebra $\overline{p} : At \longrightarrow C(P_f P_f)(At)$ sends an atom A to the coinductive tree for A.*

Proof. Applying the work of [34] to this setting, the cofree comonad is in general determined as follows: $C(P_f P_f)(At)$ is the limit of the diagram

$$\ldots \longrightarrow At \times P_f P_f(At \times P_f P_f(At)) \longrightarrow At \times P_f P_f(At) \longrightarrow At.$$

with maps determined by the projection $\pi_0 : At \times P_f P_f(At) \longrightarrow At$, with applications of the functor $At \times P_f P_f(-)$ to it.

Putting $At_0 = At$ and $At_{n+1} = At \times P_f P_f At_n$, and defining the cone

$$p_0 = id : At \longrightarrow At(= At_0)$$
$$p_{n+1} = \langle id, P_f P_f(p_n) \circ p \rangle : At \longrightarrow At \times P_f P_f At_n(= At_{n+1})$$

the limiting property of the diagram determines the coalgebra $\bar{p} : At \longrightarrow C(P_f P_f)(At)$. The image $\bar{p}(A)$ of an atom A is given by an element of the limit, equivalently a map from 1 into the limit, equivalently a cone of the diagram over 1.

To give the latter is equivalent to giving an element A_0 of At, specifically $p_0(A) = A$, together with an element A_1 of $At \times P_f P_f(At)$, specifically $p_1(A) = (A, p_0(A)) = (A, p(A))$, together with an element A_2 of $At \times P_f P_f(At \times P_f P_f(At))$, etcetera. The definition of the coinductive tree for A is inherently coinductive, matching the definition of the limit, and with the first step agreeing with the definition of p. Thus it follows by coinduction that $\bar{p}(A)$ can be identified with the coinductive tree for A.

Example 3. Let At consist of atoms A, B, C and D. Let P denote the logic program

$$A \leftarrow B, C$$
$$A \leftarrow B, D$$
$$D \leftarrow A, C$$

So $p(A) = \{\{B, C\}, \{B, D\}\}$, $p(B) = p(C) = \emptyset$, and $p(D) = \{\{A, C\}\}$.

Then $p_0(A) = A$, which is the root of the coinductive tree for A.

Then $p_1(A) = (A, p(A)) = (A, \{\{B, C\}, \{B, D\}\})$, which consists of the same information as in the first three levels of the coinductive tree for A, i.e., the root A, two or-nodes, and below each of the two or-nodes, nodes given by each atom in each antecedent of each clause with head A in the logic program P: nodes marked B and C lie below the first or-node, and nodes marked B and D lie below the second or-node, exactly as $p_1(A)$ describes.

Continuing, note that $p_1(D) = (D, p(D)) = (D, \{\{A, C\}\})$. So

$$\begin{aligned} p_2(A) &= (A, P_f P_f(p_1)(p(A))) \\ &= (A, P_f P_f(p_1)(\{\{B, C\}, \{B, D\}\})) \\ &= (A, \{\{(B, \emptyset), (C, \emptyset)\}, \{(B, \emptyset), (D, \{\{A, C\}\})\}\}) \end{aligned}$$

which is the same information as that in the first five levels of the coinductive tree for A: $p_1(A)$ provides the first three levels of $p_2(A)$ because $p_2(A)$ must map to $p_1(A)$ in the cone; in the coinductive tree, there are two and-nodes at level 3, labelled by A and C. As there are no clauses with head B or C, no or-nodes lie below the first three of the and-nodes at level 3. However, there is one or-node lying below D, it branches into and-nodes labelled by A and C, which is exactly as $p_2(A)$ tells us.

For pictures of such trees, see [23].

4 Modelling Coinductive Trees for Logic Programs Without Existential Variables

We now lift the restriction on $Var = \emptyset$ in Definition 1, and consider first-order terms and atoms in full generality, however, we restrict the definition of clauses in Definition 1 to those not containing existential variables.

The *Lawvere theory $\mathcal{L}_\Sigma$ generated by* a signature Σ is (up to isomorphism, as there are several equivalent formulations) the category defined as follows: $\mathrm{ob}(\mathcal{L}_\Sigma)$ is the set of natural numbers. For each natural number n, let $x_1, \ldots, x_n$ be a specified list of distinct variables. Define $\mathcal{L}_\Sigma(n, m)$ to be the set of m-tuples $(t_1, \ldots, t_m)$ of terms generated by the function symbols in Σ and variables $x_1, \ldots, x_n$. Define composition in $\mathcal{L}_\Sigma$ by substitution.

One can readily check that these constructions satisfy the axioms for a category, with $\mathcal{L}_\Sigma$ having strictly associative finite products given by the sum of natural numbers. The terminal object of $\mathcal{L}_\Sigma$ is the natural number 0.

Example 4. Consider ListNat. The constants 0 and nil are maps from 0 to 1 in $\mathcal{L}_\Sigma$, s is modelled by a map from 1 to 1, and cons is modelled by a map from 2 to 1. The term s(0) is the map from 0 to 1 given by the composite of the maps modelling s and 0.

Given an arbitrary logic program P with signature Σ, we can extend the set At of atoms for a variable-free logic program to the functor $At : \mathcal{L}_\Sigma^{op} \to Set$ that sends a natural number n to the set of all atomic formulae, with variables among $x_1, \ldots, x_n$, generated by the function symbols in Σ and by the predicate symbols in P. A map $f : n \to m$ in $\mathcal{L}_\Sigma$ is sent to the function $At(f) : At(m) \to At(n)$ that sends an atomic formula $A(x_1, \ldots, x_m)$ to $A(f_1(x_1, \ldots, x_n)/x_1, \ldots, f_m(x_1, \ldots, x_n)/x_m)$, i.e., $At(f)$ is defined by substitution.

As explained in the Introduction and in [20], we cannot model a logic program by a natural transformation of the form $p : At \longrightarrow P_f P_f At$ as naturality breaks down, e.g., in ListNat. So, in [21,23], we relaxed naturality to lax naturality. In order to define it, we extended $At : \mathcal{L}_\Sigma^{op} \to Set$ to have codomain Poset by composing At with the inclusion of Set into $Poset$. Mildly overloading notation, we denote the composite by $At : \mathcal{L}_\Sigma^{op} \to Poset$.

Definition 4. *Given functors $H, K : \mathcal{L}_\Sigma^{op} \longrightarrow Poset$, a lax transformation from H to K is the assignment to each object n of $\mathcal{L}_\Sigma$, of an order-preserving function $\alpha_n : Hn \longrightarrow Kn$ such that for each map $f : n \longrightarrow m$ in $\mathcal{L}_\Sigma$, one has $(Kf)(\alpha_m) \leq (\alpha_n)(Hf)$, pictured as follows:*

$$
\begin{array}{ccc}
Hm & \xrightarrow{\alpha_m} & Km \\
\downarrow{\scriptstyle Hf} & \geq & \downarrow{\scriptstyle Kf} \\
Hn & \xrightarrow[\alpha_n]{} & Kn
\end{array}
$$

Functors and lax transformations, with pointwise composition, form a locally ordered category denoted by $Lax(\mathcal{L}_\Sigma^{op}, Poset)$. Such categories and generalisations have been studied extensively, e.g., in [2,3,16,18].

Definition 5. *Define $P_f : Poset \longrightarrow Poset$ by letting $P_f(P)$ be the partial order given by the set of finite subsets of P, with $A \leq B$ if for all $a \in A$, there exists $b \in B$ for which $a \leq b$ in P, with behaviour on maps given by image. Define P_c similarly but with countability replacing finiteness.*

We are not interested in arbitrary posets in modelling logic programming, only those that arise, albeit inductively, by taking subsets of a set qua discrete poset. So we gloss over the fact that, for an arbitrary poset P, Definition 5 may yield factoring, with the underlying set of $P_f(P)$ being a quotient of the set of subsets of P. It does not affect the line of development here.

Example 5. Modelling Example 1, ListNat generates a lax transformation of the form $p : At \longrightarrow P_f P_f At$ as follows: $At(n)$ is the set of atomic formulae in *ListNat* with at most n variables.

For example, $At(0)$ consists of nat(0), nat(nil), list(0), list(nil), nat(s(0)), nat(s(nil)), list(s(0)), list(s(nil)), nat(cons(0,0)), nat(cons(0,nil)), nat(cons(nil,0)), nat(cons(nil,nil)), etcetera.

Similarly, $At(1)$ includes all atomic formulae containing at most one (specified) variable x, thus all the elements of $At(0)$ together with nat(x), list(x), nat(s(x)), list(s(x)), nat(cons(0,x)), nat(cons(x,0)), nat(cons(x,x)), etcetera.

The function $p_n : At(n) \longrightarrow P_f P_f At(n)$ sends each element of $At(n)$, i.e., each atom $A(x_1, \ldots, x_n)$ with variables among $x_1, \ldots, x_n$, to the set of sets of atoms in the antecedent of each unifying substituted instance of a clause in P with head for which a unifying substitution agrees with $A(x_1, \ldots, x_n)$.

Taking $n = 0$, nat(0) $\in At(0)$ is the head of one clause, and there is no other clause for which a unifying substitution will make its head agree with nat(0). The clause with head nat(0) has the empty set of atoms as its tail, so $p_0(\text{nat}(0)) = \{\emptyset\}$.

Taking $n = 1$, list(cons(x,0)) $\in At(1)$ is the head of one clause given by a unifying substitution applied to the final clause of ListNat, and accordingly $p_1(\text{list}(\text{cons}(x,0))) = \{\{\text{nat}(x), \text{list}(0)\}\}$.

The family of functions p_n satisfy the inequality required to form a lax transformation precisely because of the allowability of substitution instances of clauses, as in turn is required to model logic programming. The family does not satisfy the strict requirement of naturality as explained in the introduction.

Example 6. Attempting to model Example 2 by mimicking the model of ListNat as a lax transformation of the form $p : At \longrightarrow P_f P_f At$ in Example 5 fails.

Consider the clause

$$\text{connected}(x, y) \leftarrow \text{edge}(x, z), \text{connected}(z, y)$$

Modulo possible renaming of variables, the head of the clause, i.e., the atom connected(x, y), lies in $At(2)$ as it has two variables. However, the tail does not lie in $P_f P_f At(2)$ as the tail has three variables rather than two.

We dealt with that inelegantly in [21]: in order to allow $p_2(\texttt{connected}(\texttt{x}, \texttt{y}))$ to model GC in any reasonable sense, we allowed substitutions for z by any term on x, y on the basis that there is no unifying such, so we had better allow all possibilities. So, rather than modelling the clause directly, recalling that $At(2) \subseteq At(3) \subseteq At(4)$, etcetera, modulo renaming of variables, we put

$$p_2(\texttt{connected}(\texttt{x}, \texttt{y})) = \{\{\texttt{edge}(\texttt{x}, \texttt{x}), \texttt{connected}(\texttt{x}, \texttt{y})\}, \{\texttt{edge}(\texttt{x}, \texttt{y}), \texttt{connected}(\texttt{y}, \texttt{y})\}\}$$
$$p_3(\texttt{connected}(\texttt{x}, \texttt{y})) = \{\{\texttt{edge}(\texttt{x}, \texttt{x}), \texttt{connected}(\texttt{x}, \texttt{y})\}, \{\texttt{edge}(\texttt{x}, \texttt{y}), \texttt{connected}(\texttt{y}, \texttt{y})\},$$
$$\{\texttt{edge}(\texttt{x}, \texttt{z}), \texttt{connected}(\texttt{z}, \texttt{y})\}\}$$
$$p_4(\texttt{connected}(\texttt{x}, \texttt{y})) = \{\{\texttt{edge}(\texttt{x}, \texttt{x}), \texttt{connected}(\texttt{x}, \texttt{y})\}, \{\texttt{edge}(\texttt{x}, \texttt{y}), \texttt{connected}(\texttt{y}, \texttt{y})\},$$
$$\{\texttt{edge}(\texttt{x}, \texttt{z}), \texttt{connected}(\texttt{z}, \texttt{y})\}, \{\texttt{edge}(\texttt{x}, \texttt{w}), \texttt{connected}(\texttt{w}, \texttt{y})\}\}$$

etcetera: for p_2, as only two variables x and y appear in any element of $P_f P_f At(2)$, we allowed substitution by either x or y for z; for p_3, a third variable may appear in an element of $P_f P_f At(3)$, allowing an additional possible substitution; for p_4, a fourth variable may appear, etcetera.

Countability arises if a unary symbol s is added to GC, as in that case, for p_2, not only did we allow x and y to be substituted for z, but we also allowed $s^n(x)$ and $s^n(y)$ for any $n > 0$, and to do that, we replaced $P_f P_f$ by $P_c P_f$, allowing for the countably many possible substitutions.

Those were inelegant decisions, but they allowed us to give some kind of model of all logic programs.

We now turn to the relationship between the lax transformation $p : At \longrightarrow P_c P_f At$ modelling a logic program P and $\overline{p} : At \longrightarrow C(P_c P_f)At$, the corresponding coalgebra for the cofree comonad $C(P_c P_f)$ on $P_c P_f$.

We recall the central abstract result of [21], the notion of an "oplax" map of coalgebras being required to match that of lax transformation. Notation of the form H-$coalg$ refers to coalgebras for an endofunctor H, while notation of the form C-$Coalg$ refers to coalgebras for a comonad C. The subscript $oplax$ refers to oplax maps, and given an endofunctor E on $Poset$, the notation $Lax(\mathcal{L}_\Sigma^{op}, E)$ denotes the endofunctor on $Lax(\mathcal{L}_\Sigma^{op}, Poset)$ given by post-composition with E; similarly for a comonad.

Theorem 2. *For any locally ordered endofunctor E on $Poset$, if $C(E)$ is the cofree comonad on E, then there is a canonical isomorphism*

$$Lax(\mathcal{L}_\Sigma^{op}, E)\text{-}coalg_{oplax} \simeq Lax(\mathcal{L}_\Sigma^{op}, C(E))\text{-}Coalg_{oplax}$$

Corollary 1. $Lax(\mathcal{L}_\Sigma^{op}, C(P_c P_f))$ *is the cofree comonad on* $Lax(\mathcal{L}_\Sigma^{op}, P_c P_f)$.

Corollary 1 gives a bijection between lax transformations

$$p : At \longrightarrow P_c P_f At$$

and lax transformations

$$\overline{p} : At \longrightarrow C(P_c P_f)At$$

subject to the two conditions required of a coalgebra of a comonad. Subject to the routine replacement of the outer copy of P_f by P_c in the construction in Theorem 1, the same construction, if understood pointwise, extends to this setting, i.e., if one uniformly replaces At by $At(n)$ in the construction of Theorem 1, and replaces the outer copy of P_f by P_c, one obtains a description of $C(P_cP_f)At(n)$ together with the construction of $\overline{p}_n$ from p_n.

That is fine for ListNat, modelling the coinductive trees generated by ListNat, the same holding for any logic program without existential variables, but for GC, as explained in Example 6, p did *not* model the clause

$$\text{connected}(x, y) \leftarrow \text{edge}(x, z), \text{connected}(z, y)$$

directly, and so its extension *a fortiori* could *not* model the coinductive trees generated by $\text{connected}(x, y)$.

For arbitrary logic programs, $\overline{p}(A(x_1, \ldots, x_n))$ was a variant of the coinductive tree generated by $A(x_1, \ldots, x_n)$ in two key ways:

1. coinductive trees allow new variables to be introduced as one passes down the tree, e.g., with

$$\text{connected}(x, y) \leftarrow \text{edge}(x, z), \text{connected}(z, y)$$

appearing directly in it, whereas, extending Example 6, $\overline{p_1}(\text{connected}(x, y))$ does not model such a clause directly, but rather substitutes terms on x and y for z, continuing inductively as one proceeds.
2. coinductive trees are finitely branching, as one expects in logic programming, whereas $\overline{p}(A(x_1, \ldots, x_n))$ could be infinitely branching, e.g., for GC with an additional unary operation s.

5   Modelling Coinductive Trees for Arbitrary Logic Progams

We believe that our work in [21] provides an interesting model of ListNat, in particular because it agrees with the coinductive trees generated by ListNat. However, the account in [21] is less interesting when applied to GC, thus in the full generality of logic programming. Restriction to non-existential examples such as ListNat is common for implementational reasons [9,10,15,23], so [21] does allow the modeling of coinductive trees for a natural class of logic programs. Here we seek to model coinductive trees for logic programs in general, *a fortiori* doing so for GC.

In order to model coinductive trees, it follows from Example 6 that the endofunctor $Lax(\mathcal{L}_\Sigma^{op}, P_fP_f)$ on $Lax(\mathcal{L}_\Sigma^{op}, Poset)$ that sends At to P_fP_fAt, needs to be refined as $\{\{\text{edge}(x, z), \text{connected}(z, y)\}\}$ is not an element of $P_fP_fAt(2)$ as it involves three variables x, y and z. Motivated by that example, we refine our axiomatics in general so that the codomain of p_n is a superset of $P_fP_fAt(m)$ for every $m \geq n$. There are six injections of 2 into 3, inducing six inclusions

$At(2) \subseteq At(3)$, so six inclusions $P_f P_f At(2) \subseteq P_f P_f At(3)$, and one only wants to count each element of $P_f P_f At(2)$ once. So we refine $P_f P_f At(n)$ to become $(\Sigma_{m \geq n} P_f P_f At(m))/ \equiv$, where $\equiv$ is generated by the injections $i : n \longrightarrow m$. This can be made precise in abstract category theoretic terms as follows.

For any Lawvere theory L, there is a canonical identity-on-objects functor from the category Inj of injections $i : n \longrightarrow m$ of natural numbers into L^{op}. So, in particular, there is a canonical identity on objects functor $J : Inj \longrightarrow \mathcal{L}_{\Sigma}^{op}$, upon which $\Sigma_{m \geq n} P_f P_f At(m)/ \equiv$ may be characterised as the colimit (see [26] or, for the enriched version, [17])

$$\int^{m \in n/Inj} P_f P_f At J(m)$$

or equivalently, given $n \in Inj$, the colimit of the functor from n/Inj to $Poset$ that sends an injection $j : n \longrightarrow m$ to $P_f P_f At J(m)$.

This construction extends to a functor $P_{ff}(At) : \mathcal{L}_{\Sigma}^{op} \longrightarrow Poset$ by sending a map $f : n \longrightarrow n'$ in $\mathcal{L}_{\Sigma}$ to the order-preserving function

$$\int^{m \in n'/Inj} P_f P_f At J(m) \longrightarrow \int^{m \in n/Inj} P_f P_f At J(m)$$

determined by the fact that each $m \in n'/Inj$ is, up to coherent isomorphism, uniquely of the form $n' + k$, allowing one to apply $P_f P_f At$ to the map $f + k : n + k \longrightarrow n' + k = m$ in $\mathcal{L}_{\Sigma}$. This is similar to the behaviour of the monad for local state on maps [29].

It is routine to generalise the construction from At to make it apply to an arbitrary functor $H : \mathcal{L}_{\Sigma}^{op} \longrightarrow Poset$.

In order to make the construction functorial, i.e., in order to make it respect maps $\alpha : H \Rightarrow K$, we need to refine $Lax(\mathcal{L}_{\Sigma}^{op}, Poset)$ as the above colimit strictly respects injections, i.e., for any *injection* $i : n \longrightarrow m$, we want the diagram

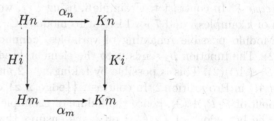

to commute.

Summarising this discussion yields the following:

Definition 6. *Let $Lax_{Inj}(\mathcal{L}_{\Sigma}^{op}, Poset)$ denote the category with objects given by functors from $\mathcal{L}_{\Sigma}^{op}$ to $Poset$, maps given by lax transformations that strictly respect injections, and composition given pointwise.*

Proposition 2. *cf [29] Let $J : Inj \longrightarrow \mathcal{L}_{\Sigma}^{op}$ be the canonical inclusion. Define*

$$P_{ff} : Lax_{Inj}(\mathcal{L}_{\Sigma}^{op}, Poset) \longrightarrow Lax_{Inj}(\mathcal{L}_{\Sigma}^{op}, Poset)$$

by $(P_{ff}(H))(n) = \int^{m \in n/Inj} P_f P_f H J(m)$, with, for any map $f : n \longrightarrow n'$ in $\mathcal{L}_{\Sigma}$,

$$(P_{ff}(H))(f) : \int^{m \in n'/Inj} P_f P_f H J(m) \longrightarrow \int^{m \in n/Inj} P_f P_f H J(m)$$

determined by the fact that each $m \in n'/Inj$ is, up to coherent isomorphism, uniquely of the form $n' + k$, allowing one to apply $P_f P_f H$ to the map $f + k : n + k \longrightarrow n' + k = m$ in $\mathcal{L}_{\Sigma}$.

Given $\alpha : H \Rightarrow K$, define $P_{ff}(\alpha)(n)$ by the fact that $m \in n/Inj$ is uniquely of the form $n + k$, and using

$$\alpha_{n+k} : H(m) = H(n + k) \longrightarrow K(n + k) = K(m)$$

Then P_{ff} is an endofunctor on $Lax_{Inj}(\mathcal{L}_{\Sigma}^{op}, Poset)$.

The proof is routine but requires lengthy calculation involving colimits. Observe that we have not required countability anywhere in the definition of P_{ff}, using only finiteness as we sought at the end of Sect. 4.

We can now model an arbitrary logic program by a map $p : At \longrightarrow P_{ff}At$ in $Lax_{Inj}(\mathcal{L}_{\Sigma}^{op}, Poset)$, modelling ListNat as we did in Example 5 but now modelling the clauses of GC directly rather than using the awkward substitution instances of Example 6.

Example 7. Except for the restriction of $Lax(\mathcal{L}_{\Sigma}^{op}, Poset)$ to $Lax_{Inj}(\mathcal{L}_{\Sigma}^{op}, Poset)$, ListNat is modelled in exactly the same way here as it was in Example 5, the reason being that no clause in ListNat has a variable in the tail that does not already appear in the head. We need only observe that, although p is not strictly natural in general, it does strictly respect injections. For example, if one views $\texttt{list}(\texttt{cons}(\texttt{x}, 0))$ as an element of $At(2)$, its image under p_2 agrees with its image under p_1.

Example 8. In contrast to Example 6, using P_{ff}, we can emulate the construction of Examples 5 and 7 for ListNat to model GC.

Modulo possible renaming of variables, $\texttt{connected}(\texttt{x}, \texttt{y})$ is an element of $At(2)$. The function p_2 sends it to the element $\{\{\texttt{edge}(\texttt{x}, \texttt{z}), \texttt{connected}(\texttt{z}, \texttt{y})\}\}$ of $(P_{ff}(At))(2)$. This is possible by taking $n = 2$ and $m = 3$ in the formula for $P_{ff}(At)$ in Proposition 2. In contrast, $\{\{\texttt{edge}(\texttt{x}, \texttt{z}), \texttt{connected}(\texttt{z}, \texttt{y})\}\}$ is not an element of $P_f P_f At(2)$, hence the failure of Example 6.

The behaviour of $P_{ff}(At)$ on maps ensures that the lax transformation p strictly respects injections. For example, if $\texttt{connected}(\texttt{x}, \texttt{y})$ is seen as an element of $At(3)$, the additional variable is treated as a fresh variable w, so does not affect the image of $\texttt{connected}(\texttt{x}, \texttt{y})$ under p_3.

Theorem 3. *The functor* $P_{ff} : Lax_{Inj}(\mathcal{L}_{\Sigma}^{op}, Poset) \longrightarrow Lax_{Inj}(\mathcal{L}_{\Sigma}^{op}, Poset)$ *induces a cofree comonad* $C(P_{ff})$ *on* $Lax_{Inj}(\mathcal{L}_{\Sigma}^{op}, Poset)$. *Moreover, given a logic progam* P *qua* P_{ff}*-coalgebra* $p : At \longrightarrow P_{ff}(At)$, *the corresponding* $C(P_{ff})$*-coalgebra* $\bar{p} : At \longrightarrow C(P_{ff})(At)$ *sends an atom* $A(x_1, \ldots, x_n) \in At(n)$ *to the coinductive tree for* $A(x_1, \ldots, x_n)$.

Proof. The construction of Theorem 1, subject to mild rephrasing, continues to work here. Specifically, $(C(P_{ff})At)(n)$ is given by the same limit as in Theorem 1 but with At replaced by $At(n)$ and with $P_f P_f$ replaced by P_{ff}: products in the category $Lax_{Inj}(\mathcal{L}_\Sigma^{op}, Poset)$ are given pointwise, so the use of projections is the same; $[Inj, Poset]$ is locally finitely presentable and P_{ff} is an accessible functor, allowing us to extend the construction of the cofree comonad pointwise to $[Inj, Poset]$. It is routine, albeit tedious, to verify functoriality of $C(P_{ff})$ with respect to all maps and to verify the universal property. The construction of $\bar{p}$ is given pointwise, with it following from its coinductive construction that it yields the coinductive trees as required.

The lax naturality in respect to general maps $f : m \longrightarrow n$ means that a substitution applied to an atom $A(x_1, \ldots, x_n) \in At(n)$, i.e., application of the function $At(f)$ to $A(x_1, \ldots, x_n)$, followed by application of $\bar{p}$, i.e., taking the coinductive tree for the substituted atom, or application of the function $(C(P_f f)At)f)$ to the coinductive tree for $A(x_1, \ldots, x_n)$ potentially yield different trees: the former substitutes into $A(x_1, \ldots, x_n)$, then takes its coinductive tree, while the latter applies a substitution to each node of the coinductive tree for $A(x_1, \ldots, x_n)$, then prunes to remove redundant branches.

Example 9. Extending Example 8, consider connected$(x, y) \in At(2)$. In expressing GC as a map $p : At \longrightarrow P_{ff}At$ in Example 8, we put

$$p_2(\text{connected}(x, y)) = \{\{\text{edge}(x, z), \text{connected}(z, y)\}\}$$

Accordingly, $\bar{p}_2(\text{connected}(x, y))$ is the coinductive tree for connected(x, y), thus the infinite tree generated by repeated application of the same clause modulo renaming of variables.

If we substitute x for y in the coinductive tree, i.e., apply the function $(C(P_{ff})At)(x, x)$ to it (see the definition of L_Σ at the start of Sect. 4 and observe that (x, x) is a 2-tuple of terms generated trivially by the variable x), we obtain the same tree but with y systematically replaced by x. However, if we substitute x for y in connected(x, y), i.e., apply the function $At(x, x)$ to it, we obtain connected$(x, x) \in At(1)$, whose coinductive tree has additional branching as the first clause of GC, i.e., connected$(x, x) \leftarrow$ may also be applied.

In contrast to this, we have strict naturality with respect to injections: for example, an injection $i : 2 \longrightarrow 3$ yields the function $At(i) : At(2) \longrightarrow At(3)$ that, modulo renaming of variables, sends connected$(x, y) \in At(2)$ to itself seen as an element of $At(3)$, and the coinductive tree for connected(x, y) is accordingly also sent by $(C(P_{ff})At)(i)$ to itself seen as an element of $(C(P_{ff})At)(3)$.

Example 9 illustrates why, although the condition of strict naturality with respect to injections holds for P_{ff}, it does not hold for $Lax(\mathcal{L}_\Sigma^{op}, P_f P_f)$ in Example 6 as we did not model the clause

$$\text{connected}(x, y) \leftarrow \text{edge}(x, z), \text{connected}(z, y)$$

directly there, but rather modelled all substitution instances into all available variables.

6   Complementing Saturated Semantics

Bonchi and Zanasi's approach to modelling logic programming in [6] was to consider $P_f P_f$ as we did in [21], sending At to $P_f P_f At$, but to ignore the inherent laxness, replacing $Lax(\mathcal{L}_\Sigma^{op}, Poset)$ by $[ob(\mathcal{L}_\Sigma), Set]$, where $ob(\mathcal{L}_\Sigma)$ is the set of objects of $\mathcal{L}_\Sigma$ treated as a discrete category, i.e., as one with only identity maps.

The central mathematical fact that supports saturated semantics is that, regarding $ob(\mathcal{L}_\Sigma)$ as a discrete category, with inclusion functor $I : ob(\mathcal{L}_\Sigma) \longrightarrow \mathcal{L}_\Sigma$, the functor

$$[I, Set] : [\mathcal{L}_\Sigma^{op}, Set] \longrightarrow [ob(\mathcal{L}_\Sigma)^{op}, Set]$$

that sends a functor $H : \mathcal{L}_\Sigma^{op} \longrightarrow Set$ to the composite functor $HI : ob(\mathcal{L}_\Sigma) = ob(\mathcal{L}_\Sigma)^{op} \longrightarrow Set$ has a right adjoint. That adjoint is given by right Kan extension. It is primarily the fact of the existence of the right adjoint, rather than its characterisation as a right Kan extension, that enabled Bonchi and Zanasi's various constructions, in particular those of saturation and desaturation.

That allows us to mimic Bonchi and Zanasi's saturation semantics, but starting from $Lax(\mathcal{L}_\Sigma^{op}, Poset)$ rather than from $[ob(\mathcal{L}_\Sigma), Set]$. We are keen to allow this as laxness is an inherent fact of the situation, as we have explained through the course of this paper. Such laxness has been valuable in related semantic endeavours, such as in Tony Hoare's pioneering work on the modelling of data refinement [13,14,18], of which substitution in logic programming can be seen as an instance.

The argument, which was originally due to Ross Street, cf [32], goes as follows.

Theorem 4. *[3] For any finitary 2-monad T on a cocomplete 2-category K, the inclusion*

$$J : T\text{-}Alg_s \longrightarrow T\text{-}Alg_l$$

of the category of strict T-algebras and strict maps of T-algebras into the category of strict T-algebras and lax maps of T-algebras has a left adjoint.

Example 10. For any Lawvere theory L, there is a finitary locally ordered monad T on $[ob(L), Poset^{op}]$ for which $[L, Poset^{op}]$ is isomorphic to $T\text{-}Alg_s$, with $T\text{-}Alg_l$ isomorphic to $Lax(L, Poset^{op})$. The monad T is given by the composite of the functor

$$[J, Poset^{op}] : [L, Poset^{op}] \longrightarrow [ob(L), Poset^{op}]$$

where $J : ob(L) \longrightarrow L$ is the inclusion, cf Bonchi and Zanasi's construction [6], with its left adjoint, which is given by left Kan extension. The fact that the functor $[J, Poset^{op}]$ also has a right adjoint, given by right Kan extension, implies that the monad T is finitary.

Corollary 2. *For any Lawvere theory L, the inclusion*

$$[L^{op}, Poset] \longrightarrow Lax(L^{op}, Poset)$$

has a right adjoint.

Proof. Poset is a complete 2-category as it is a complete locally ordered category. So $Poset^{op}$ is a cocomplete 2-category, and so $[ob(L), Poset^{op}]$ is a cocomplete 2-category. So the conditions of Theorem 4 hold for Example 10, and so the inclusion

$$[L, Poset^{op}] \longrightarrow Lax(L, Poset^{op})$$

has a left adjoint. But $[L, Poset^{op}]^{op}$ is canonically isomorphic to $[L^{op}, Poset]$, and $Lax(L, Poset^{op})^{op}$ is canonically isomorphic to $Lax(L^{op}, Poset)$, and in general, a functor $H : A \longrightarrow B$ has a right adjoint if and only if $H : A^{op} \longrightarrow B^{op}$ has a left adjoint. The combination of these facts yields the result.

With this result in hand, one can systematically work through Bonchi and Zanasi's paper, adapting their constructions for saturation and desaturation, without discarding the inherent laxness that logic programming, cf data refinement, possesses.

We have stated the results here for arbitrary lax transformations, but they apply equally to those that strictly respect injections, i.e., a subtle extension of the above argument shows that the inclusion

$$[L^{op}, Poset] \longrightarrow Lax_{Inj}(L^{op}, Poset)$$

has a right adjoint, that right adjoint being a further variant of the right Kan extension that Bonchi and Zanasi used. The argument for lax naturality from the Introduction retains its force, so in Bonchi and Zanasi's sense, this does not yield compositionality of lax semantics, but it does further refine their analysis of saturation, eliminating more double counting.

7 Conclusions

For variable-free logic programs, in [20], we used the cofree comonad on $P_f P_f$ to model the coinductive trees generated by a logic program. The notion of coinductive tree had not been isolated at the time of writing of [20], or of [21], so we did not explicitly explain the relationship in [20], hence our doing so here, but the result was effectively in [20], just explained in somewhat different terms.

Using lax transformations, we extended the result in [21], albeit again not stating it explicitly but again explained explicitly here, to arbitrary logic programs, including existential programs a leading example being GC, as studied extensively by Sterling and Shapiro [31]. The problem of existential clauses is well-known in the literature on theorem proving and within communities that use term-rewriting, TM-resolution or their variants. In TRS [33], existential variables are not allowed to appear in rewriting rules, and in type inference, the restriction to non-existential programs is common [11]. In LP, the problem of handling existential variables when constructing proofs with TM-resolution marks the boundary between the theorem-proving and problem-solving aspects, as explained in Sect. 2.

The papers [21,23] also contained a kind of category theoretic semantics for existential logic programs such as GC, but that semantics was limited, not modelling the coinductive trees generated by TM-resolution for such logic programs. Here, we have refined lax semantics, refining $Lax(\mathcal{L}_\Sigma^{op}, Poset)$ to $Lax_{Inj}(\mathcal{L}_\Sigma^{op}, Poset)$, thus insisting upon strict naturality for injections, and refining the construction $P_c P_f At$ to $P_{ff}(At)$, thus allowing for additional variables in the tail of a clause in a logic program and not introducing countability, cf the modelling of local state in [29]. This has allowed us to model coinductive trees for arbitrary logic programs.

We have further mildly refined Bonchi and Zanasi's saturation semantics for logic programming [6], showing how it may be seen to complement rather than to replace lax semantics.

References

1. Amato, G., Lipton, J., McGrail, R.: On the algebraic structure of declarative programming languages. Theor. Comput. Sci. **410**(46), 4626–4671 (2009)
2. Benabou, J.: Introduction to bicategories. In: Bénabou, J., Davis, R., Dold, A., Isbell, J., MacLane, S., Oberst, U., Roos, J.-E. (eds.) Reports of the Midwest Category Seminar. Lecture Notes in Mathematics, vol. 47, pp. 1–77. Springer, Heidelberg (1967)
3. Blackwell, R., Kelly, G.M., Power, A.J.: Two-dimensional monad theory. J. Pure Appl. Algebra **59**, 1–41 (1989)
4. Bonchi, F., Montanari, U.: Reactive systems, (semi-)saturated semantics and coalgebras on presheaves. Theor. Comput. Sci. **410**(41), 4044–4066 (2009)
5. Bruni, R., Montanari, U., Rossi, F.: An interactive semantics of logic programming. TPLP **1**(6), 647–690 (2001)
6. Bonchi, F., Zanasi, F.: Saturated semantics for coalgebraic logic programming. In: Heckel, R., Milius, S. (eds.) CALCO 2013. LNCS, vol. 8089, pp. 80–94. Springer, Heidelberg (2013)
7. Bonchi, F., Zanasi, F.: Bialgebraic semantics for logic programming. CoRR abs/1502.06095 (2015)
8. Comini, M., Levi, G., Meo, M.C.: A theory of observables for logic programs. Inf. Comput. **169**(1), 23–80 (2001)
9. Fu, P., Komendantskaya, E.: A type-theoretic approach to resolution. In: Falaschi, M., et al. (eds.) LOPSTR 2015. LNCS, vol. 9527, pp. 91–106. Springer, Heidelberg (2015). doi:10.1007/978-3-319-27436-2_6
10. Fu, P., Komendantskaya, E., Schrijvers, T., Pond, A.: Proof relevant corecursive resolution. In: Kiselyov, O., King, A., et al. (eds.) FLOPS 2016. LNCS, vol. 9613, pp. 126–143. Springer, Heidelberg (2016). doi:10.1007/978-3-319-29604-3_9
11. Jones, S.P., Jones, M., Meijer, E.: Type classes: an exploration of the design space. In: Haskell Workshop (1997)
12. Gupta, G., Costa, V.: Optimal implementation of and-or parallel prolog. In: PARLE 1992, pp. 71–92 (1994)
13. Jifeng, H., Hoare, C.A.R.: Categorical semantics for programming languages. In: Main, M., Melton, A., Mislove, M., Schmidt, D. (eds.) MFPS 1989. LNCS, vol. 442, pp. 402–417. Springer, Heidelberg (1989)
14. Jifeng, H., Hoare, C.A.R.: Data refinement in a categorical setting. Technical Monograph PRG-90. Oxford University Computing Laboratory, Programming Research Group, Oxford (1990)

15. Johann, P., Komendantskaya, E., Komendantskiy, V.: Structural resolution for logic programming. In: Technical Communications of ICLP 2015 (2015)
16. Kelly, G.M.: Coherence theorems for lax algebras and for distributive laws. In: Kelly, G.M. (ed.) Category Seminar. Lecture Notes in Mathematics, vol. 420, pp. 281–375. Spriniger, Heidelberg (1974)
17. Kelly, G.M.: Basic Concepts of Enriched Category Theory. London Math. Soc. Lecture Notes Series, vol. 64. Cambridge University Press, Cambridge (1982)
18. Kinoshita, Y., Power, A.J.: Lax naturality through enrichment. J. Pure Appl. Algebra **112**, 53–72 (1996)
19. Kinoshita, Y., Power, J.: A fibrational semantics for logic programs. In: Dyckhoff, R., Herre, H., Schroeder-Heister, P. (eds.) ELP 1996. LNCS (LNAI), vol. 1050, pp. 177–191. Springer, Heidelberg (1996)
20. Komendantskaya, E., McCusker, G., Power, J.: Coalgebraic semantics for parallel derivation strategies in logic programming. In: Johnson, M., Pavlovic, D. (eds.) AMAST 2010. LNCS, vol. 6486, pp. 111–127. Springer, Heidelberg (2011)
21. Komendantskaya, E., Power, J.: Coalgebraic semantics for derivations in logic programming. In: Corradini, A., Klin, B., Cîrstea, C. (eds.) CALCO 2011. LNCS, vol. 6859, pp. 268–282. Springer, Heidelberg (2011)
22. Komendantskaya, E., Power, J.: Coalgebraic derivations in logic programming. In: CSL. LIPIcs, pp. 352–366. Schloss Dagstuhl (2011)
23. Komendantskaya, E., Power, J., Schmidt, M.: Coalgebraic logic programming: from Semantics to Implementation. J. Log. Comput. **26**(2), 745–783 (2016)
24. Komendantskaya, E., Schmidt, M., Heras, J.: Exploiting parallelism in coalgebraic logic programming. Electr. Notes Theor. Comput. Sci. **303**, 121–148 (2014)
25. Lloyd, J.: Foundations of Logic Programming, 2nd edn. Springer, Heidelberg (1987)
26. Mac Lane, S.: Categories for the Working Mathematician. Graduate Texts in Mathematics. Springer, Heidelberg (1971)
27. Power, A.J.: An algebraic formulation for data refinement. In: Main, M., Melton, A., Mislove, M., Schmidt, D. (eds.) MFPS 1989. LNCS, vol. 442, pp. 390–401. Springer, Heidelberg (1989)
28. Plotkin, G., Power, J.: Adequacy for algebraic effects. In: Honsell, F., Miculan, M. (eds.) FOSSACS 2001. LNCS, vol. 2030, pp. 1–24. Springer, Heidelberg (2001)
29. Plotkin, G., Power, J.: Notions of computation determine monads. In: Nielsen, M., Engberg, U. (eds.) FOSSACS 2002. LNCS, vol. 2303, pp. 342–356. Springer, Heidelberg (2002)
30. Simon, L., Bansal, A., Mallya, A., Gupta, G.: Co-logic programming: extending logic programming with coinduction. In: Arge, L., Cachin, C., Jurdziński, T., Tarlecki, A. (eds.) ICALP 2007. LNCS, vol. 4596, pp. 472–483. Springer, Heidelberg (2007)
31. Sterling, L., Shapiro, E.: The Art of Prolog. MIT Press, Cambridge (1986)
32. Street, R.: The formal theory of monads. J. Pure Appl. Algebra **2**, 149–168 (1972)
33. Terese: Term Rewriting Systems. Cambridge University Press (2003)
34. Worrell, J.: Terminal sequences for accessible endofunctors. In: Proceedings of the CMCS 1999. Electronic Notes in Theoretical Computer Science, vol. 19, pp. 24–38 (1999)

Product Rules and Distributive Laws

Joost Winter[✉]

Faculty of Mathematics, Informatics, and Mechanics,
University of Warsaw, Warsaw, Poland
jwinter@mimuw.edu.pl

Abstract. We give a categorical perspective on various product rules, including Brzozowski's product rule $((st)_a = s_a t + o(s)t_a)$ and the familiar rule of calculus $((st)_a = s_a t + st_a)$. It is already known that these product rules can be represented using distributive laws, e.g. via a suitable quotient of a GSOS law. In this paper, we cast these product rules into a general setting where we have two monads S and T, a (possibly copointed) behavioural functor F, a distributive law of T over S, a distributive law of S over F, and a suitably defined distributive law $TF \Rightarrow FST$. We introduce a coherence axiom giving a sufficient and necessary condition for such triples of distributive laws to yield a new distributive law of the composite monad ST over F, allowing us to determinize FST-coalgebras into lifted F coalgebras via a two step process whenever this axiom holds.

1 Introduction

In [Brz64], Brzozowski introduced a calculus of derivatives for regular expressions, including amongst other rules the *product rule*

$$(xy)_a = x_a y + o(x)y_a$$

which can be contrasted with the familiar (Leibniz) product rule from calculus (here rephrased with a somewhat uncommon nomenclature, to highlight the correspondences and differences with Brzozowski's rule)

$$(xy)_a = x_a y + xy_a.$$

In the work of Rutten, starting with [Rut98], a picture started to emerge in which Brzozowski's calculus of derivatives was absorbed into the framework of universal coalgebra. Notably, in [Rut03, Rut05] both product rules were considered as behavioural differential equations, or coinductive definitions, possessing unique solutions, and *defining* respectively the convolution product (Brzozowski's rule) and the shuffle product (Leibniz' rule) on the final coalgebra.

J. Winter—This work was supported by the Polish National Science Centre (NCN) grant 2012/07/E/ST6/03026.

I. Hasuo (Ed.): CMCS 2016, LNCS 9608, pp. 114–135, 2016.
DOI: 10.1007/978-3-319-40370-0_8

In the work of Bonsangue and Rutten together with the present author, starting with [WBR11], a coalgebraic presentation of the context-free languages was given, in which Brzozowski's derivative rules (excluding the star rule) again play a crucial role. This work was based on a concretely presented extension of $2 \times \mathcal{P}_\omega(-^*)^A$-coalgebras into $2 \times -^A$-coalgebras, as in the following diagram:

$$
\begin{array}{ccc}
X & \xrightarrow{\ \eta_X\ } \mathcal{P}_\omega(X^*) \xrightarrow{\ [\![-]\!]\ } & \mathcal{P}(A^*) \\
{\scriptstyle (o,\delta)}\Big\downarrow \quad {\scriptstyle (\hat{o},\hat{\delta})} & & \Big\downarrow{\scriptstyle (O,\Delta)} \\
2 \times \mathcal{P}_\omega(X^*)^A & \xrightarrow{\quad 2 \times [\![-]\!]^A \quad} & 2 \times \mathcal{P}_\omega(X^*)^A
\end{array}
$$

In [BHKR13], it was shown that the above diagram could be understood in terms of a distributive law of the monad $\mathcal{P}_\omega(-^*)$ over the *cofree copointed functor* on $2 \times -^A$, that is, over $(- \times (2 \times -^A), \pi_1)$ by means of a suitable quotient of distributive laws. This firmly established the connection between the work of [WBR11] and the general categorical framework of *bialgebras* and *distributive laws*, considered in e.g. [LPW00, Bar04, Jac06a, Jac06b, Kli11, JSS12], as well as to the (closely related) *generalized powerset construction* presented in [SBBR10, SBBR13].

Moreover, in [BRW12], the approach was generalized from formal languages to formal power series, thus giving a coalgebraic characterization of the *constructively algebraic power series*.

In [Win14], the author's Ph.D. thesis, this extension was broken up into a two step process, as in the diagram,

$$
\begin{array}{cccc}
X & \xrightarrow{\ \eta_X^0\ } X^* & \xrightarrow{\ \eta_{X^*}^1\ } \mathcal{P}_\omega(X^*) \xrightarrow{\ [\![-]\!]\ } & \mathcal{P}(A^*) \\
{\scriptstyle (o,\delta)}\Big\downarrow \quad {\scriptstyle (o^\#,\delta^\#)} \quad {\scriptstyle (\hat{o},\hat{\delta})} & & & \Big\downarrow{\scriptstyle (O,\Delta)} \\
2 \times \mathcal{P}_\omega(X^*)^A & & \xrightarrow{\quad 2 \times [\![-]\!]^A \quad} & 2 \times \mathcal{P}_\omega(X^*)^A
\end{array}
$$

and it was noted that the first extension corresponded directly to the product rule (combined with the unit rule), whereas the second extension was simply an instance of ordinary determinization, applied to a coalgebra which is by itself infinite. This two-step process again easily generalizes from the setting of languages to the setting of noncommuting power series over commutative semirings; however, no general categorical presentation of this two-step process was given in the thesis.

The main aim of this paper is to give such a categorical presentation of this two-step process. This involves a setting including both distributive laws between monads $\lambda^0 : TS \Rightarrow ST$, distributive laws of a monad over a functor (or copointed functor) $\lambda^2 : SF \Rightarrow FS$, as well as a, suitably defined, distributive law of type $\lambda^1 : TF \Rightarrow FST$ (this definition will be given in Sect. 3).

In Sect. 4, we will present our main result, which states that the three laws as suggested above can be combined to form a new distributive law $\hat{\lambda}$ of the composite monad ST over F, if and only if a certain *coherence axiom* involving the three laws is satisfied. This construction works regardless whether F is an ordinary or copointed functor. The proof that satisfaction of the coherence axiom implies the multiplicative law of the distributive law of the composite monad ST over F was automatically derived using a Prolog program written for this purpose.

We then show that both the Brzozowski and Leibniz product rules, as well as the (simpler) pointwise rule defining the Hadamard product, can be cast into this setting, and that the coherence axiom is satisfied in these cases, and additionally present a counterexample where the coherence axiom is not satisfied. We conclude with a presentation of an extension of the generalized powerset construction to incorporate these two-step extensions.

Related work. All of the aforementioned references can be regarded as related work. In particular, [Che07] gives a similar coherence condition, for the existence of composite distributive laws in the context of distributive laws between monads. The relationships between [Che07] and the present paper involve some subtleties, and in Sect. 6, we will present a more detailed investigation of the relationship.

Additionally, [MMS13, Sch14] also give a compositional approach to distributive laws, however, in a setting different from our approach, not including the combination of distributive laws between monads and distributive laws of monads over functors into new distributive laws. In [Jac06b], a coinductive presentation of the shuffle product using a two-step process is also given, however, the coinductive definition of the multiplication there is not given using a distributive law.

2   Preliminaries

2.1   General Preliminaries

We assume familiar the basic notions of category theory (functors, natural transformations, monads, (Eilenberg-Moore) algebras), of (commutative) monoids and semirings, and of coalgebras for a functor. These can be found in e.g. [Awo10] or [Mac71] for general categorical notions, and in [Rut00] for the basics of coalgebra.

An important endofunctor we will use is the functor $K \times -^A$ on **Set**, representing automata with output in K over the input alphabet A, often called *Moore automata* but in this paper simply K-*automata*. The transition of coalgebras for this functor is represented as a pair (o, δ), called an *output-derivative pair*, with $\delta(x)(a)$ denoted by x_a. A final coalgebra for this functor exists, and can be given by $(K\langle\!\langle A \rangle\!\rangle, O, \Delta)$, where $K\langle\!\langle A \rangle\!\rangle$ is the function space $A^* \to K$, $O(\sigma) = \sigma(1)$ and $\sigma_a(w) = \sigma(aw)$ for all $\sigma \in K\langle\!\langle A \rangle\!\rangle$, $a \in A$, and $w \in A^*$. Here, and elsewhere in this paper, 1 denotes the empty word.

A *copointed endofunctor* is a pair (F, ϵ), where F is a **C**-endofunctor, and ϵ is a natural transformation $\epsilon : F \Rightarrow 1_{\mathbf{C}}$. A coalgebra for a copointed endofunctor (F, ϵ) consists of a coalgebra (X, δ) for the endofunctor F, satisfying the condition that $\epsilon_X \circ \delta = 1_X$.

Of special interest to us are *cofree copointed endofunctors*, which can be constructed in any category that has binary products. Given an ordinary endofunctor F, the cofree copointed endofunctor on F is the pair $(\mathrm{Id} \times F, \pi_1)$. For any functor F such that a cofree copointed endofunctor on F exists, there is an isomorphism between coalgebras for the ordinary endofunctor F, and coalgebras for the copointed endofunctor $(\mathrm{Id} \times F, \pi_1)$.

Given a monad T (we will frequently refer to a monad simply using the name of its functorial part) and an endofunctor F, a (T, F)-bialgebra is a triple (X, α, δ) such that (X, α) is an algebra for the monad T and (X, δ) is a F-coalgebra. If F is a copointed functor (here again, we often refer to the copointed functor simply using the name of the functor), we additionally require the coalgebraic part of a (T, F)-bialgebra to be a coalgebra for the copointed functor. (We will turn to λ-bialgebras involving a distributive law soon!)

Given a commutative semiring K (in this paper, we will essentially only be concerned with commutative semirings), a K-semimodule $(M, 0, +, \times)$ consists of a commutative monoid $(M, 0, +)$ and an operation $\times : K \times M \to M$ (called *scalar product*) such that, for all $k, l \in k$ and $m, n \in M$:

$$k \times (m + n) = (k \times m) + (k \times n) \quad 1_K \times m = m$$
$$(k + l) \times m = (k \times m) + (l \times m) \quad 0_K \times m = 0$$
$$(kl) \times m = k \times (l \times m) \qquad k \times 0 = 0$$

For any commutative semiring K^1, the K-semimodules are exactly the algebras for the monad $(\mathrm{Lin}_K(-), \eta, \mu)$, where $\mathrm{Lin}_K(X)$ is the set of finitely supported functions on X, and, given a function $f : X \to Y$,

$$\mathrm{Lin}_K(f) = y \mapsto \sum_{x \in f^{-1}(y)} f(x).$$

The multiplication of the monad, for any X and any $f \in \mathrm{Lin}_K(\mathrm{Lin}_K(X))$, can be given by

$$\mu_X(f) = x \mapsto \sum_{g \in \mathrm{supp}(f)} f(g) \cdot g(x)$$

and the unit of the monad, for any $x \in X$, by:

$$\eta_X(x) = y \mapsto \text{if } x = y \text{ then } 1 \text{ else } 0$$

Given some $\sigma \in \mathrm{Lin}_K(X)$ and some $x \in X$, we furthermore write $[\sigma \Downarrow x]$ for the application of σ at x.

Given a commutative semiring K, a K-algebra[2] (see e.g. [Eil76]) is a tuple $(X, 0, 1, +, \cdot, \times)$ such that $(X, 0, 1, +, \cdot)$ is a semiring, $(X, 0, +, \times)$ is a K-semimodule, such that for all $k \in K$ and $x \in X$:

$$(k \times 1) \cdot x = k \times x = x \cdot (k \times 1)$$

---

[1] For a semiring K that is not commutative, one can define two distinct monad structures on $\mathrm{Lin}_K(-)$, with a different multiplication, one corresponding to left-K-semimodules, and one corresponding to right K-semimodules. When K is commutative, these two monads are identical.

[2] In the literature often called a unital associative algebra.

An equivalent, and as it turns out for us more convenient, characterization is the following: a K-algebra is a tuple $(X, 0, 1, +, \cdot, \times)$ such that $(X, 1, \cdot)$ is a monoid, $(X, 0, +, \times)$ is a K-semimodule, furthermore satisfying the following axioms:

$$x \cdot (y + z) = x \cdot y + x \cdot z$$
$$(x + y) \cdot z = x \cdot z + y \cdot z$$
$$(k \times 1) \cdot x = k \times x = x \cdot (k \times 1)$$

(Note that the pair of semiring axioms $0 \cdot x = 0 = x \cdot 0$ can be derived from the pair of axioms on the third line.)

For some semirings K, the categories of K-semimodules and K-algebras correspond to familiar categories. We summarize some of these in the following table:

K	K-semimodules	K-algebras
$\mathbb{B}$	join-semilattices	idempotent semirings
$\mathbb{N}$	commutative monoids	semirings
$\mathbb{Z}$	abelian groups	rings

Moreover, it is easily verified that the monad $\mathrm{Lin}_{\mathbb{B}}(-)$ is naturally isomorphic to the finite powerset monad $\mathcal{P}_\omega$.

2.2   Distributive Laws Between Monads

We now summarize the definitions and some important facts about distributive laws between monads, which can be found in e.g. [Bec69, BW85].

Given monads (S, μ^S, η^S) and (T, μ^T, η^T), a *distributive law* of the monad T over the monad S is a natural transformation $\lambda : TS \Rightarrow ST$ such that the four diagrams of natural transformations

$$
\begin{array}{ccc}
S \xLongrightarrow{\eta^T S} TS & & TTS \xLongrightarrow{\mu^T S} TS \\
\quad\searrow_{S\eta^T} \quad \downarrow \lambda & & T\lambda \downarrow \qquad \qquad \downarrow \lambda \\
\qquad\qquad ST & & TST \xLongrightarrow{\lambda T} STT \xLongrightarrow{S\mu^T} ST
\end{array}
$$

$$
\begin{array}{ccc}
T \xLongrightarrow{T\eta^S} TS & & TSS \xLongrightarrow{T\mu^S} TS \\
\quad\searrow_{\eta^S T} \quad \downarrow \lambda & & \lambda S \downarrow \qquad \qquad \downarrow \lambda \\
\qquad\qquad ST & & STS \xLongrightarrow{S\lambda} SST \xLongrightarrow{\mu^S T} ST
\end{array}
$$

commute.

In the presence of a distributive law of a monad T over another monad S, we obtain, amongst other things, the following:

1. A monad structure on the composite functor ST, given by:

$$\eta^{ST} = \eta^S T \circ \eta^T \quad \text{and} \quad \mu^{ST} = S\mu^T \circ \mu^S TT \circ S\lambda T$$

2. A lifting of the monad S to a monad $\hat{S}$ in the category $\mathbf{C}^T$ of T-algebras, with the functorial part of $\hat{S}$ given by:

$$\hat{S}(X, \alpha) = (SX, S\alpha \circ \lambda_X)$$

3. An isomorphism of categories between the categories of algebras for the monad ST and algebras for the monad $\hat{S}$. Given an $\hat{S}$-algebra (X, α^T, α^S), the corresponding ST-algebra can be given by $(X, \alpha^S \circ S\alpha^T)$, and given a ST-algebra (X, α), the corresponding $\hat{S}$-algebra can be given by $(X, \alpha \circ \eta_{TX}^S, \alpha \circ S\eta_X^T)$. These constructions are mutually inverse, as shown in [Bec69, Section 2].

In the case where $T = -^*$ and $S = \mathrm{Lin}_K(-)$ (where K is a commutative semiring), there is a distributive law $\lambda : (\mathrm{Lin}_K(-))^* \Rightarrow \mathrm{Lin}_K(-^*)$ of T over S, given by:

$$\lambda_X \left(\prod_{i=1}^{n} \sum_{j=1}^{m_i} k_{ij} \times x_{ij} \right) = \sum_{j_1=1}^{m_1} \cdots \sum_{j_n=1}^{m_n} \left(\prod_{i=1}^{n} k_{ij_i} \times \prod_{i=1}^{n} x_{ij_i} \right)$$

(recall that $\times$ is the scalar product, and thus, the k_{ij} here are scalars of K!) Here the product symbol $\prod$ denotes the operation on the monoid structure, and the summation symbol $\sum$ and times $\times$ denote the (sum and scalar product) on the semimodule structure, respectively.

It is well-known, at least[3] in the cases of the semirings $\mathbb{B}$, $\mathbb{N}$, and $\mathbb{Z}$ (and probably in the remaining cases as well—nevertheless we give a full proof in the appendix), that this is a distributive law of $-^*$ over $\mathrm{Lin}_K(-)$:

Proposition 1. *The natural transformation λ as given above is a distributive law of the monad $\mathrm{Lin}_K(-)$ over the monad $-^*$.*

We let $K\langle - \rangle = \mathrm{Lin}_K(-^*)$ denote the resulting composite monad. Elements of $K\langle X \rangle$ can be identified with noncommuting polynomials in variables from X. Moreover, the algebras for this monad are precisely the K-algebras: by definition, a K-algebra is both a monoid and a semimodule additionally satisfying

$$x \cdot (y + z) = x \cdot y + x \cdot z$$
$$(x + y) \cdot z = x \cdot z + y \cdot z$$
$$(k \times 1) \cdot x = k \times x = x \cdot (k \times 1)$$

---

[3] See e.g. [Bec69] for $\mathbb{Z}$ and [Jac06a] for $\mathbb{N}$ and $\mathbb{B}$.

for elements x, y, z and scalars k and an algebra for the monad $K\langle-\rangle$ is both a monoid and a semimodule additionally satisfying

$$\prod_{i=1}^{n}\sum_{j=1}^{m_i} k_{ij} \times x_{ij} = \sum_{j_1=1}^{m_1} \cdots \sum_{j_n=1}^{m_n} \left(\prod_{i=1}^{n} k_{ij_i} \times \prod_{i=1}^{n} x_{ij_i} \right)$$

It is easily verified that these two conditions are equivalent.

2.3   Distributive Laws of a Monad over a (Copointed) Endofunctor

We now turn to a brief summary of distributive laws of monads over endofunctors and copointed endofunctors. The material which now follows can all be found in either [Bar04] or [JSS12]; supplementary presentations can be found in e.g. [Jac06a, Jac06b, Kli11, LPW00] (in which the notions of distributive laws of monads over endofunctors and copointed endofunctors were first considered).

Given a monad (T, μ, η) and an endofunctor F on any category $\mathbf{C}$, a *distributive law* of the monad T over F is a natural transformation $\lambda : TF \Rightarrow FT$ such that the two diagrams of natural transformations

$$
\begin{array}{ccc}
F \stackrel{\eta F}{\Longrightarrow} TF & & TTF \stackrel{\mu F}{\Longrightarrow} TF \\
\quad \searrow{\scriptstyle F\eta} \quad \Downarrow{\lambda} & \text{and} & T\lambda \Downarrow \qquad \qquad \Downarrow \lambda \\
FT & & TFT \stackrel{\lambda T}{\Longrightarrow} FTT \stackrel{F\mu}{\Longrightarrow} FT
\end{array}
$$

commute.

Given a distributive law λ of T over F, a λ-bialgebra is a (T, F)-bialgebra for which the following diagram commutes:

$$
\begin{array}{ccccc}
TX & \stackrel{\alpha}{\longrightarrow} & X & \stackrel{\delta}{\longrightarrow} & FX \\
\downarrow{\scriptstyle T\delta} & & & & \uparrow{\scriptstyle F\alpha} \\
TFX & & \stackrel{\lambda_X}{\longrightarrow} & & FTX
\end{array}
$$

A distributive law of a monad (T, η, μ) over a copointed endofunctor (F, ϵ) is a distributive law of the monad over the (ordinary) endofunctor F, such that additionally the diagram

$$
\begin{array}{ccc}
TF & \stackrel{\lambda}{\Longrightarrow} & FT \\
& \searrow{\scriptstyle T\epsilon} \quad & \Downarrow{\epsilon T} \\
& T &
\end{array}
$$

commutes.

For distributive laws of a monad over a copointed endofunctor, bialgebras can again be defined: the only modification needed here is that we additionally require (X, δ) to be a coalgebra for the copointed endofunctor F. Note that any distributive law λ of a monad T over an ordinary functor F extends to a distributive law λ' of T over the cofree copointed functor $(\text{Id} \times F)$, such that the λ-bialgebras and λ'-bialgebras are isomorphic as categories.

In the presence of a distributive law λ of T over F, we obtain, amongst other things, the following (both for ordinary and copointed endofunctors, see e.g. [Bar04] for proofs):

1. A lifting of F to a functor $\hat{F}$ in the category of algebras for the monad T, and of T to a functor $\hat{T}$ the category of F-coalgebras, together with an iso-morphic correspondence between the categories of λ-bialgebras, $\hat{F}$-coalgebras, and algebras for the monad $\hat{T}$.
2. If F has a final coalgebra (Ω, ω), there is a unique algebra ξ for the monad T on Ω such that (Ω, ξ, ω) is a λ-bialgebra, which moreover is a final λ-bialgebra.

An important instance of a distributive law, considered in e.g. [JSS12], is the law of the functor $K \times -^A$ over the monad $\text{Lin}_K(-)$ where K is any semiring[4], given by:

$$\lambda_X \left(\sum_{i=1}^{n} k_i \times (o_i, a \mapsto d_{ia}) \right) = \left(\sum_{i=1}^{n} k_i o_i, a \mapsto \sum_{i=1}^{n} k_i \times d_{ia} \right)$$

This distributive law can be obtained canonically by means of the strength operation (see e.g. [JSS12]), using the obvious K-semimodule structure on K itself. The bialgebras for this distributive law are precisely the structures that are both K-automata and K-semimodules, such that the output and derivative (with respect to an arbitrary alphabet symbol) are both linear mappings. A concrete formulation is the following: a K-*linear automaton*[5] is a tuple $(X, 0, +, \times, o, \delta)$ such that $(X, 0, +, \times)$ is a K-semimodule, (X, o, δ) is a K-automaton, and o and $\delta(-)(a)$ (for each $a \in A$) are linear mappings.

2.4   The Generalized Powerset Construction

We now turn to a summary of some of the main observations from [SBBR10, SBBR13] pertaining to the *generalized powerset construction* (many instances of which can be found in these two papers as examples). Beyond that, we also sketch how the 'generalized powerset construction' framework can be extended to distributive laws over (cofree) copointed functors, and highlight some of the (additional) subtleties that arise in this case.

Given a (possibly copointed) endofunctor F such that a final F-coalgebra exists, a monad T (on the same category), and a distributive law of T over F, any FT-coalgebra (X, δ) (if F is copointed, we additionally require that

---

[4] If K is not commutative, we can consider left K-semimodules, but for this paper, this is not relevant.

[5] called *linear weighted automata* in e.g. [BBB+12].

$\epsilon_{TX} \circ \delta = \eta_X$) can be extended canonically to an F-coalgebra $(TX, \hat{\delta})$, where $\hat{\delta}$ can be specified as:

$$\hat{\delta} = F\mu_X \circ \lambda_{TX} \circ T\delta$$

In fact, it follows that $(TX, \mu_X, \hat{\delta})$ is a λ-bialgebra (and, hence, that FTX has the structure of an algebra for the monad T). If F has a final coalgebra (Ω, ω), we thus obtain

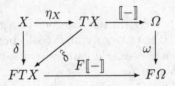

where $[\![-]\!]$ is a morphism of λ-bialgebras to the unique λ-bialgebra on the final F-coalgebra. If F is copointed, note that

$$\epsilon_{TX} \circ \hat{\delta} = \epsilon_{TX} \circ F\mu_X \circ \lambda_{TX} \circ T\delta = \mu_X \circ \epsilon_{TTX} \circ \lambda_{TX} \circ T\delta$$
$$= \mu_X \circ T\epsilon_X \circ T\delta = \mu_X \circ T\eta_X = 1_{TX}$$

so that $\hat{\delta}$ is indeed a coalgebra for the copointed functor.

When the functor F is a cofree copointed functor of the form $(\mathrm{Id} \times G, \pi_1)$, note that δ has to be of the form (η_X, ζ), where ζ is an (ordinary) GT-coalgebra, and because $\hat{\delta}$ is a coalgebra for the cofree copointed functor, it has to be of the form $(1_{TX}, \zeta^\flat)$ for some $\zeta^\flat$. Similarly, ω has to be of the form $(1_\Omega, \psi)$, where (Ω, ψ) is the final G-coalgebra. Now note that there is a correspondence of commuting diagrams of the form

$$
\begin{array}{ccccc}
X & \xrightarrow{\eta_X} & TX & \xrightarrow{[\![-]\!]} & \Omega \\
{\scriptstyle (\eta_X, \zeta)}\downarrow & {\scriptstyle (1_{TX}, \zeta^\flat)} & & & \downarrow{\scriptstyle (1_\Omega, \psi)} \\
TX \times GTX & \xrightarrow{[\![-]\!] \times G[\![-]\!]} & & & \Omega \times G\Omega
\end{array}
$$

and:

$$
\begin{array}{ccccc}
X & \xrightarrow{\eta_X} & TX & \xrightarrow{[\![-]\!]} & \Omega \\
{\scriptstyle \zeta}\downarrow & {\scriptstyle \hat{\zeta}} & & & \downarrow{\scriptstyle \psi} \\
GTX & & \xrightarrow{G[\![-]\!]} & & G\Omega
\end{array}
\tag{1}
$$

However, in general we are *not* guaranteed that the second diagram occurs as the result of an ordinary distributive law of T over G. Hence, in general, although we are guaranteed that $TX \times GTX$ has the structure of an algebra for the monad T, we do not have this guarantee of GTX itself (this may in fact fail, e.g. for GSOS rules that cannot be presented using the simple SOS format). On the other hand, we are still guaranteed that $[\![-]\!]$ is a T-algebra morphism, because this follows from the upper diagram.

3 Distributive Laws of a Monad over a (Copointed) Endofunctor into a Composite Monad

In this section, we present a definition of a new (as far as the author is aware) type of distributive law, which allows us to formulate the 'two-step' generalized powerset construction. We start by giving the general formulation, before turning to the main examples.

Given two monads (S, μ^S, η^S), (T, μ^T, η^T) such that a distributive law $\lambda^0 : TS \Rightarrow ST$ exists, and an endofunctor F, on any category $\mathbf{C}$, a *distributive law* of the monad T over F into the composite monad ST is a natural transformation $\lambda : TF \Rightarrow FST$ such that the two diagrams of natural transformations

$$
\begin{array}{ccc}
F \xrightarrow{\eta^T F} TF & & TTF \xrightarrow{\mu^T F} TF \\
\underset{F\eta^{ST}}{\searrow} \quad \Big\downarrow \lambda & \text{and} & T\lambda \Big\downarrow \qquad\qquad\qquad \Big\downarrow \lambda \\
FST & & TFST \xrightarrow{\lambda ST} FSTST \xrightarrow{F\mu^{ST}} FST
\end{array}
$$

commute.

We can again define a suitable notion of a λ-bialgebra for a distributive law of T over F into the composite monad ST. For such a λ, a λ-bialgebra (X, α, δ) is a (ST, F)-bialgebra such that the diagram

$$
\begin{array}{ccccccc}
TX & \xrightarrow{\eta^S_{TX}} & STX & \xrightarrow{\alpha} & X & \xrightarrow{\delta} & FX \\
T\delta \Big\downarrow & & & & & & \Big\uparrow F\alpha \\
TFX & & & \xrightarrow{\lambda_X} & & & FSTX
\end{array}
$$

commutes (note that we can here replace $\alpha \circ \eta^S_{TX}$ with α^T).

A distributive law λ of a monad (T, η, μ) over a copointed endofunctor (F, ϵ) into a composite monad is a distributive law of the monad over the (ordinary) endofunctor F into the same composite monad, such that additionally the diagram

$$
\begin{array}{ccc}
TF & \xRightarrow{\lambda} & FST \\
T\epsilon \Big\Downarrow & & \Big\Downarrow \epsilon ST \\
T & \xRightarrow{\eta S} & ST
\end{array}
$$

commutes. Again, the corresponding notion of a λ-bialgebra is obtained by adding the additional requirement that (X, δ) is a coalgebra for the copointed functor F.

The following result is unspectacular, but will be useful for us later on:

Lemma 2. *Given monads (S, η^S, μ^S) and (T, η^T, μ^T), and a (possibly copointed) endofunctor F, and a distributive law $\lambda^0 : TS \Rightarrow ST$, if a natural transformation $\lambda : TF \Rightarrow FT$ is a distributive law of T over F, then*

$\hat{\lambda} = F\eta^S T \circ \lambda$ *is a distributive law of T over F into the composite monad* ST. *If η is pointwise monic and F preserves monos, the converse also holds.*

3.1   Product Rules as Distributive Laws

Instantiating $S = \text{Lin}_K(-)$, $T = -^*$, and $F = K \times -^A$, we can model each of the three product rules as distributive laws of the monad (T, η^T, μ^T) over the copointed functor $(\text{Id} \times F, \pi_1)$ into the composite monad $(ST, \eta^{ST}, \mu^{ST})$ as follows:

1. The Brzozowski product rule can be specified as

$$\lambda_X\left(\prod_{i=1}^n (x_i, o_i, a \mapsto d_{ia})\right)$$
$$= \left(\prod_{i=1}^n x_i, \prod_{i=1}^n o_i, a \mapsto \sum_{i=1}^n \left(\prod_{k=1}^{i-1} o_i\right) \times \left(d_{ia} \cdot \prod_{k=i+1}^n x_i\right)\right)$$

or equivalently inductively as:

$$\lambda_X(1) = (1, 1, a \mapsto 0)$$
$$\lambda_X(x, o, a \mapsto d_a)w = (x\pi_0(\lambda_X(w)),$$
$$o\pi_1(\lambda_X(w)), a \mapsto d_a\pi_0(\lambda_X(w)) + o\pi_2(\lambda_X(w))(a))$$

The latter pair of equations can be more conveniently represented using the output-derivative notation as:

$$o(1) = 1 \qquad\qquad 1_a = 0$$
$$o(xw) = o(x)o(w) \quad (xw)_a = x_a w + o(x)w_a$$

2. The Leibniz product rule can be given as

$$\lambda_X\left(\prod_{i=1}^n (x_i, o_i, a \mapsto d_{ia})\right)$$
$$= \left(\prod_{i=1}^n x_i, \prod_{i=1}^n o_i, a \mapsto \sum_{i=1}^n \left(\left(\prod_{k=1}^{i-1} x_i\right) \cdot d_{ia} \cdot \prod_{k=i+1}^n x_i\right)\right)$$

or inductively and using the output-derivative notation as:

$$o(1) = 1 \qquad\qquad 1_a = 0$$
$$o(xw) = o(x)o(w) \quad (xw)_a = x_a w + x w_a$$

3. Finally, the pointwise (Hadamard) product rule is given as

$$\lambda_X\left(\prod_{i=1}^n (o_i, a \mapsto d_{ia})\right) = \left(\prod_{i=1}^n o_i, a \mapsto \prod_{k=1}^n d_{ia}\right)$$

or inductively and using the output-derivative notation as:

$$o(1) = 1 \qquad 1_a = 1$$
$$o(xw) = o(x)o(w) \quad (xw)_a = x_a w_a$$

In the case of the Brzozowski and Leibniz product rules, it can be verified that they yield distributive laws using a lengthy verification.

Proposition 3. *The Brzozowski and Leibniz product rule both are distributive laws of $-^*$ over the copointed endofunctor $(\mathrm{Id} \times F, \pi_1)$ into the composite monad $K\langle - \rangle$.*

In the case of the Hadamard product, everything is a bit easier.

Proposition 4. *The pointwise product rule is a distributive law of $-^*$ over the endofunctor F into the composite monad $K\langle - \rangle$.*

Proof. This follows from the fact that the Hadamard product can be given as a pointwise distributive law of $-^*$ over $K \times -^A$ obtained using the strength operation and multiplication over K giving its monoid structure (see, e.g. [Jac06b] for a detailed discussion of distributive laws obtained using strength). By Lemma 2, this distributive law yields another distributive law into the composite monad. □

We will henceforth refer to the three above distributive laws as λ^{brz}, λ^{lei}, and λ^{had}, respectively.

From the above propositions and the definition of λ-bialgebras for distributive laws of a monad over a functor into another monad, it follows directly that any λ-bialgebra for any of the above distributive laws satisfies the corresponding product rule for arbitrary elements. For example, it is easy to see from the definitions that the λ-bialgebras for the Brzozowski product rule are precisely the (T, F)-bialgebras that satisfy the unit and product rule

$$o(1) = 1 \qquad o(xy) = o(x)o(y)$$
$$1_a = 0 \qquad (xy)_a = x_a y + o(x) y_a$$

for arbitrary elements x and y of the bialgebra and alphabet symbols a. *Mutatis mutandis*, the same holds for the two other product rules.

4   Composite Distributive Laws

In this section we will introduce a *coherence axiom*, comparable to but different from the Yang-Baxter condition presented in [Che07]. The relation with the work in [Che07] will be discussed in Sect. 6. We consider a setting, where we are given monads S and T, a (possibly copointed) endofunctor F, on an arbitrary category $\mathbf{C}$, together with three distributive laws, as follows:

– $\lambda^0 : TS \Rightarrow ST$, a distributive law of T over S.

- $\lambda^1 : TF \Rightarrow FST$, a distributive law of T over F into the composite monad ST.
- $\lambda^2 : SF \Rightarrow FS$, a distributive law of S over F.

For such triples[6], we can compose the distributive laws as follows, yielding a new natural transformation $\hat{\lambda} : (ST)F \Rightarrow F(ST)$ given by:

$$STF \xRightarrow{S\lambda^1} SFST \xRightarrow{\lambda^2 ST} FSST \xRightarrow{F\mu^S T} FST$$

A triple of distributive laws $(\lambda^0, \lambda^1, \lambda^2)$ as above is said to satisfy the *coherence axiom* whenever the following diagram commutes:

$$
\begin{array}{ccccccc}
TSF & \xRightarrow{T\lambda^2} & TFS & \xRightarrow{\lambda^1 S} & FSTS & \xRightarrow{FS\lambda^0} & FSST \\
\Big\Vert{\lambda^0 F} & & & & & & \Big\downarrow{F\mu^S T} \\
STF & \xRightarrow{S\lambda^1} & SFST & \xRightarrow{\lambda^2 ST} & FSST & \xRightarrow{F\mu^S T} & FST
\end{array}
\tag{2}
$$

As it turns out, the coherence axiom is a sufficient and necessary condition for $\hat{\lambda}$ to be a distributive law:

Theorem 5. *Given three distributive laws as above for monads S and T and a functor (or copointed functor) F, the natural transformation $\hat{\lambda}$ is a distributive law of the composite monad ST over the functor (or copointed functor) F if and only if the coherence axiom is satisfied.*

Perhaps interesting to note, after initially finding myself unable to prove one of the directions of the above theorem by hand (the initial aim was to prove that a triple of distributive laws *always* yields a composite distributive law), I wrote a Prolog program generating all the possible ways to go from $STSTF$ to STF, together with equivalences determined by naturality squares, the multiplicative laws for the monad, and the various distributive laws (however excluding the unit laws), as well as all of these equivalences with any functor applied to it. Using this program, the source of which can be found at

http://www.mimuw.edu.pl/~jwinter/distlaws.prolog

the following facts were found:

1. There are 784 different ways of going from $STSTF$ to FST, using compositions of arbitrary natural transformations satisfying the regular expression:

$$(S|T|F)^* (\mu^S | \mu^T | \lambda^0 | \lambda^1 | \lambda^2)(S|T|F)^*$$

2. Without the coherence axiom, and using the equivalence relation named above, these divide up into two different equivalence classes (of sizes 218 and 566, respectively).

---

[6] In some of the literature, monads are called triples, but in this paper it simply means a list of three elements.

3. With the coherence axiom added, the two equivalence classes collapse into one.

After the proof was *found* automatically, it was, however, easily *verified* by hand, and transformed into a (rather large) commuting diagram.

Later, it was found (by hand) that the coherence axiom can also be derived from the existence of a composite distributive law, turning the theorem into one giving a sufficient and necessary condition, and indeed a counterexample (soon to be presented) to the coherence axiom was found. (Note that neither the theorem, nor the falling apart of the paths into two equivalence classes is *by itself* a proof that there are counterexamples to the coherence axiom, but only a proof that the coherence axiom cannot be derived using one particular method.)

Furthermore, it turns out that each of the distributive laws from Sect. 3.1 satisfies the coherence axiom, when combined with the distributive laws presented in Sects. 2.2 and 2.3:

Proposition 6. *The three distributive laws for the Brzozowski, Leibniz, and Hadamard product rules satisfy the coherence axiom, when instantiating λ^0 with the distributive law between monads presented in Sect. 2.2, and instantiating λ^2 with the distributive law of S over F from Sect. 2.3.*

We will henceforth refer to the extensions of the three distributive laws earlier named as $\hat{\lambda}^{\mathrm{brz}}$, $\hat{\lambda}^{\mathrm{lei}}$, and $\hat{\lambda}^{\mathrm{had}}$, respectively.

We now turn to the counterexample, consisting of a distributive law of T over F into the composite monad ST, which, together with the same λ^2 and λ^0 as before, does not satisfy the coherence axiom.

Counterexample 1. Consider the natural transformation given by $\lambda^1 : (\mathbb{N} \times -^A)^* \Rightarrow \mathbb{N} \times \mathbb{N}\langle-\rangle^A$ given by

$$\lambda^1_X\left(\prod_{i=1}^{n}(x_i, o_i, a \mapsto^* d_{ia})\right) = \left(\prod_{i=1}^{n} x_i, \sum_{i=1}^{n} o_i, a \mapsto \prod_{i=1}^{n} d_{ia}\right)$$

This distributive law can again be given by means of a (pointwise) distributive law from $-^*$ over $\mathbb{N} \times -^A$ using the strength operator, this time by considering the additive monoid structure on $\mathbb{N}$ (as opposed to the multiplicative monoid structure, which yields the Hadamard product).

Now consider

$$((x, 1, a \mapsto x) + (x, 1, a \mapsto x)) \cdot (x, 1, a \mapsto x)$$

which is an element of $TSF\{x\}$. It now follows that by applying

$$F\mu^S T \circ FS\lambda^0 \circ \lambda^1 S \circ T\lambda^2$$

we obtain an element $(y, o, d) \in FST\{x\}$ such that $o = 3$, whereas by applying

$$F\mu^S T \circ \lambda^2 ST \circ S\lambda^1 \circ \lambda^0 F$$

we obtain an element $(y', o', d') \in FST\{x\}$ such that $o' = 4$. This proves that the coherence axiom is not satisfied, and hence, by Theorem 5, also that $\hat{\lambda}$, although a natural transformation, is not a distributive law of the composite monad ST over F.

Proposition 7. *Given three distributive laws $\lambda^0, \lambda^1, \lambda^2$ as above, such that $\hat{\lambda}$ is a distributive law over ST over F, a (ST, F)-bialgebra (X, α, δ) is a $\hat{\lambda}$-bialgebra if and only if (X, α^S, δ) is a λ^2-bialgebra and (X, α^T, δ) is a λ^1-bialgebra.*

Proof. We will establish the result by establishing the following implications (in the diagrams, each of the (internal) faces commutes by either one of the assumed algebras or bialgebras, one of the axioms for monads or distributive laws, a naturality square, or a functor applied to any of these types of diagrams):

– If (X, α, δ) is a $\hat{\lambda}$-bialgebra, then (X, α^T, δ) is a λ^1-bialgebra. This is established by the following diagram:

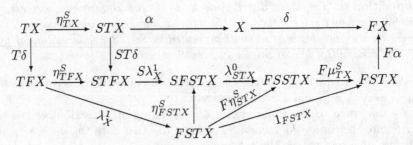

– If (X, α, δ) is a $\hat{\lambda}$-bialgebra, then (X, α^S, δ) is a λ^2-bialgebra. This is established by the following diagram:

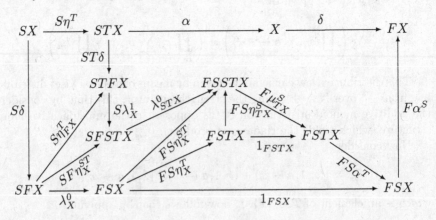

– If (X, α^T, δ) is a λ^1-bialgebra and (X, α^S, δ) is a λ^2-bialgebra, then (X, α, δ) is a $\hat{\lambda}$-bialgebra. This is established by the following diagram:

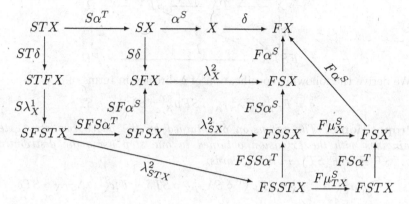

$\square$

From the preceding proposition, we can now directly conclude that, given a commutative semiring K, the corresponding $\hat{\lambda}^{\mathrm{brz}}$-bialgebras are precisely the structures $(X, 0, 1, +, \cdot, \times, o, \delta)$ such that $(X, 0, 1, +, \cdot, \times)$ is a K-algebra, that $(X, 0, +, \times, o, \delta)$ is a K-linear automaton, and such that additionally the unit and product rule are satisfied for arbitrary elements. Again, *mutatis mutandis*, the same holds for the Leibniz and pointwise product rules. It may be worthwhile to note that (in the context of a singleton alphabet A and a field as underlying semiring) the bialgebras for $\hat{\lambda}^{\mathrm{lei}}$, are precisely *differential algebras* (as defined in e.g. [RR08]) together with output functions that are K-algebra morphisms.

Another corollary of this proposition is that the final $\hat{\lambda}$-bialgebra has a structure compatible with the final λ^2-bialgebra, because the final $\hat{\lambda}$-bialgebra is a λ^2-bialgebra (by the preceding proposition) compatible with the final F-coalgebra (as a consequence of [Bar04, Corollary 3.4.19]), and again by [Bar04, Corollary 3.4.19], there is a *unique* λ^2-bialgebra compatible with the final F-coalgebra. By similar reasoning, the final $\hat{\lambda}$-bialgebra also has a structure with the final λ^1-bialgebra. So, the final $\hat{\lambda}$-bialgebra can now be seen to naturally extend both the final λ^1-bialgebra and the final λ^2-bialgebra, inheriting the coalgebraic structure from the final F-coalgebra, and with algebraic structure compatible to that of the final bialgebras for λ^1 and λ^2.

5 . The Generalized Powerset Construction for Composite Distributive Laws

We can now formulate a two-step version of the generalized powerset construction, in the setting where we are given three distributive laws $\lambda^0, \lambda^1, \lambda^2$ as in the previous section. Here the extensions can be specified as

$$\delta^{\sharp} = F\mu_X^{ST} \circ \lambda_{STX}^1 \circ T\delta \qquad \text{and} \qquad \hat{\delta} = F\mu_{TX}^S \circ \lambda_{STX}^2 \circ S\delta^{\sharp}.$$

giving the following diagram if a final F-coalgebra exists:

$$
\begin{array}{ccccccc}
X & \xrightarrow{\;\eta_X^T\;} & TX & \xrightarrow{\;\eta_{TX}^S\;} & STX & \xrightarrow{\;[\![-]\!]\;} & \Omega \\
\downarrow{\scriptstyle\delta} & \swarrow{\scriptstyle\hat{\delta}^\sharp} & \downarrow{\scriptstyle\hat{\delta}^\flat} & & & & \downarrow{\scriptstyle\omega} \\
FSTX & & & \xrightarrow{\quad F[\![-]\!]\quad} & & & F\Omega
\end{array}
$$

We derive the following specification of $\hat{\delta}$ directly in terms of δ:

$$
\hat{\delta} = F\mu_{TX}^S \circ \lambda_{STX}^2 \circ SF\mu_X^{ST} \circ S\lambda_{STX}^1 \circ ST\delta
$$

Proposition 8. *The extension $\hat{\delta}$ obtained by the above two step extension coincides with the extension obtained in one step using the distributive law $\hat{\lambda} : (ST)F \Rightarrow F(ST)$. In other words:*

$$
F\mu_{TX}^S \circ \lambda_{STX}^2 \circ SF\mu_X^{ST} \circ S\lambda_{STX}^1 \circ ST\delta = F\mu_X^{ST} \circ \hat{\lambda}_{STX} \circ ST\delta
$$

Proof. The following commuting diagram

$$
\begin{array}{ccccccc}
FSSTSTX & \xrightarrow{FSS\lambda_{TX}^0} & FSSSTTX & \xrightarrow{FS\mu_{TTX}^S} & FSSTTX & \xrightarrow{FSS\mu_X^T} & FSSTX \\
\downarrow{\scriptstyle F\mu_{TSTX}^S} & & \downarrow{\scriptstyle F\mu_{STTX}^S} & & \downarrow{\scriptstyle F\mu_{TTX}^S} & & \downarrow{\scriptstyle F\mu_{TX}^S} \\
FSTSTX & \xrightarrow{FS\lambda_{TX}^0} & FSSTTX & \xrightarrow{F\mu_{TTX}^S} & FSTTX & \xrightarrow{FS\mu_X^T} & FSTX
\end{array}
$$

establishes that $F\mu_{TX}^S \circ FS\mu_X^{ST} = F\mu_X^{ST} \circ F\mu_{TSTX}^S$.
Next, observe that

$$
\begin{aligned}
& F\mu_{TX}^S \circ \lambda_{STX}^2 \circ SF\mu_X^{ST} \circ S\lambda_{STX}^1 \circ ST\delta \\
={} & F\mu_{TX}^S \circ FS\mu_X^{ST} \circ \lambda_{STSTX}^2 \circ S\lambda_{STX}^1 \circ ST\delta \quad \text{(naturality)} \\
={} & F\mu_X^{ST} \circ F\mu_{TSTX}^S \circ \lambda_{STSTX}^2 \circ S\lambda_{STX}^1 \circ ST\delta \quad \text{(result of above diagram)} \\
={} & F\mu_X^{ST} \circ \hat{\lambda}_{STX} \circ ST\delta
\end{aligned}
$$

completing the proof. $\qquad\square$

As a result of this proposition, we can directly conclude that, in the two-step diagram above, $FSTX$ can again be assigned the structure of an algebra for the monad ST, and the mappings $\hat{\delta}$ and $[\![-]\!]$ are ST-algebra morphisms.

Finally note that, if F is a cofree copointed functor $(\mathrm{Id} \times G)$, the same considerations as those surrounding the diagrams in (1) again apply.

Remark 9. Note that, when instantiating the above construction with the laws λ^0 from Sect. 2.2, λ^{brz}, and λ^2 from 2.3, the resulting composite distributive law is a law between the monad $S\langle-\rangle$ and the cofree copointed endofunctor on $K \times -^A$. This law differs from (but is closely related to) the laws obtained in [WBR13, MPW16], in which a distributive law is defined between a suitably defined monad on $K\langle A + -\rangle$ and the endofunctor $K \times -^A$ (without copoint). However, the resulting law $\hat{\lambda}$ still has the property, that a language is context-free (resp. a power series is constructively algebraic) iff it is the final coalgebra semantics of the determinization of a finite $K \times K\langle-\rangle^A$-coalgebra.

6   Comparison to the Coherence Condition from [Che07]

In [Che07], a similar coherence condition to the one from this paper is presented, providing a way of iterating distributive laws between which a coherence condition similar to (but simpler than) the one presented in this paper is used. We will concentrate on the specific case of just three monads with interacting distributive laws, as the more general iterated cases are less directly relevant to the present paper. In this case, the main result from [Che07] instantiates as:

Proposition 10. *Given monads* (S, η^S, μ^S), (T, η^T, μ^T), *and* (U, η^U, μ^U), *together with distributive laws* $\lambda^0 : UT \Rightarrow TU$, $\lambda^1 : US \Rightarrow SU$, *and* $\lambda^2 : TS \Rightarrow ST$, *if the diagram*

$$
\begin{array}{ccccc}
UTS & \overset{U\lambda^2}{\Longrightarrow} & UST & \overset{\lambda^1 T}{\Longrightarrow} & SUT \\
{\scriptstyle \lambda^0 S}\Big\Downarrow & & & & \Big\Downarrow{\scriptstyle S\lambda^0} \\
TUS & \underset{T\lambda^1}{\Longrightarrow} & TSU & \underset{\lambda^2 U}{\Longrightarrow} & STU
\end{array}
$$

(called the coherence condition or Yang-Baxter condition*) commutes, the natural transformations*

$$\lambda^2 U \circ T\lambda^1 : (TU)S \Rightarrow S(TU)$$

and

$$S\lambda^0 \circ \lambda^1 T : U(ST) \Rightarrow (ST)U$$

are distributive laws between monads (of the composite monad TU *over* S, *and of* U *over the composite monad* ST*), respectively). Moreover both of these laws yield the same composite monad on* STU.

A difference between this result and the corresponding result from our paper is that the result in [Che07] is only presented in one direction, instead of as an equivalence. Thus, one can wonder if in this case, too, the result can be turned into an equivalence. In fact, it can be stated in the form of the following three-way equivalence:

Proposition 11. *Given monads* S, T, *and* U *and distributive laws* λ^0, λ^1, *and* λ^2 *as in the previous proposition, the following are equivalent:*

1. *The Yang-Baxter condition holds.*
2. $\lambda^2 U \circ T\lambda^1$ *is a distributive law of the composite monad* TU *over* S.
3. $S\lambda^0 \circ \lambda^1 T$ *is a distributive law of* U *over the composite monad* ST.

As a variant on this result, bridging the gap between it and our result, we will move to a setting where, rather than three monads, we have two monads, (S, η^S, μ^S) and (T, η^T, μ^T) and an endofunctor F, together with a distributive law between monads $\lambda^0 : TS \Rightarrow ST$, and two distributive laws of the monad-over-endofunctor type, $\lambda^1 : TF \Rightarrow FT$, and $\lambda^2 : SF \Rightarrow FS$. We now obtain the following result, in much the same manner as before:

Proposition 12. *Given monads* (S, η^S, μ^S), (T, η^T, μ^T), *and an endofunctor* F, *together with distributive laws* $\lambda^0 : TS \Rightarrow ST$, $\lambda^1 : TF \Rightarrow FT$, *and* $\lambda^2 : SF \Rightarrow FS$, *the following are equivalent:*

1. *The Yang-Baxter condition holds, as follows:*

$$
\begin{array}{ccccc}
TSF & \xRightarrow{T\lambda^2} & TFS & \xRightarrow{\lambda^1 S} & FTS \\
\lambda^0 F \Big\Downarrow & & & & \Big\Downarrow F\lambda^0 \\
STF & \xRightarrow{S\lambda^1} & SFT & \xRightarrow{\lambda^2 T} & FST
\end{array} \tag{3}
$$

2. $\lambda^2 T \circ S\lambda^1$ *is a distributive law of the composite monad* ST *over* F.

(Note that this result is very similar to, but different from, one direction of [BMSZ15, Proposition 7.1], which is concerned with two monads, an endofunctor, and one distributve law between these monads, and two Kleisli-type laws, i.e. laws of the endofunctor over both of the monads.)

In fact, the result can be related to the results from Sect. 4 by means of Lemma 2 and the following proposition:

Proposition 13. *Given monads* (S, η^S, μ^S) *and* (T, η^T, μ^T), *an endofunctor* F, *and distributive laws* $\lambda^0 : TS \Rightarrow ST$, $\lambda^1 : TF \Rightarrow FT$, *and* $\lambda^2 : SF \Rightarrow FS$, *the Yang-Baxter condition* (3) *holds if and only if the law* $\hat{\lambda}^1 = F\eta^S T \circ \lambda^1 : TF \Rightarrow FST$ *makes diagram* (2) *(Sect. 4) commute.*

Connecting this result to the three product rules mentioned in this paper, we recall that the Hadamard product rule can also be given directly as a distributive law of type $TF \Rightarrow FT$ (with $T = -^*$, $F = K \times -^A$) using the strength operator. This law can be given concretely by

$$
\lambda_X \left(\prod_{i=1}^{n}(o_i, a \mapsto d_{ia}) \right) = \left(\prod_{i=1}^{n} o_i, a \mapsto \prod_{k=1}^{n} d_{ia} \right)
$$

with the difference that the right hand side is now regarded as an element of FT, instead of an element of FST (with $S = \mathrm{Lin}_K(-)$). The Leibniz and Brzozowski product rules, however, do not fit into this simpler scheme, as a direct result of the occurrence of a summation on the right hand side.

It thus turns out that, while the results from [Che07] are easily modified to a coalgebraic setting with laws $\lambda^0 : TS \Rightarrow ST$, $\lambda^1 : TF \Rightarrow FT$, and $\lambda^2 : SF \Rightarrow FS$, the usage of distributive laws into a composite monad (where λ^1 is of the form $TF \Rightarrow FST$) allows us to specify product rules of the type $(xy)_a = x_y a + o(x)y_a$, where a summation occurs in the right hand side, that are not possible to specify in a two-step mechanism using the simple variant of the result in [Che07].

Additionally, in the case of distributive laws into composite monads, we require an octagonal coherence condition, rather than the hexagonal condition from [Che07].

7 Further Directions

One interesting direction for future work consists of making a further generalization from copointed functors to *comonads*. Another possible direction, owing inspiration to [Wor09] (where it is shown that K-algebras are precisely the monoids in the monoidal category of K-semimodules), consists of investigating whether we can reverse the order of the two-step process in the concrete setting of product rules, going from X via $\text{Lin}_K(X)$ to $K\langle X \rangle$ instead of via X^*. Finally, it may be very worthwhile to further develop the Prolog-program, developed here in an *ad hoc* manner to prove a part of Theorem 5, into a more general tool able to (at least in certain contexts) prove commutativity of diagrams.

Acknowledgements. For various discussions, comments, and criticism (hopefully in all cases taken as constructively as possible), I would like to thank Henning Basold, Marcello Bonsangue, Helle Hvid Hansen, Bart Jacobs, Bartek Klin, Jurriaan Rot, Jan Rutten, and Alexandra Silva, as well as to Charles Paperman for suggesting automation of the proofs (which in the end was carried out for only a part of just one of the propositions). Additionally I would like to thank the anonymous referees (as well as the anonymous refees of the earlier, rejected, submission) for a large amount of corrections and constructive comments.

References

[Awo10] Awodey, S.: Category Theory. Oxford University Press, Oxford (2010)

[Bar04] Bartels, F.: On generalized coinduction and probabilistic specification formats. Ph.D. thesis, Vrije Universiteit Amsterdam (2004)

[BBB+12] Bonchi, F., Bonsangue, M.M., Boreale, M., Rutten, J., Silva, A.: A coalgebraic perspective on linear weighted automata. Inf. Comput. **211**, 77–105 (2012)

[Bec69] Beck, J.: Distributive laws. In: Eckmann, B. (ed.) Seminar on Triples and Categorical Homology Theory, pp. 119–140. Springer, Heidelberg (1969)

[BHKR13] Bonsangue, M.M., Hansen, H.H., Kurz, A., Rot, J.: Presenting distributive laws. In: Heckel, R., Milius, S. (eds.) CALCO 2013. LNCS, vol. 8089, pp. 95–109. Springer, Heidelberg (2013)

[BMSZ15] Bonchi, F., Milius, S., Silva, A., Zanasi, F.: Killing epsilons with a dagger: a coalgebraic study of systems with algebraic label structure. Theoret. Comput. Sci. **604**, 102–126 (2015). doi:10.1016/j.tcs.2015.03.024

[BRW12] Bonsangue, M.M., Rutten, J., Winter, J.: Defining context-free power series coalgebraically. In: Pattinson and Schröder [PS12], pp. 20–39

[Brz64] Brzozowski, J.A.: Derivatives of regular expressions. J. ACM **11**, 481–494 (1964)

[BW85]   Barr, M., Wells, C.: Toposes, Triples, and Theories. Grundlehren der mathematischen Wissenschaften. Springer, New York (1985)

[Che07] Cheng, E.: Iterated distributive laws (2007). http://arxiv.org/pdf/0710. 1120v1.pdf

[Eil76] Eilenberg, S.: Automata, Languages, and Machines. Academic Press Inc., Orlando (1976)

[Jac06a] Jacobs, B.: A bialgebraic review of deterministic automata, regular expressions and languages. In: Futatsugi, K., Jouannaud, J.-P., Meseguer, J. (eds.) Algebra, Meaning, and Computation. LNCS, vol. 4060, pp. 375–404. Springer, Heidelberg (2006)

[Jac06b] Jacobs, B.: Distributive laws for the coinductive solution of recursive equations. Inf. Comput. **204**(4), 561–587 (2006)

[JSS12] Jacobs, B., Silva, A., Sokolova, A.: Trace semantics via determinization. In: Pattinson and Schröder [PS12], pp. 109–129

[Kli11] Klin, B.: Bialgebras for structural operational semantics: an introduction. Theoret. Comput. Sci. **412**(38), 5043–5069 (2011)

[LPW00] Lenisa, M., Power, J., Watanabe, H.: Distributivity for endofunctors, pointed and co-pointed endofunctors, monads and comonads. Electr. Notes Theor. Comput. Sci. **33**, 230–260 (2000)

[Mac71] MacLane, S.: Categories for the Working Mathematician. Springer, New York (1971)

[MMS13] Milius, S., Moss, L.S., Schwencke, D.: Abstract GSOS rules and a modular treatment of recursive definitions. Logical Methods Comput. Sci. **9**(3) (2013)

[MPW16] Milius, S., Pattinson, D., Wißmann, T.: A new foundation for finitary corecursion. CoRR, abs/1601.01532 (2016)

[PS12] Pattinson, D., Schröder, L. (eds.): CMCS 2012. LNCS, vol. 7399. Springer, Heidelberg (2012)

[RR08] Rosenkranz, M., Regensburger, G.: Integro-differential polynomials and operators. In: Proceedings of the Twenty-First International Symposium on Symbolic and Algebraic Computation, ISSAC 2008, pp. 261–268. ACM, New York (2008)

[Rut98] Rutten, J.J.M.M.: Automata and coinduction (an exercise in coalgebra). In: Sangiorgi, D., de Simone, R. (eds.) CONCUR 1998. LNCS, vol. 1466, pp. 194–218. Springer, Heidelberg (1998)

[Rut00] Rutten, J.: Universal coalgebra: a theory of systems. Theoret. Comput. Sci. **249**(1), 3–80 (2000)

[Rut03] Rutten, J.: Behavioural differential equations: a coinductive calculus of streams, automata, and power series. Theoret. Comput. Sci. **308**(1–3), 1–53 (2003)

[Rut05] Rutten, J.: A coinductive calculus of streams. Math. Struct. Comput. Sci. **15**(1), 93–147 (2005)

[SBBR10] Silva, A., Bonchi, F., Bonsangue, M.M., Rutten, J.: Generalizing the powerset construction, coalgebraically. In: Lodaya, K., Mahajan, M. (eds.) FSTTCS. LIPIcs, vol. 8, pp. 272–283. Schloss Dagstuhl–Leibniz-Zentrum für Informatik (2010)

[SBBR13] Silva, A., Bonchi, F., Bonsangue, M., Rutten, J.: Generalizing determinization from automata to coalgebras. Logical Methods Comput. Sci. **9**(1) (2013)

[Sch14] Schwencke, D.: Compositional and effectful recursive specification formats. Ph.D. thesis, Technische Universität Braunschweig (2014)

[WBR11] Winter, J., Bonsangue, M.M., Rutten, J.: Context-free languages, coalgebraically. In: Corradini, A., Klin, B., Cîrstea, C. (eds.) CALCO 2011. LNCS, vol. 6859, pp. 359–376. Springer, Heidelberg (2011)

[WBR13] Winter, J., Bonsangue, M.M., Rutten, Jan J. M. M : Coalgebraic characterizations of context-free languages. Logical Methods Comput. Sci. **9**(3:14) (2013)

[Win14] Winter, J.: Coalgebraic characterizations of automata-theoretic classes. Ph.D. thesis, Radboud Universiteit Nijmegen (2014)

[Wor09] Worthington, J.: Automata, representations, and proofs. Ph.D. thesis, Cornell University (2009)

On the Logic of Generalised Metric Spaces

Octavian Babus[✉] and Alexander Kurz[✉]

University of Leicester, Leicester, UK
octavianbabus@yahoo.com, ak155@le.ac.uk

Abstract. The aim of the paper is to work towards a generalisation of coalgebraic logic enriched over a commutative quantale. Previous work has shown how to dualise the coalgebra type functor $T : \Omega\text{-Cat} \longrightarrow \Omega\text{-Cat}$ in order to obtain the modal operators and axioms describing transitions of type T. Here we give a logical description of the dual of Ω-Cat.

1 Introduction

Recently, following on work of Rutten [18] and Worrell [24], the interest in coalgebras enriched over posets or, more generally, enriched over a commutative quantale has attracted some attention. In particular, the question of a coalgebraic logic in this setting has been asked [2].

In the non-enriched situation we start with a functor $T : \mathsf{Set} \longrightarrow \mathsf{Set}$ and ask for a logic that allows us to completely describe T-coalgebras up to bisimilarity. More specifically, we would like to ensure *strong expressivity* in the sense that for any property $p \subseteq X$ of any T-coalgebra (X, ξ) there is a formula ϕ such that p coincides with the semantics $[\![\phi]\!]_{(X,\xi)}$ of ϕ on (X, ξ). Moreover, we would like to have *completeness* in the sense that if $[\![\phi]\!]_{(X,\xi)} \subseteq [\![\phi]\!]_{(X,\xi)}$ then $\phi \leq \psi$ in the initial algebra of formulas.

To achieve the above, the **first step** is to let $LA = [T([A, 2]), 2]$ in

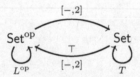

and to declare the initial L-algebra, if it exists, as the "Lindenbaum-algebra" of T. This terminology is justified in sofar as the adjoint transpose

$$\delta : L([-, 2]) \longrightarrow [T-, 2]$$

of the iso $L \longrightarrow [T([-, 2]), 2]$ allows us to define the semantics $[\![\,]\!]_{(X,\xi)}$ wrt a coalgebra (X, ξ) as the unique arrow from the initial L-algebra $\iota : LI \longrightarrow I$ as in

© IFIP International Federation for Information Processing 2016
Published by Springer International Publishing Switzerland 2016. All Rights Reserved
I. Hasuo (Ed.): CMCS 2016, LNCS 9608, pp. 136–155, 2016.
DOI: 10.1007/978-3-319-40370-0_9

$$LI \xrightarrow{\quad \iota \quad} I$$

$$L\llbracket \rrbracket_{(X,\xi)} \downarrow \qquad\qquad \downarrow \llbracket \rrbracket_{(X,\xi)}$$

$$L([X,2]) \xrightarrow[\delta_X]{} [TX,2] \xrightarrow[[\xi,2]]{} [X,2]$$

But the reason why, at this stage, we cannot truly speak of $\iota : LI \longrightarrow I$ as a Lindenbaum algebra is that it lives in $\mathsf{Set}^{\mathrm{op}}$ and is not (yet) an algebra over Set with elements and operations in the usual sense.

The **second step**, then, consists in using the well-known fact that $[-,2]$: $\mathsf{Set}^{\mathrm{op}} \longrightarrow \mathsf{Set}$ is monadic and, therefore, $\mathsf{Set}^{\mathrm{op}}$ is equivalent to a category of algebras defined by operations and equations. In particular, we know that $\mathsf{Set}^{\mathrm{op}}$ is equivalent to the category of complete atomic Boolean algebras, which now allows us to consider (L, ι) as the Lindenbaum algebra of infinitary T-logic.

The aim of the paper is to carry out these steps in the case where we replace Set by the category $\Omega\text{-}\mathsf{Cat}$ of categories enriched over Ω for a commutative quantale Ω. It is based on the Ω-generalisations of the downset monad $\mathcal{D}$ and the upset monad $\mathcal{U}$. We will define algebras for operations $\Sigma_{\mathcal{DU}}$ and equations $E_{\mathcal{DU}}$ and will argue via (3), Theorems 16, 22, and 49 that $\langle \Sigma_{\mathcal{DU}}, E_{\mathcal{DU}} \rangle$-algebras complete the table

$\mathsf{Set}^{\mathrm{op}}$	complete atomic Boolean algebras
$\Omega\text{-}\mathsf{Cat}^{\mathrm{op}}$	$\langle \Sigma_{\mathcal{DU}}, E_{\mathcal{DU}} \rangle$-algebras

Importantly, this includes an equation for generalised distributivity, Axiom (13), and will allow us to derive a logic for $\Omega\text{-}\mathsf{Cat}$ very much in the same way as we can say that Boolean logic is the logic of Set.

Related Work. The results in Sect. 3 generalize the results in [13] by Marmolejo, Rosebrugh, and Wood to arbitrary commutative quantales Ω but the proofs remain the same as they work in any 3-category where 3-cells form a poset. Theorem 16 is also a special case of a theorem of Stubbe [19].

Section 4 generalises the well-known dual adjunction,

$$\mathsf{Pre}^{\mathrm{op}} \underset{\longleftarrow}{\overset{\longrightarrow}{\rightleftarrows}} \mathsf{CDL}$$

between preorders and completely distributive lattices to categories enriched over a commutative quantale. This is similar in spirit to the work in Hofmann [5], where a generalisation from preorders to topological spaces and to approach spaces can be found.

The category of distributive complete Ω-lattices of Lai and Zhang [9] coincides with what we denote CCD in Definition 18. Compared to their work, we add the argument of how to obtain CCD from the monad $[[-,\Omega],\Omega]$ and we show that the CCD is isomorphic to the category of (ordinary, set-based) $\langle \Sigma_{\mathcal{DU}}, E_{\mathcal{DU}} \rangle$-algebras.

In Pu and Zhang [14] it is shown, amongst other things, that the category of anti-symmetric CCD's is monadic over Set, but the proof proceeds by Beck's monadicity theorem whereas we give the operations and equations $\langle \Sigma_{\mathcal{D}\mathcal{U}}, E_{\mathcal{D}\mathcal{U}} \rangle$ explicitly.

The double powerset monad $\mathcal{D}\mathcal{U}$ is investigated in detail, in the case $\Omega = 2$, by Vickers in [20–23].

2   Preliminaries and Related Work

We are interested in categories enriched over commutative quantales [15].

Definition 1. *By a quantale $\Omega = ((\Omega, \leq), e, \otimes)$ we understand a complete lattice with a binary operation $\otimes : \Omega \times \Omega \longrightarrow \Omega$ with unit e, such that $\otimes$ preserves colimits in both arguments. We call a quantale commutative if the operation $\otimes$ is commutative.*

Since $\otimes$ preserves joins, a commutative quantale Ω can be considered as a symmetric monoidal closed category and one can enrich over Ω, see [7]. A category enriched over Ω is also called a Ω-category and the 2-category of Ω-categories, Ω-functors and Ω-natural transformations is denoted by Ω-Cat. The interpretation of such enriched categories as metric spaces is due to [10] and recalled in the following examples:

Example 2. 1. $\Omega = 2 = ((2, \leq), 1, \wedge)$. Categories enriched over 2 are preorders and the corresponding functors are monotone maps. The closed structure is implication.

2. $\Omega = (([0, \infty], \geq_{\mathbb{R}}), 0, +)$, the non-negative reals with infinity and the opposite of the natural order. That the top element is 0 formalises the idea that the elements of Ω measure distance from 'truth'. A Ω-category is called a *generalised metric space*. The corresponding functors are non-expansive maps. The closed structure is given by

$$[0, \infty](r, s) = s \dotminus r = \text{ if } s \leq_{\mathbb{R}} r \text{ then } 0 \text{ else } s - r$$

Examples inlcude:
   (a) $[0, \infty]$ itself.
   (b) The real numbers $(R, \leq_{\mathbb{R}})$ with the metric given by $R(a, b) = $ if $a \leq_{\mathbb{R}} b$ then 0 else $a - b$
   (c) Any metric space.

3. $\Omega = (([0, 1], \leq_{\mathbb{R}}), 1, \cdot)$ where $x \cdot y$ is the usual multiplication. Then

$$x \Rightarrow y = \text{ if } x \leq y \text{ then } 1 \text{ else } \frac{y}{x}$$

The exponential map $x \mapsto \exp(-x)$ induces an isomorphism from the Ω of the previous item, so that we can think of both representing two views of the same mathematics, one in terms of distances and the other in terms of truth-values.

4. $\Omega = (([0,1], \geq_{\mathbb{R}}), 0, \max)$. This is example is in the same spirit as above, but this time Ω-categories are *generalised ultrametric spaces* [17]. The closed structure is given by

$$[0,1](x,y) = \text{ if } x \geq_{\mathbb{R}} y \text{ then } 0 \text{ else } y$$

Examples include:
(a) $[0,1]$ itself, as well as $[0,1]^{\mathrm{op}}$ with 1 and min.
(b) Let A^{∞} be the finite and infinite words over A. Define $A^{\infty}(v,w) = 0$ if v is a prefix of w and $A^{\infty}(v,w) = 2^{-n}$ otherwise where $n \in \mathbb{N}$ is the largest number such that $v_n = w_n$ (where v_n is the prefix of v consisting of n letters from A).

Whenever we talk about limits or colimits in a Ω-category we understand a weighted limit or weighted colimit and we will use the same notations as in [7, Chapter 3].

Note that every Ω-category X is equipped with a preorder $x \leq y \Leftrightarrow X(x,y) \geq e$.

We call a Ω-category X *anti-symmetric* if $x \leq y$ and $y \leq x$ implies $x = y$. In the examples above, this order coincides with the expected one. For example, in Example 2b the induced order on R is the natural one and in Example 3b it is the prefix order.

Proposition 3. *1. The order $x \leq y \Leftrightarrow \Omega(x,y) \geq e$ is the order of Ω.*
2. $[X, \Omega]$ is anti-symmetric for any Ω-category X.
3. $[X, Y]$ is anti-symmetric iff Y is anti-symmetric.

We already said that Ω-categories form a category Ω-Cat of small Ω-categories. Ω-Cat is Ω-Cat enriched, with the distance between two Ω-functors $f, g : A \longrightarrow B$ given by Ω-Cat$(A,B)(f,g) = [A,B](f,g) = \bigwedge_{a \in A} B(fa, ga)$. Hence Ω-Cat is an object of $(\Omega$-Cat)-Cat. The category $(\Omega$-Cat)-Cat of Ω-Cat-categories, Ω-Cat-functors, and Ω-Cat-natural transformations is a 3-category in which natural transformations $\alpha, \beta : F \longrightarrow G : \mathcal{A} \longrightarrow \mathcal{B}$ are pre-ordered since $(\Omega$-Cat)-Cat$(\mathcal{A}, \mathcal{B})(F, G)$ is a Ω-category.

The reason to insist on pre-ordered natural transformations is that we can make use of the following notion due to [8] and reformulated by [12].

Definition 4. *By a KZ-doctrine M on Ω-Cat we understand a monad (M, η, μ) such that we have the adjunctions $M\eta \dashv \mu \dashv \eta M$. Dually a co-$KZ$-doctrine is a monad where $\eta M \dashv \mu \dashv M\eta$.*

The following proposition is Kock's definition of a KZ-doctrine simplified to the pre-ordered setting, see [8].

Proposition 5. *(M, η, μ) is a KZ-doctrine if and only if there exists a natural transformation $\delta : M\eta \longrightarrow \eta M$ and $\mu \circ \eta M = \mu \circ M\eta = \mathrm{id}$. Dually (M, η, μ) is a co-KZ-doctrine if there exists a natural transformation $\lambda : \eta M \longrightarrow M\eta$ and if $\mu \circ \eta M = \mu \circ M\eta = \mathrm{id}$.*

If one has two monads for their composite to be again a monad one needs to have a distributive law between them, as in [1].

Definition 6. *A distributive law between two monads $\mathcal{D}$ and $\mathcal{U}$ is a natural transformation $r : \mathcal{U}\mathcal{D} \Rightarrow \mathcal{D}\mathcal{U}$ subject to the commutativity of*

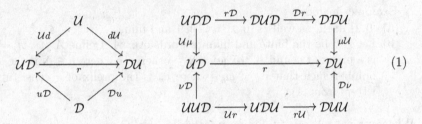

$$\tag{1}$$

Definition 7. *Let $D = (D, \eta, \mu)$ be a monad. By a D-algebra A we understand a pair $A = (A, \alpha)$, where A is category and $\alpha : DA \longrightarrow A$ is a functor such that*

$$
\begin{array}{ccc}
DDA & \xrightarrow{\ \mu_A\ } & DA \\
{\scriptstyle D\alpha}\Big\downarrow & & \Big\downarrow{\scriptstyle \alpha} \\
DA & \xrightarrow[\ \alpha\ ]{} & A
\end{array}
\qquad
\begin{array}{ccc}
A & \xrightarrow{\ \eta_A\ } & DA \\
& {\scriptstyle \mathrm{id}_A} \searrow & \Big\downarrow{\scriptstyle \alpha} \\
& & A
\end{array}
\tag{2}
$$

The next proposition is due to [8].

Proposition 8. *Let $M = (M, \eta, \mu)$ be any KZ-doctrine, then $A = (A, \alpha)$ is an M-algebra if and only if the structure map α is a left adjoint of η_M.*

The following two propositions are stated in the case $\Omega = 2$ in [13] and their proof transfers unchanged to our setting (because Ω is anti-symmetric).

Proposition 9. *If at least one of the monads D or U is either a KZ or a co-KZ-doctrines then there is at most one distributive law $r : UD \longrightarrow DU$.*

Proposition 10. *For monads D, U and a natural transformation $r : UD \longrightarrow DU$*

1. *if (D, d, μ) is KZ and (U, u, ν) is either KZ or co-KZ then $r : UD \longrightarrow DU$ is a distributive law if it satisfies $r \circ Ud = dU$ and $r \circ uD \leq Du$;*
2. *if (U, u, ν) is co-KZ and (D, d, μ) is either KZ or co-KZ then $r : UD \longrightarrow DU$ is a distributive law if it satisfies $r \circ uD = Du$ and $r \circ Ud \leq dU$.*

3   Monads and algebras

Ω being a symmetric monoidal closed category, we have the contravariant adjunction where $U = [-, \Omega] : \Omega\text{-}\mathsf{Cat} \longrightarrow \Omega\text{-}\mathsf{Cat}^{\mathrm{op}}$ is a left adjoint to $D = [-, \Omega] : \Omega\text{-}\mathsf{Cat}^{\mathrm{op}} \longrightarrow \Omega\text{-}\mathsf{Cat}$. We want to study the algebras for the monad $M = DU$ generated by it. For that we will prove that this monad is equivalent to the composite monad $\mathcal{D}\mathcal{U}$ where $\mathcal{D}, \mathcal{U} : \Omega\text{-}\mathsf{Cat} \longrightarrow \Omega\text{-}\mathsf{Cat}$ are the $\Omega\text{-}\mathsf{Cat}$ analogues of the downset and the upset monad as defined in Sect. 3.1. From there we will obtain two different sets of operations, one for each monad, and a distributive law between them. In the end we will give a categorical description for the category of algebras.

As Ω is symmetric monoidal closed, we have from [7, Chapter 1.5]

Proposition 11. $[-,\Omega]$: $\Omega\text{-Cat} \longrightarrow \Omega\text{-Cat}^{\mathrm{op}}$ is left adjoint to $[-,\Omega]$: $\Omega\text{-Cat}^{\mathrm{op}} \longrightarrow \Omega\text{-Cat}$.

As explained in the introduction, we want to consider $\Omega\text{-Cat}^{\mathrm{op}}$ as the category of algebras of a 'Ω-Cat-logic'. Since $[-,\Omega]$: $\Omega\text{-Cat}^{\mathrm{op}} \longrightarrow \Omega\text{-Cat}$ need not be monadic itself, we are going to study instead its monadic closure. That is, we let $M = DU$ and work with the category $\Omega\text{-Cat}^M$ of algebras for the monad M. We show that there is an adjunction relating them to $\Omega\text{-Cat}^{\mathrm{op}}$, as in the following picture, which will guide us through this section.

$$
\begin{array}{c}
\Omega\text{-Cat}^{\mathrm{op}} \underset{K}{\overset{AT}{\underset{\perp}{\rightleftarrows}}} \Omega\text{-Cat}^M \\
[-,\Omega] \Big(\dashv \Big) [-,\Omega] \\
\Omega\text{-Cat}
\end{array}
\tag{3}
$$

3.1   Doctrines

The aim of this subsection is to describe two monads $\mathcal{D},\mathcal{U}$: $\Omega\text{-Cat} \longrightarrow \Omega\text{-Cat}$ such that $DU = \mathcal{DU}$. Furthermore, $\mathcal{D}$ will be a KZ-doctrine, and $\mathcal{U}$ will be a co-KZ-doctrine, which in turn will help us to describe the distributive law relating them.

Recall that for any category X, one has two Yoneda embeddings dX : $X \longrightarrow [X^{\mathrm{op}},\Omega]$ given by $x \mapsto X(-,x)$ and uX : $X \longrightarrow [X,\Omega]^{\mathrm{op}}$ given by $x \mapsto X(x,-)$.

On objects, $\mathcal{D}$ maps X to $[X^{\mathrm{op}},\Omega]$ and on arrows it constructs the left Kan extension along Yoneda, while $\mathcal{U}$ maps an object X to $[X,\Omega]^{\mathrm{op}}$ and an arrow to the right Kan extension along Yoneda. Thus for any $f : X \longrightarrow Y$ in $\Omega\text{-Cat}$, let $\mathcal{D}f$ be defined as $\mathrm{Lan}_{dX}dY \circ f = \mathrm{Lan}_{dX}Y(-,f)$ and $\mathcal{U}f = \mathrm{Ran}_{dX}uY \circ f = \mathrm{Ran}_{dX}Y(f,-)$ as in

$$
\begin{array}{ccc}
\mathcal{D}X \xrightarrow[\mathrm{Lan}_{dX}(dY\circ f)]{\mathcal{D}f=} \mathcal{D}Y & \qquad & \mathcal{U}X \xrightarrow[\mathrm{Ran}_{uX}(uY\circ f)]{\mathcal{U}f=} \mathcal{U}Y \\
dX \Big\uparrow \qquad \Big\uparrow dY & & uX \Big\uparrow \qquad \Big\uparrow uY \\
X \xrightarrow[f]{} Y & & X \xrightarrow[f]{} Y
\end{array}
\tag{4}
$$

Writing down the formula for left and right Kan extensions, see [7, Chapter 4.2], we obtain for $\varphi : X^{\mathrm{op}} \longrightarrow V$ and $\psi : X \longrightarrow \Omega$

$$\mathcal{D}f(\varphi) = Lan_{dX}(dY \circ f)(\varphi) = \int^{x \in X} [X^{\mathrm{op}}, \Omega](X(-,x), \varphi) \otimes Y(-, f(x))$$

$$= \int^{x \in X} \varphi(x) \otimes Y(-, f(x)) = \varphi * (dY \circ f),$$

and

$$\mathcal{U}f(\psi) = \int_{x \in X} \mathcal{U}X(\psi, X(x,-)) \pitchfork Y(f(x), -)$$

$$= \int_{x \in X} \psi(x) \pitchfork Y(f(x), -).$$

But considering that we calculate this end in $[Y, \Omega]^{\mathrm{op}}$, in $[Y, \Omega]$ it becomes

$$\mathcal{U}f(\psi) = \int^{x \in X} \psi(x) \otimes Y(f(x), -) = \psi * (uY \circ f)$$

Because $\mathcal{D}d$ and $u\mathcal{U}$ are, respectively, left and right Kan extensions, their universal properties yield

Proposition 12. *There exist natural transformations* $\lambda : \mathcal{D}d \longrightarrow d\mathcal{D}$ *and* $\delta :$ $u\mathcal{U} \longrightarrow \mathcal{U}u.$

We want $\mathcal{D}$ to be a KZ-doctrine, so the multiplication $\mu : \mathcal{D}\mathcal{D} \longrightarrow \mathcal{D}$ has to be a left adjoint of $d\mathcal{D}$. As $d\mathcal{D}$ preserves all limits and the right Kan extension of $\mathrm{id}_\mathcal{D}$ along $d\mathcal{D}$ exists, using [7, Theorem 4.81], we know that the left adjoint of $d\mathcal{D}$ exists and is expressed by $Ran_{d\mathcal{D}} \, \mathrm{id}_\mathcal{D}$. Dually, the right adjoint of $u\mathcal{U}$ exists and is expressed by $Lan_{u\mathcal{U}} \, \mathrm{id}_\mathcal{U}$.

$$
\begin{array}{ccc}
\mathcal{D}\mathcal{D} \xrightarrow[Ran_{d\mathcal{D}} \, \mathrm{id}]{\mu=} \mathcal{D} & \qquad & \mathcal{U}\mathcal{U} \xrightarrow[Lan_{u\mathcal{U}} \, \mathrm{id}_\mathcal{U}]{\nu=} \mathcal{U} \\
d\mathcal{D} \uparrow \quad \nearrow \mathrm{id}_\mathcal{D} & & u\mathcal{U} \uparrow \quad \nearrow \mathrm{id}_\mathcal{U} \\
\mathcal{D} & & \mathcal{U}
\end{array}
\tag{5}
$$

$$\mu G = \int_{\varphi \in \mathcal{D}X} \mathcal{D}\mathcal{D}X(G, \mathcal{D}X(-, \varphi)) \pitchfork \varphi \qquad \nu F = \int_{\psi \in \mathcal{U}X} [\mathcal{U}X, \Omega](G, u\mathcal{U}(\psi)) \pitchfork \psi \tag{6}$$

Furthermore as $d\mathcal{D}$ and $u\mathcal{U}$ are fully faithful, one has $\mu \circ d\mathcal{D} = \mathrm{id}_\mathcal{D}$ and $\nu \circ u\mathcal{U} = \mathrm{id}_\mathcal{U}$. Following Proposition 5 to show that $\mathcal{D}$ is a KZ-doctrine we just have to prove that $\mu \circ \mathcal{D}d = \mathrm{id}$ as well. For that we know that μ is a left adjoint so it preserves left Kan extensions, so $\mu X \circ \mathcal{D}dX = \mu X \circ Lan_{dX}(d\mathcal{D}X \circ dX) = Lan_{dX}(\mu X \circ \mathcal{D}dX \circ dX) = Lan_{dX}(\mathrm{id}_{\mathcal{D}X} \circ d) = Lan_{dX}dX = \mathrm{id}_{\mathcal{D}X}$

$$
\begin{array}{ccc}
\mathcal{D}X \xrightarrow{\mathcal{D}dX} \mathcal{D}\mathcal{D}X \xrightarrow{\mu} \mathcal{D}X \\
dX \uparrow \qquad d_{\mathcal{D}X} \uparrow \quad \nearrow \mathrm{id}_{\mathcal{D}X} \\
X \xrightarrow{dX} \mathcal{D}X
\end{array}
\tag{7}
$$

Similarly, $\nu \circ \mathcal{U}u = \text{id}_\mathcal{U}$, so $\mathcal{U}$ is a co-KZ-doctrine. Thus we have proved

Proposition 13. $(\mathcal{D}, d, \mu)$ *is a* KZ-*doctrine and* $(\mathcal{U}, u, \nu)$ *is a co-*KZ *doctrine.*

3.2 Distributive Laws and Equivalence of $\mathcal{D}\mathcal{U}$ with $[[-, \Omega], \Omega]$

In the previous section we constructed two monads, but in order for their composite to be a monad, one needs a distributive law between them.

Verifying that a natural transformation is indeed a distributive law may not be easy, but, thanks to [13], for KZ-doctrines, we just have to check the conditions of Proposition 10. To construct $\mathcal{D}$ and $\mathcal{U}$, we have used Kan extensions, thus it make sense that a distributive law between them is a Kan extension as well. Looking at the diamond above and as both $u\mathcal{D}$ and $\mathcal{U}d$ are fully faithful, a Kan extension along any of them would make that triangle commute, so intuitively, it should make no difference from which triangle one starts. So if one calculates all four Kan extensions one obtains

1. $r_\mathcal{D}^r = \text{Ran}_{u\mathcal{D}} \mathcal{D}u = \mathcal{U}\mathcal{D}(\mathcal{D}u, -)$
2. $r_\mathcal{U}^l = \text{Lan}_{\mathcal{U}d} d\mathcal{U} = \mathcal{U}\mathcal{D}(\mathcal{U}d, -)$
3. $r_\mathcal{D}^l = \text{Lan}_{u\mathcal{D}} \mathcal{D}u = \mathcal{U}\mathcal{D}(d\mathcal{D} \circ d, -) * d\mathcal{U} \circ u$
4. $r_\mathcal{U}^r = \text{Ran}_{\mathcal{U}d} d\mathcal{U} = \{\mathcal{U}\mathcal{D}(-, \mathcal{U}d), d\mathcal{U}\}$

Now as for any X and any $\varphi \in \mathcal{D}X$ and any $\psi \in \mathcal{U}X$ one has $\mathcal{D}uX(\varphi)(\psi) = \mathcal{U}dX(\varphi)(\psi)$ it follows $\text{Ran}_{u\mathcal{D}} \mathcal{D}u = \text{Lan}_{\mathcal{U}d} d\mathcal{U}$.

Proposition 14. *The natural transformation* $r = \text{Ran}_{u\mathcal{D}} \mathcal{D}u = \text{Lan}_{\mathcal{U}d} d\mathcal{U} :$ $\mathcal{U}\mathcal{D} \longrightarrow \mathcal{D}\mathcal{U}$ *is a distributive law between* $\mathcal{D}$ *and* $\mathcal{U}$.

In a similar way one has a distributive law $l = \text{Ran}_{\mathcal{D}u} u\mathcal{D} = \text{Lan}_{d\mathcal{U}} \mathcal{U}d :$ $\mathcal{D}\mathcal{U} \longrightarrow \mathcal{U}\mathcal{D}$, given by $l = \mathcal{D}\mathcal{U}(-, \mathcal{D}u)$.

Proposition 15. *With the notations from above we have* $l \dashv r$.

Next, we state that the monad $\mathcal{D}\mathcal{U}$ is equivalent to the double dualisation monad DU, a result due to [13] and generalised in [19].

Theorem 16. *For a commutative quantale* Ω, *the composite monad* $\mathcal{D}\mathcal{U}$ *is equivalent to the monad generated by the adjunction* $[-, \Omega] \dashv [-, \Omega] :$ $\Omega\text{-Cat} \longrightarrow \Omega\text{-Cat}^{\text{op}}$.

3.3 CCD: complete and completely distributive algebras

In this section we discuss the algebras of the two monads defined above. As $\mathcal{D}$ is a KZ-doctrine, following [8], a $\mathcal{D}$-algebra A is a tuple $A = (A, \alpha)$ such that $\alpha : \mathcal{D}A \longrightarrow A$ is a left adjoint to d_A, and since $\mathcal{U}$ is a co-KZ-doctrine a $\mathcal{U}$-algebra B is a tuple $B = (B, \beta)$ such that $\beta : \mathcal{U}B \longrightarrow B$ is a right adjoint to u_B.

Proposition 17. *The carrier A of a $\mathcal{D}$-algebra $A = (A, \alpha_A)$ is co-complete, and the carrier C of an $\mathcal{U}$-algebra $C = (C, \beta_C)$ is complete. Moreover, $f : (A, \alpha_A) \longrightarrow (B, \alpha_B)$ is $\mathcal{D}$-morphism if and only if f preserves all weighted colimits, and it is a $\mathcal{U}$-morphism if and only if it preserves all weighted limits.*

The following transfers the notion of complete distributivity of [4] from 2 to a commutative quantale Ω.

Definition 18. *A $\mathcal{D}$ algebra (A, α) is called ccd if the structure map α has a left adjoint. We denote with CCD the subcategory of $\mathcal{D}$-alg such that the objects are ccd and the arrows preserves weighted limits and colimits. Dually, a $\mathcal{U}$-algebra for which the structure map has a right adjoint is called ^{op}ccd.*

Example 19. In the case $\Omega = 2$, a poset A equipped with a $\mathcal{D}$-algebra structure α is a join semi-lattice. Moreover, A is ccd in the sense of the definition above iff it is completely distributive in the usual order-theoretic sense.

Definition 20. *A $\mathcal{DU}$-algebra is a $\mathcal{U}$-algebra (A, β) which has a $\mathcal{D}$-structure $\alpha : \mathcal{D}A \longrightarrow A$ such that α is a $\mathcal{U}$-homomorphism, i.e. the following diagram commutes.*

$$\begin{array}{ccc} \mathcal{U}\mathcal{D}A \xrightarrow{\ r_A\ } \mathcal{D}\mathcal{U}A \xrightarrow{\ \mathcal{D}\beta\ } \mathcal{D}A \\ \\ {}_{\mathcal{U}\alpha}\searrow \qquad \downarrow{\scriptstyle\alpha} \\ \mathcal{U}A \xrightarrow{\ \beta\ } A \end{array} \qquad (8)$$

For any two $\mathcal{DU}$-algebras (A, α_A, β_A) and (B, α_B, β_B) a $\mathcal{DU}$-morphism from A to B is a map $f : A \longrightarrow B$ such that it is simultaneously $\mathcal{D}$ and $\mathcal{U}$ morphism.

Lemma 21. *The carrier A of a ccd-algebra (A, α) is complete and cocomplete.*

The following result is due to [13].

Theorem 22. *$\mathcal{DU}$-alg $\cong$ CCD, and $\mathcal{UD}$-alg $\cong$ ^{op}CCD.*

Whereas naturally occurring metric spaces, such as Euclidean spaces, are typically not ccd, the spaces of many-valued *predicates* over metric spaces are ccd:

Example 23. For any X in Ω-Cat,

1. $(\mathcal{D}X, \mu X)$ is ccd.
2. $(\mathcal{U}X, \nu X)$ is ^{op}ccd.

4   The comparison functor Ω-Cat$^{\mathrm{op}} \to \mathcal{DU}$-alg

Following [11], let the comparison functor $K : \Omega$-Cat$^{\mathrm{op}} \longrightarrow \mathcal{DU}$-alg be given by $KX = (X, \mathcal{D}\epsilon X)$, for the adjunction $U \dashv D$. As Ω-Cat^{op} is cocomplete, K has a left adjoint. In order to describe it we first define the concept of atoms, also known as tiny or small projective objects, see [6] and [7, Chapter 5.5].

4.1   The Left Adjoint of the Comparison Functor

Definition 24. *An atom in a category $\mathcal{C}$ is an object C such that $\mathcal{C}(C, -)$ preserves all colimits.* $\mathsf{At}(\mathcal{C})$ *is the full subcategory of $\mathcal{C}$ whose objects are atoms.*

Before we continue let us give some example of atoms.

Example 25.

1. In posets atoms are known as completely prime elements. In a completely distributive lattice being an atom is equivalent to being completely join irreducible.
2. The category $[0, \infty]$ seen as a generalized metric space has only one atom 0.
3. Let $[X^{\mathrm{op}}, \Omega]$ be a functor category, then using Yoneda and the definition of a colimit in a functor category, see [7, Chapter 3.3], one has that any representable is an atom. Moreover, see [7, Chapter 5.5], one has that $[X^{\mathrm{op}}, \Omega] \cong [\mathsf{At}(X)^{\mathrm{op}}, \Omega]$. In general one has $X \subseteq \mathsf{At}([X^{\mathrm{op}}, \Omega])$.

We define a functor $\mathsf{AT} : \mathcal{DU}\text{-alg} \longrightarrow \Omega\text{-Cat}^{\mathrm{op}}$ on objects by $\mathsf{AT}(A, \alpha) = (\mathsf{At}(A))^{\mathrm{op}}$. In order to define AT on maps we need some additional lemmas.

First note that for any $H : A \longrightarrow B$ in $\mathcal{DU}\text{-alg}$, since A is complete and H preserves all limits, there exists a left adjoint $L : B \longrightarrow A$ in $\Omega\text{-Cat}$.

Lemma 26. *For all $A, B \in \mathcal{A}$ and $H : A \longrightarrow B$ with left adjoint L, there exists $f : \mathsf{At}(B) \longrightarrow \mathsf{At}(A)$ such that $L \circ i_B = i_A \circ f$, where $i_A : \mathsf{At}(A) \longrightarrow A$ and $i_B : \mathsf{At}(B) \longrightarrow B$ are the atom-inclusion maps.*

We can now define $\mathsf{AT}(H) = f^{\mathrm{op}}$ with f as in the lemma. This defines a functor because composition of adjoints is again an adjoint. We are ready to prove

Theorem 27. *For any $X \in \Omega\text{-Cat}$ and $A \in \mathcal{DU}\text{-alg}$, we have a natural isomorphism of categories $\Omega\text{-Cat}(X^{\mathrm{op}}, \mathsf{At}(A))^{\mathrm{op}} \cong \mathcal{DU}\text{-alg}(A, [X, \Omega])$. Moreover this is isomorphism also an isomorphism of Ω-categories.*

Proof. We sketch the construction of the isomorphism.
We have to define the functors

$$\phi_{XA} : \Omega\text{-Cat}(X^{\mathrm{op}}, \mathsf{At}(A))^{\mathrm{op}} \longrightarrow \mathcal{DU}\text{-alg}(A, [X, \Omega]),$$

$$\psi_{XA} : \mathcal{DU}\text{-alg}(A, [X, \Omega]) \longrightarrow \Omega\text{-Cat}(X^{\mathrm{op}}, \mathsf{At}(A))^{\mathrm{op}}$$

and show that they are inverse to each other. First define ϕ_{XA} on objects. For all $h : X^{\mathrm{op}} \longrightarrow \mathsf{At}(A)$ let

$$\phi_{XA}(h) = A(h-, -) : A \longrightarrow [X, \Omega].$$

Now define ψ_{XA} on objects. Let $H : A \longrightarrow [X, \Omega]$ and let $L : [X, \Omega] \longrightarrow A$ be its left adjoint, and also let $uX : X^{\mathrm{op}} \longrightarrow [X, \Omega]$, $x \mapsto X(x, -)$ the Yoneda embedding.

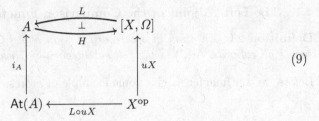

$$(9)$$

Since $L \circ uX(x)$ is an atom for all x in X, we let $\psi_{XA}(H) = L \circ uX$. In order to define ϕ_{XA} and ψ_{XA} on arrows, one uses the concept of conjugate natural transformation [11, Chapter 4.7].

Theorem 28. *The functor* $\mathsf{AT} : \mathcal{DU}\text{-alg} \longrightarrow \Omega\text{-Cat}^{\mathrm{op}}$ *is a left adjoint to the functor* $K : \Omega\text{-Cat}^{\mathrm{op}} \longrightarrow \mathcal{DU}\text{-alg}$.

Proof. Let $X \in \Omega\text{-Cat}$ and $A \in \mathcal{DU}\text{-alg}$. We have to show that $\Omega\text{-Cat}^{\mathrm{op}}(\mathsf{AT}(A), X) \cong \mathcal{DU}\text{-alg}(A, KX)$ which is equivalent to $\Omega\text{-Cat}(X, \mathsf{At}(A)^{\mathrm{op}}) \cong \mathcal{DU}\text{-alg}(A, [X, \Omega])$, and as $\Omega\text{-Cat}(X, \mathsf{At}(A)^{\mathrm{op}}) \cong \Omega\text{-Cat}(X^{\mathrm{op}}, \mathsf{At}(A))^{\mathrm{op}}$, see [7, 2.28], we have to prove that there is a natural isomorphism between

$$\Omega\text{-Cat}(X^{\mathrm{op}}, \mathsf{At}(A))^{\mathrm{op}} \cong \mathcal{DU}\text{-alg}(A, [X, \Omega]),$$

which is Theorem 27.

After having constructed a left adjoint AT of K, we next ask when $\Omega\text{-Cat}^{\mathrm{op}}$ is a full reflective subcategory of $\mathcal{DU}\text{-alg}$, that is, we ask when K is fully faithful. We also want to characterise the image of K and describe the subcategories of $\Omega\text{-Cat}^{\mathrm{op}}$ and $\mathcal{DU}\text{-alg}$ on which the adjunction restricts to an equivalence.

4.2   A Fully Faithfulness of the Comparison and Its Image

In the case of $\Omega = 2$ the comparison K is fully faithful, but this is not true for all commutative quantales Ω. In this subsection, we give necessary and sufficient conditions for K to be fully faithful and describe its image.

Using Proposition 3 we notice that K is faithful on $\Omega\text{-Cat}^{\mathrm{op}}(X, Y)$ if and only if X is anti-symmetric. Indeed, if X is not anti-symmetric let $g_1, g_2 : Y \longrightarrow X$ be two distinct equivalent maps. Then as Ω is we have that $Kg_1 = Kg_2$.

For K to be full we need that for any two categories $X, Y \in \Omega\text{-Cat}$ and every map $H : KX \longrightarrow KY$ there exists a map $h : Y \longrightarrow X$ such that $Kh = H$. Using the adjunction, we have $K \circ \mathsf{AT}(H) = H$ so if one can make sure that $\mathsf{At}(KX) \cong X$ and $\mathsf{At}(KY) \cong Y$ then the functor K will be full. For that we need the following definition [6,10].

Definition 29. *We say that* $X \in \Omega\text{-Cat}$ *is Cauchy complete if* $X \simeq \mathsf{At}([X, \Omega])^{\mathrm{op}}$. *We denote by* $\Omega\text{-Cat}_{\mathrm{cc}}$ *the full subcategory of* $\Omega\text{-Cat}$ *spanned by the antisymmetric Cauchy complete categories.*

Remark 30. 1. Let $\Omega = [0, \infty]$ and let Q and R be the rational and real numbers, respectively, with the usual Euclidean metric. Then the map in $H : [Q, \Omega] \longrightarrow [R, \Omega]$ given by $H(f)(r) = \lim_n f(q_n)$ where (q_n) is a Cauchy sequence with limit r, is in $\mathcal{DU}$-alg and cannot be restricted to a map $\mathsf{At}(H) : R \longrightarrow Q$. So K is not full in general.

2. Any poset is Cauchy complete, see [16].

3. As shown in [10], a generalised metric space X is isomorphic to $\mathsf{At}([X^{\mathrm{op}}, \Omega])$ if it is Cauchy complete in the usual sense of metric spaces.

Theorem 31. *The comparison functor for the adjunction $[-, \Omega] \dashv [-, \Omega]$: $\Omega\text{-Cat} \longrightarrow \Omega\text{-Cat}_{\mathsf{cc}}{}^{\mathrm{op}}$ is full and faithful.*

This result is conceptually important to us. When we started out from the basic picture (3), we were guided by the example $\Omega = 2$, in which $\Omega\text{-Cat}_{\mathsf{cc}} = \Omega\text{-Cat}$. Therefore we could as well have chosen $\Omega\text{-Cat}_{\mathsf{cc}}{}^{\mathrm{op}}$ instead of $\Omega\text{-Cat}^{\mathrm{op}}$ in (3). From this point of view, the theorem confirms that we are free to consider K in (3) to by fully faithful.

To characterise the image of K, we use the description of full reflective subcategories by orthogonality, see [3, Chapter 5.4]. First we need again some definitions.

Definition 32. *A functor $F : A \longrightarrow C$ is called dense if $c = C(F-, c) * F$ for all $c \in C$.*

For more equivalent descriptions of dense functors see [7, Chapter 5].

Definition 33. *A category A is called atomic if the atom-inclusion functor i_A : $\mathsf{At}(A) \longrightarrow A$ is dense.*

Let us give some example of atomic categories.

Example 34. 1. Any finite distributive lattice is atomic.

2. Any presheaf category is atomic as every functor is a colimit of representables.

3. The category $[0, \infty]$ is atomic if seen as a generalised metric space but not if seen as a poset.

We will need the following property of dense functors.

Lemma 35. *If A is cocomplete and the atom-inclusion functor $i_A : \mathsf{At}(A) \longrightarrow A$ is dense then $A \cong [\mathsf{At}(A)^{\mathrm{op}}, \Omega]$.*

Proof. Let $A \in \mathcal{A}$ such that $i : \mathsf{At}(A) \longrightarrow A$ is dense. According to [7, Theorem 5.1] if i is dense then $\tilde{i} : A \longrightarrow [\mathsf{At}(A)^{\mathrm{op}}, \Omega]$, defined by $\tilde{i}a = A(i-, a)$, is fully faithful. So we just have to show that it is essentially surjective. Let $H : \mathsf{At}(A)^{\mathrm{op}} \longrightarrow \Omega$, as A is cocomplete $H * i$ exists, then $\tilde{i}(H * i) \cong H * \tilde{i}i \cong H * d\mathsf{At}(A) \cong H$ thus $\tilde{i}$ is essentially surjective and so $A \cong [\mathsf{At}(A)^{\mathrm{op}}, \Omega]$.

Theorem 36. *An algebra A in $\mathcal{DU}$-alg is isomorphic to an algebra in the image of K if and only if it is atomic.*

Proof. We shall use orthogonality [3, Chapter 5.4]. First let us take X in Ω-Cat^{op} and show that it is atomic. Let us denote by $\theta : \mathrm{id} \longrightarrow KAT$ the unit of the adjunction $\mathsf{AT} \dashv K$. From orthogonality we obtain that for every $B \in \mathcal{D}\mathcal{U}$-alg and any $f : B \longrightarrow X$ we have a unique factorisation through θ_B, so let us take $B = X$ and $f = \mathrm{id}_X$. There exists $g : [\mathsf{At}^{\mathrm{op}}(X), \Omega] \longrightarrow X$ such that g preserves limits and colimits and such that $g \circ \theta_X = \mathrm{id}_X$. Thus, for every $x \in X$ one has

$$g(\theta_X(x)) = x.$$

Now $\theta_X(x) = X(-, x) : \mathsf{At}^{\mathrm{op}}(X) \longrightarrow \Omega$ and as every presheaf is a colimit of representables one has

$$X(-, x) = X(-, x) * d_{\mathsf{At}(X)}.$$

Thus one also has

$$
\begin{aligned}
x = g(X(-, x)) &= g(X(-, x) * d_{\mathsf{At}(X)}) \\
&= g\left(\int^{x' \in \mathsf{At}(X)} X(x', x) \otimes \mathsf{At}(X)(-, x') \right) = \int^{x' \in \mathsf{At}(X)} X(x', x) \otimes g(\mathsf{At}(X)(-, x'))) \\
&= \int^{x' \in \mathsf{At}(X)} X(x', x) \otimes g(X(-, x'))) = \int^{x' \in \mathsf{At}(X)} X(x', x) \otimes x')) \\
&= X(i_X -, x) * i_X
\end{aligned}
$$

So, in conclusion, X is atomic as $i_X : \mathsf{At}(X) \longrightarrow X$ is dense. The converse follows from Lemma 35 because $X \cong [\mathsf{At}^{\mathrm{op}}(X), \Omega] = \mathcal{D}(\mathsf{At}(X))$ which is ccd. $\qed$

Remark 37. In the case $\Omega = 2$, we have $^{\mathrm{op}}\mathsf{CCD} = \mathsf{CCD}$ (since the dual of a completely distributive lattice is a completely distributive lattice). But this is not true for general Ω. Using results from [4] and reproving them for the enriched case we can show that the categories of $\mathcal{D}\mathcal{U}$-algebras and $\mathcal{U}\mathcal{D}$-algebras are isomorphic if $\Omega \cong \Omega^{\mathrm{op}}$ in Ω-Cat.

5   Algebras for Operations and Equations

We will show that the categories of algebras for the monads $\mathcal{D}, \mathcal{U}$, and $\mathcal{D}\mathcal{U}$ are isomorphic to categories of algebras given by operations and equations over Set.

5.1   Syntactic $\mathcal{D}$-algebras and $\mathcal{U}$-algebras

Definition 38. *By a $\langle \Sigma_\mathcal{D}, E_\mathcal{D} \rangle$-algebra we understand a set A together with a family of unary operations $(v \star \_)_{v \in \Omega} : A \longrightarrow A$ indexed by Ω, and a family of operations $\bigsqcup_K : A^K \longrightarrow A$, where K ranges over all sets, satisfying the following 7 axioms. Dually the notions of a $\langle \Sigma_\mathcal{U}, E_\mathcal{U} \rangle$-algebra is given by a set B together with a family of unary operations $(v \triangleright \_)_{v \in \Omega} : B \longrightarrow B$ and for each set K an operation $\bigsqcap_K : B^K \longrightarrow B$ satisfying the following 7 axioms.*

1. $e \star - = \mathrm{id}_A$ $\qquad\qquad\qquad\qquad\qquad\qquad$ $e \triangleright \_ = \mathrm{id}_A$
2. For all $a \in A$, $b \in B$ and $v, w \in \Omega$

$$v \star (w \star a) = (v \otimes w) \star a \qquad\qquad v \triangleright (w \triangleright b) = (v \otimes w) \triangleright b$$

3. For all $v \in \Omega$ and $a_k \in [K, A]$, $b_k \in [K, B]$

$$v \star \bigsqcup_K a_k = \bigsqcup_K (v \star a_k) \qquad\qquad v \triangleright \bigsqcap_K b_k = \bigsqcap_K (v \triangleright b_k)$$

4. For all $a \in A$, $b \in B$ and $v_k \in [K, \Omega]$

$$\left(\int^K v_k\right) \star a = \bigsqcup_K (v_K \star a) \qquad\qquad \left(\int^K v_k\right) \triangleright b = \bigsqcap_K (v_K \triangleright b)$$

5. For a set K and function $J : K \longrightarrow \mathrm{Set}$ let us denote with $\bar{J} = \coprod_{k \in K} Jk$. For each $k \in K$ let $a_k : J(k) \longrightarrow A$ and let $a : \bar{J} \longrightarrow A$ be the map induced by the coproduct. For each $k \in K$ let $b_k : J(k) \longrightarrow B$ and let $b : \bar{J} \longrightarrow B$ be the map induced by the coproduct.

$$\bigsqcup_K \left(\bigsqcup_{Jk} a_k\right) = \bigsqcup_{\bar{J}} a \qquad\qquad \bigsqcap_K \left(\bigsqcap_{Jk} b_k\right) = \bigsqcap_{\bar{J}} b$$

6. Let Δ be the diagonal functor then for any set K and for all $a \in A$ and $b \in B$ we have

$$\bigsqcup_K \Delta a = a \qquad\qquad\qquad \bigsqcap_K \Delta b = b$$

7. For any two sets J, K and any bijective function $f : J \longrightarrow K$ one has

$$\bigsqcup_J \circ A^f = \bigsqcup_K \qquad\qquad\qquad \bigsqcap_J \circ A^f = \bigsqcap_K$$

Before we continue let us fix some notations and give some examples. If the set K is 2 then we put $\bigsqcup_K = \sqcup$ and $\bigsqcap_K = \sqcap$ and use infix notation. For any set K by an element a_K of A^K we understand any function $a_K : K \longrightarrow A$. If K is finite a_K can be represented as a tuple $a_K = (a_1, a_2, ..., a_k)$ where $k = |K|$.

Example 39. 1. For any quantale Ω, the Ω-category Ω is a $\langle \Sigma_{\mathcal{D}}, E_{\mathcal{D}} \rangle$-algebra, with $\bigsqcup$ given by $\bigvee$ and $v \star -$ given by $v \otimes -$. The fact that this satisfies all the axioms is trivial. In a similar way Ω is also a $\langle \Sigma_{\mathcal{U}}, E_{\mathcal{U}} \rangle$-algebra with $\bigsqcap$ given by $\bigwedge$ and $v \triangleright -$ given by $\Omega(v, -)$.
2. Any cocomplete Ω-category A is a $\langle \Sigma_{\mathcal{D}}, E_{\mathcal{D}} \rangle$-algebra. For any $v \in \Omega$ and $a \in A$ we define $v \star a$ as the colimit of a weighted by v. And for every set K and any $a_K \in A^K$ we define $\bigsqcup_K a_K$ as the colimit of a_K weighted by constant Ω-functor $e_K : K \longrightarrow \Omega$ given by $e_K(k) = e$ for all $k \in K$. That is equivalent to saying that $\bigsqcup_K$ is a coend.
3. Any complete Ω-category A is a $\langle \Sigma_{\mathcal{U}}, E_{\mathcal{U}} \rangle$–algebra
4. For any quantale Ω and any Ω-category X the functor category $[X, \Omega]$ is a $\langle \Sigma_{\mathcal{D}}, E_{\mathcal{D}} \rangle$-algebra and the functor category $[X, \Omega]^{\mathrm{op}}$ is a $\langle \Sigma_{\mathcal{U}}, E_{\mathcal{U}} \rangle$-algebra. If $\Omega \cong \Omega^{\mathrm{op}}$ then any functor category is both a $\langle \Sigma_{\mathcal{D}}, E_{\mathcal{D}} \rangle$ and a $\langle \Sigma_{\mathcal{U}}, E_{\mathcal{U}} \rangle$-algebra.

As any $\langle \Sigma_\mathcal{D}, E_\mathcal{D} \rangle$-algebra A has a preorder structure on it, given by $a \leq b \Leftrightarrow a \sqcup b = b$. We now show that A also carries a Ω-category structure.

Proposition 40. *Any $\langle \Sigma_\mathcal{D}, E_\mathcal{D} \rangle$-algebra A has a Ω-category structure given by*

$$A(a,b) = \bigvee \{v \in \Omega \mid (v \star a) \leq b\}, \tag{10}$$

for all $a, b \in A$. Also any $\langle \Sigma_\mathcal{U}, E_\mathcal{U} \rangle$-algebra B has a Ω-category structure given by

$$B(b,b') = \bigvee \{v \in \Omega \mid b \leq (v \rhd b')\}, \tag{11}$$

for all $b, b' \in B$.

One could ask why we do not define $A(a,b)$ as that $v \in \Omega$ such that $v \star a = b$, and the answer is because $\star$ is not injective in general. For example, take $\Omega = [0, \infty]$ and note that $w \star \infty = \infty$ for all $w \in \Omega$, thus there is no unique $w \in \Omega$ to define $[0, \infty](\infty, \infty)$.

Example 41. Let us look at $\Omega = (([0, \infty] \geq), 0, +)$. Define $v \star a = v + a$ and $\bigsqcup_K(v_1, ..., v_k) = \inf_\mathbb{R}(v_1, .. v_k)$, thus Ω is a $\langle \Sigma_\mathcal{D}, E_\mathcal{D} \rangle$-algebra. Let us check that the Ω-category structure given by Proposition 40 is the usual one. Let $a, b \in [0, \infty]$, then one has

$$\{v \in \Omega \mid v + a \geq_\mathbb{R} b\} = \{v \in \Omega \mid v \geq_\mathbb{R} b - a\}$$

Now obviously $[0, \infty](a,b) = b \dotminus a = \inf\{v \in \Omega \mid v \geq_\mathbb{R} b - a\} = \bigvee\{v \in \Omega \mid v \geq_\mathbb{R} b - a\}$. Also let us note that $\bigwedge\{v \in \Omega \mid v \geq_\mathbb{R} b - a\} = \infty$.

One has two equivalent definitions of a semi-lattice, one using operations and equations, and one saying that a semi-lattice is a complete/cocomplete poset. The Ω-Cat analogue is as follows.

Theorem 42. *Let A be a $\langle \Sigma_\mathcal{D}, E_\mathcal{D} \rangle$-algebra and B a $\langle \Sigma_\mathcal{U}, E_\mathcal{U} \rangle$-algebra.*

1. *For any $v \in \Omega$ and $a, b \in A$ we have $A(v \star a, b) = \Omega(v, A(a,b))$. Thus $v \star a$ is the colimit of a weighted by v.*
2. *The operation $\bigsqcup_K$ is a coend, in the sense that for any set K one has $A(\bigsqcup_K a_k, b) = \int_{k \in K} A(a_k, b)$.*
3. *For any $v \in \Omega$ and $a, b \in B$ we have $B(a, v \rhd b) = \Omega(v, A(a,b))$. Thus $v \rhd b$ is the limit of b weighted by v.*
4. *The operation $\bigsqcap_K$ is an end, in the sense that for any set K one has $B(a, \bigsqcap_K b_k) = \int_{k \in K} B(a, b_k)$.*

Thus any $\langle \Sigma_\mathcal{D}, E_\mathcal{D} \rangle$-algebra is co-complete as a Ω-category, and any $\langle \Sigma_\mathcal{U}, E_\mathcal{U} \rangle$-algebra is complete as a Ω-category.

Definition 43. *If $(A, (v \star^A \_)_{v \in \Omega}, \bigsqcup_K^A)$ and $(B, (v \star^B \_)_{v \in \Omega}, \bigsqcup_K^B)$ are $\langle \Sigma_\mathcal{D}, E_\mathcal{D} \rangle$-algebras, a map $f : A \longrightarrow B$ is a morphism if f preserves all operations, that is if the following diagrams commute.*

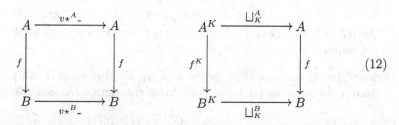

$$(12)$$

Theorem 44. *The category $\langle \Sigma_{\mathcal{D}}, E_{\mathcal{D}} \rangle$-alg of $\langle \Sigma_{\mathcal{D}}, E_{\mathcal{D}} \rangle$-algebras and their morphisms is isomorphic to the category of $\mathcal{D}$-algebras, and the category of $\langle \Sigma_{\mathcal{U}}, E_{\mathcal{U}} \rangle$-algebras and their morphisms is isomorphic to the category of $\mathcal{U}$-algebras.*

5.2 Syntactic $\mathcal{DU}$-algebras

In order to make the definition of a $\langle \Sigma_{\mathcal{DU}}, E_{\mathcal{DU}} \rangle$-algebra more readable we need some preliminary results. First let us recall the following known fact about lattices.

Lemma 45. *Let $(A, (v \star -)_{(v \in \Omega)}, (\bigsqcup_K)_K)$ be a $\langle \Sigma_{\mathcal{D}}, E_{\mathcal{D}} \rangle$-algebra and $(A, (v \triangleright -)_{(v \in \Omega)}, (\bigsqcap_K)_K)$ be a $\langle \Sigma_{\mathcal{U}}, E_{\mathcal{U}} \rangle$-algebra. In particular A is a meet-semi lattice and join semi-lattice, so the order given by these is compatible if and only if we have the following two absorption axioms:*

1. $a \sqcap (a \sqcup b) = a$ for all $a, b \in A$
2. $a \sqcup (a \sqcap b) = a$ for all $a, b \in A$

Proposition 46. *Let A be simultaneously a $\langle \Sigma_{\mathcal{D}}, E_{\mathcal{D}} \rangle$ and a $\langle \Sigma_{\mathcal{U}}, E_{\mathcal{U}} \rangle$-algebra, which satisfies the absorbtion rules defined in the previous lemma, then the Ω-category structures given by A being a $\langle \Sigma_{\mathcal{D}}, E_{\mathcal{D}} \rangle$-algebra and a $\langle \Sigma_{\mathcal{U}}, E_{\mathcal{U}} \rangle$-algebra are compatible, that is for all $a, b \in A$ we have $\bigvee \{v \in \Omega \mid v \star a \leq_{\sqcup} b\} = \bigvee \{v \in \Omega \mid a \leq_{\sqcap} v \triangleright b\}$*

Now we can formulate the following definition.

Definition 47. *By a $\langle \Sigma_{\mathcal{DU}}, E_{\mathcal{DU}} \rangle$-algebra we understand, a set A together with two unary family of operations $(v \star -)_{(v \in \Omega)} : A \longrightarrow A$ and $(v \triangleright -)_{(v \in \Omega)} : A \longrightarrow A$, and for each set K two K-arity operations $\bigsqcup_K : A^K \longrightarrow A$ and $\bigsqcap_K : A^K \longrightarrow A$, such that $(A, (v \star -)_{(v \in \Omega)}, (\bigsqcup_K)_K)$ is a $\langle \Sigma_{\mathcal{D}}, E_{\mathcal{D}} \rangle$-algebra and $(A, (v \triangleright -)_{(v \in \Omega)}, (\bigsqcap_K)_K)$ is a $\langle \Sigma_{\mathcal{U}}, E_{\mathcal{U}} \rangle$-algebra satisfying the following equations:*

1. $a \sqcap (a \sqcup b) = a$ for all $a, b \in A$
2. $a \sqcup (a \sqcap b) = a$ for all $a, b \in A$
3. *for any $v \in \Omega$ and any $a, b \in A$ one has $(v \star a) \leq b \Leftrightarrow a \leq (v \triangleright b)$*
4. *for any set K and any functions $\varphi : K \longrightarrow \Omega$ and $G : K \times A \longrightarrow \Omega$*

$$\prod_K \varphi(k) \triangleright (\bigsqcup_A G(k)(a) \star a) = \bigsqcup_A \{\varphi, \downarrow G(-, a)\} \star a, \qquad (13)$$

where $\{\varphi, \downarrow G(-, a)\}$ is a limit computed in Ω with $\downarrow G(k) : A^{op} \longrightarrow \Omega$ given by $\downarrow G(k) = \text{Lan}_i G = \int^{b \in A} A(-, i(b)) \otimes G(k)(b)$ for $i : |A| \longrightarrow A^{op}$ the object inclusion functor.

Remark 48. 1. As we will see below a $\langle \Sigma_{\mathcal{DU}}, E_{\mathcal{DU}} \rangle$-algebra A can be translated into a $\mathcal{DU}$-algebras (A, α, β). Under this translation, Axiom (13) becomes

$$\{\varphi, \alpha \circ \downarrow G\} = \alpha(\{\varphi, \downarrow G\})$$

stating that α preserves limits.

2. In the case that φ and G are crisp, that is, they take values in $\{\bot, e\} \subseteq \Omega$, we can identify K with the extension of φ and $G(k)$ with subsets of A so that Axiom (13) becomes

$$\bigsqcap \{\bigsqcup G(k) \mid k \in K\} = \bigsqcup \bigsqcap \{\downarrow G(k) \mid k \in K\}$$

Note that this coincides with the specialisation of Axiom (13) to the case $\Omega = 2$. It expresses that joins perserve meets. As observed in [4] this is equivalent (under the axiom of choice) to the usual distributive law using choice functions. Choice functions allow us to replace the intersection $\bigsqcap \{\downarrow G(k) \mid k \in K\}$, which is a meet in $\mathcal{D}A$, by a collection of meets in A. Moreover, in case that K and the $G(k)$ are finite, the join $\bigsqcup \bigsqcap \{\downarrow G(k) \mid k \in K\}$ can be replaced by a join indexed over *finitely* many choice functions, even if A is infinite. Under what circumstances the distributive law (13) can be restricted to a finite one is a question we do not pursue in this paper.

Now let us show that $\langle \Sigma_{\mathcal{DU}}, E_{\mathcal{DU}} \rangle$ algebras are indeed $\mathcal{DU}$-algebras.

Theorem 49. *The category of $\langle \Sigma_{\mathcal{DU}}, E_{\mathcal{DU}} \rangle$-algebras is isomorphic to the category of $\mathcal{DU}$-algebras.*

Proof. Let $(A, v\star-, \bigsqcup_K, v\triangleright-, \bigsqcap_K)$ be a $\langle \Sigma_{\mathcal{DU}}, E_{\mathcal{DU}} \rangle$ algebra. By Propositions 40 and 46, A has a Ω-category structure. Define $\alpha : \mathcal{D}A \longrightarrow A$ and $\beta : \mathcal{U}A \longrightarrow A$, by $\alpha(\varphi) = \bigsqcup_{|A|} \varphi(a) \star a$ and $\beta(\psi) = \bigsqcap_{|A|} \psi(a) \triangleright a$. Using all axioms of Definition 38 and Theorem 42 one shows that α is a $\mathcal{D}$-algebra and β is a $\mathcal{U}$-algebra. Moreover, A is complete and cocomplete. Axiom (13) implies that α preserves all limits, hence has a left adjoint and thus is ccd. By Theorem 22, (A, α, β) is a $\mathcal{DU}$-algebra.

For the converse, define $\star$, $\triangleright$, $\bigsqcup$, and $\bigsqcap$ as the respective (co)limits and show that ccd implies Axiom (13).

5.3 Ω-**Cat**-logic

For any (commutative) quantale Ω, we have a propositional Ω-logic given by $\langle \Sigma_{\mathcal{DU}}, E_{\mathcal{DU}} \rangle$. The language is given by

$$\mathcal{L} : p \mid v \star - \mid v \triangleright - \mid \bigsqcup \mid \bigsqcap,$$

where p are atomic propositions; $v \star -$ and $v \rhd -$ are unary operations; $\bigsqcup$ and $\bigsqcap$ are K-ary operations for each set K. These operations satisfy the equations listed in Definitions 38 and 47.

The semantics of these operations wrt a Ω-category X is as follows. If F_P is the free algebra[1] over atomic propositions $p \in P$, then any interpretation of the atomic propositions as many-valued upsets $X \longrightarrow \Omega$ induces a morphism $[\![-]\!] : F_P \longrightarrow [X, \Omega]$, which is nothing but a many-valued relation (bimodule)

$$\Vdash : X \otimes F_P \longrightarrow \Omega.$$

The values of Ω measure how well a state $x \in X$ satisfies a specification $\phi \in F_P$. We say "x satisfies ϕ up to r" if $(x \Vdash \phi) = r$ and "x satisfies ϕ" if r is top. The operations $\bigsqcup$ and $\bigsqcap$ are join and meet and we have

$$[\![v \star \phi]\!] = (x \mapsto v \otimes [\![\phi]\!](x)) \qquad [\![v \rhd \phi]\!] = (x \mapsto [v, [\![\phi]\!](x)]).$$

Reasoning in the Ω-valued setting, we are interested in judgements

$$\phi \vdash_r \psi$$

which are interpreted as $F_P(\phi, \psi) \geq r$, the latter being equivalent to

$$[x \Vdash \phi, x \Vdash \psi] \geq r$$

for all Ω-categories X and all $x \in X$.

Example 50. 1. In the case of $\Omega = 2$ the operations $\star$ and $\rhd$ are redundant and we obtain the equational theory of complete and completely distributive lattices. $\vdash_0$ is redundant and $\phi \vdash_1 \psi$ means $(x \Vdash \phi) \Rightarrow (x \Vdash \psi)$ for all X and $x \in X$.

2. In the case of $\Omega = [0, \infty]$, we write inf for $\bigsqcup$ and sup for $\bigsqcap$ (note the reversal of the order to reflect that distance 0 is top and ∞ is bottom). If x satsifies ϕ, then x satsifies $v \star \phi$ up to v. If x satisfies ϕ up to v, then x satisfies $v \rhd \phi$. That $[X, \Omega]$ is atomic means that arbitrary predicates can be built from "singletons", $v \star -$ and $\bigsqcup$. A judgement $\phi \vdash_r \psi$ means $(x \Vdash \psi) - (x \Vdash \phi) \leq_\mathbb{R} r$ for all X and $x \in X$.

6   Conclusions

We have shown in Theorem 49 that for any commutative quantale Ω the category Ω-Cat of Ω-categories, or, in other words, the category of Ω-valued generalised metric spaces, is isomorphic to a category of algebras for operations and equation in the usual sense, if we admit operations of infinite unbounded arity.

Moreover, due to the duality underlying our approach, these operations have a logical interpretation and the equations can be seen as logical axioms.

---

[1] Here we make use of the fact that due to complete distributivity free algebras exist, even though the signature has operations of unbounded arity.

The value of Theorem 49 resides not only in its statement but even more so in how we proved it: We didn't guess $\langle \Sigma_{\mathcal{DU}}, E_{\mathcal{DU}} \rangle$ and then proved the theorem, but we derived $\langle \Sigma_{\mathcal{DU}}, E_{\mathcal{DU}} \rangle$ in a systematic fashion from the functor $[-, \Omega]$. We started from the aim to derive the logic of Ω-valued predicates, that is, the logic given implicitly by the structure of the categories $[X, \Omega]$. To extract this logical structure, we considered $[X, \Omega]$ as algebras for the monad induced by $[-, \Omega]$. We then employed a result linking that monad to the 'semi-lattice' monads $\mathcal{D}$ and $\mathcal{U}$. The algebraic structure of these monads computes limits and colimits and an equational description of these was given as $\langle \Sigma_{\mathcal{DU}}, E_{\mathcal{DU}} \rangle$.

It lies in the nature of this method that the logic $\langle \Sigma_{\mathcal{DU}}, E_{\mathcal{DU}} \rangle$ we derived from Ω is not purely syntactic but still depends on Ω. The operations are infinitary and the laws contain side conditions depending on Ω. We can think of Ω as an oracle that we need to consult in our reasoning. Restricting to particular, syntactically given Ω and then describing $\langle \Sigma_{\mathcal{DU}}, E_{\mathcal{DU}} \rangle$ fully syntactically, so that consulting the oracle can be replaced by asking an automated theorem prover, is a task of future research.

In future work, finitary versions of the $\langle \Sigma_{\mathcal{DU}}, E_{\mathcal{DU}} \rangle$ will be investigated. Extension with tensor and implication will also be of interest. These should be linked with the theory of MV-algebras. Properties of $\langle \Sigma_{\mathcal{DU}}, E_{\mathcal{DU}} \rangle$ and their finitary versions should be linked with properties of Ω. Moerover, it needs to be investigated how to integrate the propositional Ω-logics with the modalities arising from coalgebraic type functors.

Acknowledgement. The second author acknowledges the influence of J. Velebil through a long-standing collaboration on enriched coalgebraic logic and his deep insights into the subject. In particular, our derscription of Ω-Cat^{op} by operations and equations confirms his suggestion that the propositional logic of Ω-Cat should have operations corresponding the the categorical (co)limits of tensor and cotensor.

References

1. Beck, J.: Distributive laws. In: Eckmann, B. (ed.) Seminar on Triples and Categorical Homology Theory. Lecture Notes in Mathematics, vol. 80, pp. 119–140. Springer, Berlin (1969)
2. Bílková, M., Kurz, A., Petrisan, D., Velebil, J.: Relation lifting, with an application to the many-valued cover modality. Log. Methods Comput. Sci. **9**(4) (2013)
3. Borceaux, F.: Handbook of Categorical Algebra 1, vol. 1. Cambridge University Press, Cambridge (1994)
4. Fawcett, B.W., Wood, R.J.: Constructive complete distributivity I. In: Mathematical Proceedings of the Cambridge Philosophical Society (1990)
5. Hofmann, D.: Duality for distributive space (2010). arXiv:1009.3892v1
6. Kelly, G.M., Schmitt, V.: Notes on enriched categories with colimits of some class. Theor. Appl. Categ. **14**, 399–423 (2005)
7. Kelly, M.: Basic Concepts of Enriched Category Theory
8. Kock, A.: Monads for which structures are adjoint to units. J. Pure Appl. Algebra **104**(1), 41–59 (1995)
9. Lai, H., Zhang, D.: Many-Valued Complete Distributivity (2006)

10. Lawvere, F.: Metric spaces, generalized logic and closed categories. Rendiconti del Seminario Matematico e Fisico di Milano, XLIII. Republished in Reprints in Theory Appl. Categ. (1973)
11. Mac Lane, S.: Categories for the Working Mathematician. Graduate Texts in Mathematics. Springer, New York (1971)
12. Marmolejo, F.: Doctrines whose structure forms a fully faithful adjoint string. Theor. Appl. Categ. **3**, 24–44 (1997)
13. Marmolejo, F., Rosebrugh, R., Wood, R.J.: A basic distributive law. J. Pure Appl. Algebra **168**(2), 209–226 (2002)
14. Pu, Q., Zhang, D.: Categories enriched over a quantaloid: algebras. Theor. Appl. Categ. **30**, 751–774 (2015)
15. Rosenthal, K.I.: Quantales and Their Applications. Pitman Research Notes in Mathematics Series, vol. 234. Longman Scientific & Technical, Harlow (1990)
16. Rosolini, G.: A note on cauchy completeness for preorders
17. Rutten, J.J.M.M.: Elements of generalized ultrametric domain theory. Theoret. Comput. Sci. **170**(1–2), 349–381 (1996)
18. Rutten, J.J.M.M.: Relators and metric bisimulations (extended abstract). In: CMCS 1998, vol. 11 (1998)
19. Stubbe, I.: The double power monad is the composite power monad. Technical report, LMPA, Université du Littoral-Côte d'Opale (2013)
20. Vickers, S.: The double powerlocale and exponentiation: a case study in geometric reasoning. Theor. Appl. Categ. **12**, 372–422 (2004)
21. Vickers, S.: Localic completion of generalized metric spaces I. Theor. Appl. Categ. **14**, 328–356 (2005)
22. Vickers, S.: Localic completion of generalized metric spaces II: powerlocales. J. Log. Anal. **1**(11), 1–48 (2009)
23. Vickers, S., Townsend, C.: A universal characterization of the double powerlocale. Theoret. Comput. Sci. **316**, 297–321 (2004)
24. Worrell, J.: Coinduction for recursive data types: partial order, metric spaces and omega-categories. In: CMCS 2000, vol. 33 of ENTCS (2000)

A Complete Logic for Behavioural Equivalence in Coalgebras of Finitary Set Functors

David Sprunger[✉]

Indiana University, Bloomington, USA
dasprung@indiana.edu

Abstract. This paper presents a sound and complete sequent-style deduction system for determining behavioural equivalence in coalgebras of finitary set functors preserving weak pullbacks. We select finitary set functors because they are quotients of polynomial functors: the polynomial functor provides a ready-made signature and the quotient provides necessary additional axioms. We also show that certain operations on functors can be expressed with uniform changes to the presentations of the input functors, making this system compositional for a range of widely-studied classes of functors, such as the Kripke polynomial functors. Our system has roots in the FLR_0 proof system of Moschovakis et al., particularly as used by Moss, Wennstrom, and Whitney for non-wellfounded sets. Similarities can also be drawn to expression calculi in the style of Bonsangue, Rutten, and Silva.

1 Introduction

In this paper, we propose a logic for detecting bisimilar states in coalgebras of finitary Set-endofunctors. We focus on finitary functors because they have presentations whereby they can be represented as the quotient of a signature functor by a collection of equations.[1] The signature provides a syntax in which the coalgebras can be expressed, and the equations add the axioms necessary to distinguish reasoning among functors of similar syntax.

In particular, we will consider *specifications* on sets of variables in the signature of finitary functors. These are total assignments of variables to terms which serve as definitions, and may be considered a recasting of the longstanding tradition of systems of simultaneous equations going back to Kahn, Manna and Vuillemin, and Lawvere. We show every coalgebra of a finitary functor has at least one corresponding specification.

Our system is comparable to FLR_0, as considered by Moss in [2] and Moss et al. in [3]. FLR_0 has distinctive terms of the form x_i **where** $\{x_1 = A_1, \ldots, x_n = A_n\}$. This **where** operator binds variables, but also allows terms to carry their context.

---

[1] Finitary functors in finitely presentable categories outside of Set may also have finitary presentations, see [1].

Published by Springer International Publishing Switzerland 2016. All Rights Reserved
I. Hasuo (Ed.): CMCS 2016, LNCS 9608, pp. 156–173, 2016.
DOI: 10.1007/978-3-319-40370-0_10

As a consequence, many of the FLR_0 rules concern moving a definition in and out of a subscope or evolving the term before the **where** clause. We avoid these issues by fixing a specification sending x_i to A_i, roughly, and considering terms with this context backgrounded. Additionally, FLR_0 and the full FLR language are intended as general languages of recursion with semantics of various flavors. The application here to coalgebras distinguishes our version somewhat.

We might also compare this work to that of Bonsangue et al. in [4,5] or Milius' related work in the setting of vector spaces [6], where a μ operator provides a similar variable binding. The work of Bonsangue et al. feature an inductive class of functors, the so-called Kripke polynomial functors, and a syntax of expressions based on the inductive class. They build a sound and complete axiomatization for these expressions which is compositional, meaning the laws involved are built in parallel with the definition of the functor and the expressions. We show that the presentations involved in our setting enjoy similar compositional properties.

In Chap. 5 of Silva's PhD thesis [7] and the related paper [8], she gives an extension of this μ calculus to finitary functors, demonstrating that the expressions of this calculus exactly coincide with the behaviours of locally finite coalgebras. However, at the end of this work, questions regarding axiomatization and uniform proofs of soundness and completeness for the system are left open.

We are able to prove soundness and completeness for our logic for the finitary functors which preserve weak pullbacks, a common condition with numerous pleasant coalgebraic consequences including that bisimilarity and behavioural equivalence coincide, see Rutten [9]. In particular, polynomial functors and the finite powerset functor preserve weak pullbacks, so the functors in our setting properly include those of Bonsangue et al. and Moss et al.

Outline. In Sect. 2, we briefly recall some background on coalgebras, signatures, and finitary functors. This section introduces the interplay between coalgebras of a finitary functor and coalgebras of its related signature functor that are of central importance to later sections. In Sect. 3, we introduce bisimulation up-to-presentation, a novel up-to technique which permits expressing bisimulations for finitary functors in terms of bisimulations for their signatures. In Sect. 4, we present a formal proof system capturing the notion of bisimulation-up-to. We show this system is sound and complete in the sense that it detects the bisimilarity of states in coalgebras for finitary functors preserving weak pullbacks. In Sect. 5, we note that signatures for previously studied inductive classes of these functors–including the Kripke polynomial functors–can be constructed compositionally. This allows the proof system developed in Sect. 4 to be constructed compositionally as well.

2 Background

In this section, we recall definitions and basic results about coalgebras, finitary signatures, finitary functors, and introduce the notion of a specification. Our setting is the category Set, and all functors are assumed to be Set-endofunctors. Additionally, we will often assume functors preserve weak pullbacks, but make a special note when this assumption is needed.

2.1 Coalgebras

Given a Set-endofunctor F, an F-*coalgebra* is a set X together with a map $f : X \to FX$. The set is often called the *carrier* of the coalgebra, while f gives its *structure* or *dynamics*.

A *coalgebra morphism* from an F-coalgebra (X, f) to another F-coalgebra (Y, g) is a map $\varphi : X \to Y$ such that the following diagram commutes:

$$
\begin{array}{ccc}
X & \xrightarrow{\ f\ } & FX \\
{\scriptstyle \varphi}\downarrow & & \downarrow{\scriptstyle F\varphi} \\
Y & \xrightarrow{\ g\ } & FY
\end{array}
$$

F-coalgebras together with coalgebra morphisms between them form a category, often denoted $\mathbf{Coalg}_F$. Of particular interest are the F where $\mathbf{Coalg}_F$ has a final object. This final coalgebra has a natural interpretation as a semantic object: since there is a unique coalgebra morphism from any F-coalgebra into it, points in coalgebras which have the same image in the final coalgebra can be considered (behaviourally) equivalent to one another.

A related notion is that of an (Aczel-Mendler) F-*bisimulation* on a coalgebra. An F-bisimulation is a relation $R \subseteq X \times X$ such that there is an F-coalgebra structure on R, ρ, such that the following diagram holds:[2]

$$
\begin{array}{ccccc}
X & \xleftarrow{\ \pi_1\ } & R & \xrightarrow{\ \pi_2\ } & X \\
{\scriptstyle f}\downarrow & & {\scriptstyle \rho}\downarrow & & \downarrow{\scriptstyle f} \\
FX & \xleftarrow{\ F\pi_1\ } & FR & \xrightarrow{\ F\pi_2\ } & FX
\end{array}
$$

Roughly speaking, a relation is a bisimulation if two points in a coalgebra being related implies their structures are also related. This gives a different notion of equivalence, which is known to coincide with the behavioural equivalence for weak pullback preserving functors. For more details, we refer the reader to Rutten [9].

2.2 Finitary Signatures and Functors

A *finitary signature* is a set Σ with a map $ar : \Sigma \to \omega$. In the sequel, we often abbreviate "finitary signature" to "signature" and refer to the signature as Σ instead of (Σ, ar) when there is no risk of confusion. The elements of Σ are the *symbols* of the signature, and each symbol $f \in \Sigma$ has the *arity* $ar(f)$. The collection of all symbols with arity n is denoted Σ_n.

Each finitary signature has an associated *signature functor*, H_Σ, given by $\coprod_n \Sigma_n \times X^n$. We denote a typical element of $H_\Sigma X$ by $f(x_1, \ldots, x_{ar(f)})$ if $x_i \in X$ or $f(\boldsymbol{x})$ if $\boldsymbol{x} : ar(f) \to X$. $H_\Sigma X$ is often referred to as the set of all "flat terms" using symbols from Σ with variables from X.

---

[2] Throughout this paper we will write π_i for the more cumbersome $\pi_i|_R$.

F is a *finitary functor* if there is a finitary signature Σ together with a (pointwise) epic natural transformation $\epsilon : H_\Sigma \twoheadrightarrow F$.[3] If F is a finitary functor, we say (Σ, ϵ) is a *presentation* of F. Finitary functors have a number of alternate characterizations, including functors which preserve ω-filtered colimits [1].

Example 1. For each set A, the constant functor $FX = A$ is finitary with signature $\Sigma = \Sigma_0 = A$ and the transformation ϵ with components $\epsilon_X : a() \mapsto a$.

Example 2. The identity functor is finitary with signature $\Sigma = \Sigma_1 = \{*\}$ and the transformation ϵ with components $\epsilon_X : *(x) \mapsto x$.

Example 3. The finite powerset functor $\mathcal{P}_\omega$ is finitary with signature $\Sigma_n = \{\sigma_n\}$ and the transformation $\epsilon_X : \sigma_n(x_1, \ldots, x_n) \mapsto \{x_1, \ldots, x_n\}$. Note that unlike the previous two transformations, this ϵ is not an isomorphism.

Example 4. The 3 powerset functor $\mathcal{P}_3$, which assigns each set to the set of its subsets of cardinality < 3, is finitary with signature $\Sigma_n = \{\sigma_n\}$ for $0 \leq n < 3$ and the same ϵ as in P_ω, restricted to the smaller set of terms.

Example 5. The functor $ZX = \{0, 1\} \times X \times X$ is finitary with signature $\Sigma = \Sigma_2 = \{0 : \text{zip}, 1 : \text{zip}\}$ and transformation given by

$$\epsilon_X(i : \text{zip}(x_1, x_2)) = (i, x_1, x_2).$$

We call this the "zip functor" in the sequel, though this is not standard terminology.

A *specification in the signature* Σ is an H_Σ-coalgebra, a set X together with a function $d : X \to H_\Sigma X$. Elements of X are called *variables*, and d gives their *definition*. Every specification in the signature of a finitary functor F gives rise to an F-coalgebra: given $d : X \to H_\Sigma X$ postcomposing with ϵ_X yields $\epsilon_X \circ d : X \to FX$.

A single F-coalgebra (X, f) will correspond to potentially many specifications in its signature. For each section s of ϵ_X, the composition $s \circ f : X \to H_\Sigma X$ is a specification in Σ. The F-coalgebra related to each of these specifications will be, not surprisingly, (X, f). Note at least one section of ϵ_X is guaranteed to exist since Set has split epis.

At a broad level, this paper could be seen as an attempt to use the quotient relationship $\epsilon : H_\Sigma \twoheadrightarrow F$ between the functors H_Σ and F to understand the relationships between H_Σ- and F-coalgebras and particularly the relationships between H_Σ- and F-bisimulations. Rather than constantly clarifying which functor we are considering, we hereafter reserve "specification" to mean an H_Σ-coalgebra, and the undecorated "coalgebra" to mean F-coalgebra.

Since we can readily recast specifications and coalgebras for finitary Set functors, we translate standard notions from coalgebras to specifications. For

---

[3] That is, we are assuming each component ϵ_X is epic. For natural transformations between functors into Set, pointwise epic and epic in the functor category coincide. [10, p. 91].

example, R is an F-bisimulation on the specification (X, d) when it is an F-bisimulation on the coalgebra $(X, \epsilon_X \circ d)$, the *standard semantics* for a variable in a specification is its image in a given final F-coalgebra, and two variables are *behaviourally equivalent* when they have the same standard semantics. Note that though the standard semantics of a variable depends on the final coalgebra under consideration, whether two variables are behaviourally equivalent is independent of this choice.

We write $\vDash_{(X,d)} x = y$ when x and y are behaviourally equivalent states in the specification. When the specification is clear from context, we write $\vDash x = y$. (This notation will be relevant mostly in Sect. 4.)

Example 6. We can give a specification for the zip functor with $X = \{x, y, z, w\}$ and

$$d(x) = 0 : \text{zip}(y, z) \qquad d(y) = 1 : \text{zip}(x, w)$$
$$d(z) = 1 : \text{zip}(z, w) \qquad d(w) = 0 : \text{zip}(w, z)$$

As shown by Kupke and Rutten in [11] and Grabmayer et al. in [12], a final coalgebra for this functor is the set of streams in $\{0, 1\}$.[4] With this final coalgebra in mind the standard semantics for x is the Thue-Morse sequence.

Example 7. Another zip specification for $Y = \{x, y, z, w, u, v, q\}$ is given by

$$d(x) = 0 : \text{zip}(y, z) \quad d(y) = 1 : \text{zip}(x, v) \qquad \quad d(v) = 0 : \text{zip}(w, u)$$
$$d(w) = 0 : \text{zip}(w, z) \quad d(z) = 1 : \text{zip}(z, w) \qquad \quad d(u) = 1 : \text{zip}(u, v)$$
$$d(q) = 0 : \text{zip}(y, u)$$

In this specification, the states x and q are behaviourally equivalent. Our goal is to give a uniform account for detecting this behavioural equivalence.

Note also x in this example and x in Example 6 are behaviourally equivalent. We could consider the problem of showing the equivalence of two variables in two separate specifications, but by taking the disjoint union of the two specifications and determining equivalence within this single joint specification we get the same effect.

3   Bisimulation Up to Presentation

In this section, we introduce the notion of bisimulation up to presentation. Roughly, bisimulations up to presentation are H_Σ-bisimulations relaxed up to the kernel of ϵ in such a way that they correspond nicely to F-bisimulations. This allows us to detect F-bisimulations using the more syntactic H_Σ-bisimulations and so-called ϵ laws. We also give an alternate characterization of bisimulation up to presentation and several related sufficient criteria to conclude that a relation is a subset of the bisimilarity relation.

---

[4] Kupke and Rutten actually considered a slight variation on **Coalg**$_Z$ which has the same final coalgebra but a slightly different final map.

Since bisimulation up to presentation provides an alternate criterion which suffices for detecting bisimulations, we have intentionally named this type of relation in the style of enhanced coalgebraic bisimulations studied recently by Rot et al. [13] with veins of research going back to Milner, Park, Sangiorgi, and others. We are also struck by the similarities between the results in Sect. 3.2 and the flavor of standard up-to results. However, we are unaware of a formal connection between these bodies of work since our setting relates bisimulations of two related functors, and the standard literature deals with bisimulations of a single functor. We would be very glad to learn of a connection, though.

Recall the standard (Aczel-Mendler) bisimulation diagram for an H_Σ-bisimulation on X:

$$
\begin{array}{ccccc}
X & \xleftarrow{\ \pi_1\ } & R & \xrightarrow{\ \pi_2\ } & X \\
{\scriptstyle d}\downarrow & & {\scriptstyle \rho}\downarrow & & \downarrow{\scriptstyle d} \\
H_\Sigma X & \xleftarrow[H_\Sigma \pi_1]{} & H_\Sigma R & \xrightarrow[H_\Sigma \pi_2]{} & H_\Sigma X
\end{array}
$$

We say $R \subseteq X \times X$ is a *bisimulation up to the presentation* (Σ, ϵ) if there is a $\rho : R \to H_\Sigma R$ such that $\epsilon_X \circ d \circ \pi_i = \epsilon_X \circ H_\Sigma \pi_i \circ \rho$ for $i \in \{1, 2\}$. That is, ρ nearly gives R a H_Σ-coalgebra structure except that the paths in the diagram above are coequalized by ϵ_X instead of commuting outright.

$$
\begin{array}{ccccc}
R & \xrightarrow{\ \pi_i\ } & X & \xrightarrow{\ d\ } & H_\Sigma X \\
{\scriptstyle \rho}\downarrow & & & & \downarrow{\scriptstyle \epsilon_X} \\
H_\Sigma R & \xrightarrow[H_\Sigma \pi_i]{} & H_\Sigma X & \xrightarrow[\epsilon_X]{} & FX
\end{array}
$$

Theorem 1. *For all specifications (X, d) for a finitary set functor with presentation (Σ, ϵ), a relation $R \subseteq X \times X$ is an F-bisimulation if and only if it is a bisimulation up to the presentation (Σ, ϵ).*

Proof. ($\Leftarrow$) Let ρ give R the structure of a bisimulation up to the presentation (Σ, ϵ). Then $\epsilon_R \circ \rho$ gives R an F-coalgebra structure such that

$$
\begin{array}{ccccc}
X & \xleftarrow{\ \pi_1\ } & R & \xrightarrow{\ \pi_2\ } & X \\
{\scriptstyle \epsilon_X \circ d}\downarrow & & {\scriptstyle \epsilon_R \circ \rho}\downarrow & & \downarrow{\scriptstyle \epsilon_X \circ d} \\
FX & \xleftarrow[F\pi_1]{} & FR & \xrightarrow[F\pi_2]{} & FX
\end{array}
$$

commutes, so R is an F-bisimulation.

($\Rightarrow$) Given an F-bisimulation structure ϕ on R, we claim $s \circ \phi$ gives a bisimulation up to presentation structure to R where s is any section of ϵ_R. To see this, note $\epsilon_X \circ H_\Sigma \pi_i \circ s \circ \phi = F\pi_i \circ \epsilon_R \circ s \circ \phi = F\pi_i \circ \phi = \epsilon_X \circ d \circ \pi_i$

We note that this statement is not that the F-bisimulation structures are in 1-1 correspondence with the bisimulation up to presentation structures, but that the carriers are in a 1-1 correspondence. Much like the correspondence between coalgebras and specifications, there is a possibly distinct bisimulation up to presentation structure for each section of ϵ_R.

Corollary 1. *The biggest F-bisimulation on a coalgebra is the biggest bisimulation up to the presentation (Σ, ϵ) on that coalgebra.*

We denote bisimilarity, the biggest F-bisimulation on a coalgebra, by $\sim$. The bisimilarity relation is known to be an equivalence relation on all coalgebras for all Set functors preserving weak pullbacks, see e.g. Rutten [9].

3.1   An Explicit Characterization

We defined a bisimulation up to presentation using a variation of the Aczel-Mendler diagram, but there is a more concrete characterization for bisimulations up to presentation which we describe in this section.

For flat terms $\alpha, \beta \in H_\Sigma X$ we write $\alpha =_\epsilon \beta$ to mean $\epsilon_X(\alpha) = \epsilon_X(\beta)$. If $\alpha =_\epsilon \beta$ we say this is an ϵlaw, or that α may be rewritten to β using ϵ laws. Note the $=_\epsilon$ relation on $H_\Sigma X$ is an equivalence relation.

Definition 1 $(c(R))$. The *flat contextual closure* of a relation $R \subseteq X \times X$ is the relation $c(R) \subseteq H_\Sigma X \times H_\Sigma X$ defined by $f(x_1, \ldots x_{ar(f)}) \; c(R) \; f(y_1, \ldots, y_{ar(f)})$ if and only if $x_i \, R \, y_i$ for all $1 \leq i \leq ar(f)$.

We denote the pointwise composition of relations R and S by $R \bullet S$. That is, $x(R \bullet S)z$ iff there exists a y such that xRy and ySz.

Definition 2 $(\sim_R)$. Given a relation R on X, we define $\sim_R$ to be $=_\epsilon \bullet c(R) \bullet =_\epsilon$, a relation on $H_\Sigma X$.

Since $\sim_R$ also depends on the transformation ϵ it would be more proper to denote it $\sim_{R,\epsilon}$, but since ϵ is standard for each functor we elide it from the notation.

Here we also emphasize the distinction between two very similar symbols: $\sim$ denotes bisimilarity on X, and has no direct relationship with the symbol $\sim_R$ just defined.

Theorem 2. *Given a finitary functor F and a specification (X, d) in that functor's signature, $R \subseteq X \times X$ is a bisimulation up to the presentation (Σ, ϵ) if and only if xRy implies $d(x) \sim_R d(y)$. More explicitly, for each $(x, y) \in R$, there is an $f \in \Sigma_n$ and $(x_1, y_1), \ldots, (x_n, y_n) \in R$ such that:*

1. $d(x) =_\epsilon f(x_1, \ldots, x_n)$
2. $d(y) =_\epsilon f(y_1, \ldots, y_n)$

Proof. ($\Rightarrow$) Suppose we have $\rho : R \to H_\Sigma R$ such that $\epsilon_X \circ d \circ \pi_i = \epsilon_X \circ H_\Sigma \pi_i \circ \rho$. Let $(x, y) \in R$ and write $\rho(x, y) = f((x_1, y_1), \ldots, (x_n, y_n))$ where $f \in \Sigma$ and $(x_i, y_i) \in R$. Then $(H_\Sigma \pi_1 \circ \rho)(x, y) = f(x_1, \ldots, x_n)$, so by the hypothesis on ρ, $d(x) =_\epsilon f(x_1, \ldots, x_n)$, as desired. Similarly considering $(H_\Sigma \pi_2 \circ \rho)(x, y)$ yields item 2.

($\Leftarrow$) Suppose we have a relation satisfying the latter condition, and define $\rho : R \to H_\Sigma R$ by $\rho(x, y) = f((x_1, y_1), \ldots, (x_n, y_n))$. Then by item 1, $\epsilon_X \circ d \circ \pi_1 = \epsilon_X \circ H_\Sigma \pi_1 \circ \rho$, and similarly for item 2.

Theorem 2 gives an explicit characterization for bisimulations up to presentation. To check that a relation is a bisimulation up to presentation, for each pair (x, y) in the relation we need to rewrite $d(x)$ and $d(y)$ using ϵ laws so that they have the same symbol and all corresponding variables are related.

We can now show x and q from Example 7 are related by a bisimulation up to presentation.

Example 8. Recall that the zip functor has function symbols $\Sigma = \Sigma_2 = \{0 : \text{zip}, 1 : \text{zip}\}$ with no nontrivial ϵ laws. Then for the specification

$$d(x) = 0 : \text{zip}(y, z) \qquad d(y) = 1 : \text{zip}(x, v) \qquad d(v) = 0 : \text{zip}(w, u)$$
$$d(w) = 0 : \text{zip}(w, z) \qquad d(z) = 1 : \text{zip}(z, w) \qquad d(u) = 1 : \text{zip}(u, v)$$
$$d(q) = 0 : \text{zip}(y, u)$$

we propose $R = \{(x, q), (z, u), (w, v)\} \cup \Delta_X$ as a bisimulation up to presentation. The diagonal part clearly satisfies the required properties. Then

- $0 : \text{zip}(y, z) \sim_R 0 : \text{zip}(y, u)$ since yRy and zRu.
- $1 : \text{zip}(z, w) \sim_R 1 : \text{zip}(u, v)$ since zRu and wRv.
- $0 : \text{zip}(w, z) \sim_R 0 : \text{zip}(w, u)$ since wRw' and zRu.

Since the zip functor has no nontrivial ϵ laws this is just an ordinary H_Σ bisimulation. Matters are more complicated for non-polynomial functors.

Example 9. Recall that the functor $\mathcal{P}_3$ from Example 4 has a presentation with three function symbols, $\{\sigma_i\}_{i<3}$, each with arity i and ϵ laws of the forms $\sigma_2(x, y) =_\epsilon \sigma_2(y, x)$ and $\sigma_2(x, x) =_\epsilon \sigma_1(x)$.

An example specification in this signature for $X = \{x, y, z\}$ might be

$$d(x) = \sigma_2(x, y) \qquad d(y) = \sigma_1(z) \qquad d(z) = \sigma_2(z, z)$$

All of these are behaviourally equivalent, so they should be related by a bisimulation up to presentation. We propose $R = \{(x, y), (y, z), (x, z), (z, z)\}$. For this we need to check four things:

- $\sigma_2(x, y) \sim_R \sigma_1(z)$: we use $\sigma_1(z) =_\epsilon \sigma_2(z, z)$ to rewrite the RHS and note xRz and yRz.
- $\sigma_1(z) \sim_R \sigma_2(z, z)$: uses the same rewrite and zRz twice.
- $\sigma_2(x, y) \sim_R \sigma_2(z, z)$: immediate from xRz and yRz.
- $\sigma_2(z, z) \sim_R \sigma_2(z, z)$: immediate from zRz.

Therefore all three of these variables are related by a bisimulation up to presentation.

3.2 Enhanced Bisimulation Up to Presentation

In Example 8, we showed two variables in a specification were related by a bisimulation up to presentation, but in the course of this proof we added in the diagonal relation to make the bisimulation hypothesis go through. This is reminiscent of other combination bisimulation up to techniques, such as those studied by Rot et al. in [13]. In this section, we provide several enhancements to the bisimulation up to presentation technique which will be useful in the sequel.

First we note bisimulation up to presentation interacts well with union of bisimulations.

Lemma 1. *Suppose (X, d) is a specification for a finitary functor presented by (Σ, ϵ). Let S be any bisimulation on X, T be a relation containing S, and R be a relation on X such that xRy implies $d(x) \sim_T d(y)$. Then $R \cup S$ also has the property that $(x, y) \in R \cup S$ implies $d(x) \sim_T d(y)$.*

Proof. Since S is a bisimulation on X, it is also a bisimulation up to presentation by Theorem 1. Then by Theorem 2, xSy implies $d(x) \sim_S d(y)$. Since $S \subseteq T$, we have $c(S) \subseteq c(T)$ and therefore $\sim_S \subseteq \sim_T$. Hence xSy implies $d(x) \sim_T d(y)$. Combining this with the hypothesis on R, we have the desired result.

Corollary 2. *Recall that $\sim$ is the biggest F-bisimulation on X. If R is a relation on a specification such that any of the following hold:*

- *$xRy \rightarrow d(x) \sim_{R \cup \Delta_X} d(y)$*
- *$xRy \rightarrow d(x) \sim_{R \cup \sim} d(y)$*

then $R \subseteq \sim$.

Proof. Both Δ_X and $\sim$ are bisimulations so by Lemma 1, taking $T = R \cup B$ where $B \in \{\Delta_X, \sim\}$, we get a relation T such that $xTy \rightarrow d(x) \sim_T d(y)$. Then T is a bisimulation up to presentation by Theorem 2, so $R \subseteq T \subseteq \sim$.

Bisimulation up to presentation also behaves well with respect to symmetric closures:

Lemma 2. *Suppose (X, d) is a specification for a finitary functor presented by (Σ, ϵ). Let T be any symmetric relation and R be any relation on X such that xRy implies $d(x) \sim_T d(y)$. Then $s(R)$, the symmetric closure of R, also has the property $x \, s(R) \, y$ implies $d(x) \sim_T d(y)$.*

Proof. It is easy to check that T symmetric implies $c(T)$ symmetric, which in turn implies $\sim_T$ symmetric. Then if $x \, s(R) \, y$, either xRy or yRx by definition of $s(R)$. The hypothesis on R yields $d(x) \sim_T d(y)$ or $d(y) \sim_T d(x)$, respectively. Then $\sim_T$ symmetric allows us to conclude that in either case $d(x) \sim_T d(y)$, as desired.

Corollary 3. *If R is a relation such that $xRy \to d(x) \sim_{s(R)} d(y)$, then $R \subseteq \sim$.*

Bisimulation up to presentation for functors *preserving weak pullbacks* also plays well with equivalence closures. Preservation of weak pullbacks is a critical assumption here. We recall the following definition and theorem from [14, p. 14], slightly recast to use our notation:

Definition 3. A presentation is *dominated* if for every ϵ law $f(\boldsymbol{x}) =_\epsilon g(\boldsymbol{y})$ where $f \in \Sigma_n$, $g \in \Sigma_m$, $\boldsymbol{x} : n \to X$, $\boldsymbol{y} : m \to X$, there is a symbol $h \in \Sigma_k$ and functions $u : k \to n$ and $v : k \to m$ such that $h(u) =_\epsilon f(id_n)$, $h(v) =_\epsilon g(id_m)$, and $\boldsymbol{x} \circ u = \boldsymbol{y} \circ v$.

To give some intuition for dominated presentations, consider the presentation for $\mathcal{P}_\omega$ from Example 3. We could say the symbol σ_4 dominates the symbol σ_2 since there is a $u : 4 \to 2$, namely $u(i) = \lfloor \frac{i}{2} \rfloor$, such that $\sigma_4(u) = \sigma_4(0,0,1,1) =_\epsilon \sigma_2(0,1) = \sigma_2(id_2)$. Then any time we have a term using σ_2, we could replace σ_2 with σ_4, following the substitution scheme hinted at by u, and remain in the same component of the kernel of ϵ. So, for example, this domination would imply $\sigma_2(x,y) =_\epsilon \sigma_4(x,x,y,y)$.

Note σ_3 is also dominated by σ_4, for example by $\sigma_3(0,1,2) = \sigma_4(0,2,1,1)$. This would allow us to rewrite $\sigma_3(x,y,x) =_\epsilon \sigma_4(x,x,y,y)$. Combining with the previous paragraph would allow us to derive $\sigma_3(x,y,x) = \sigma_2(x,y)$ via rewrites to σ_4. We can then say that $\sigma_3(x,y,x) =_\epsilon \sigma_2(x,y)$ is a consequence of the joint domination of σ_2 and σ_3 by σ_4.[5]

The verbiage for all this notation then is that a presentation is dominated means for every ϵ law $f(\boldsymbol{x}) =_\epsilon g(\boldsymbol{y})$ there is a dominating symbol h with two variable substitutions u and v such that the ϵ law is a consequence of the joint domination of f and g by h via the substitutions u and v. We will be relying on facts about dominated presentations only in the proof of Lemma 3.

Theorem 3 (Adámek, Gumm, Trnková). *A finitary functor weakly preserves pullbacks if and only if it has a dominated presentation.*

As a result of this theorem, for the next lemmas we can assume our presentation is dominated without loss of generality.

Lemma 3. *Suppose T is an equivalence relation and (Σ, ϵ) is a dominated presentation. Then $\sim_T$ is an equivalence relation.*

Proof. Reflexivity and symmetry are straightforward. T is reflexive and symmetric, therefore $c(T)$ is reflexive and symmetric, and so $=_\epsilon \bullet c(T) \bullet =_\epsilon$ is reflexive and symmetric.

---

[5] Obviously, the joint domination is not unique in the case of $\mathcal{P}_\omega$. σ_2 and σ_3 are jointly dominated by σ_i for all $i \geq 3$ and even for each dominating symbol there may be many different substitutions which yield the desired equation as a consequence of the joint domination.

Transitivity requires the dominated presentation. Suppose $\alpha \sim_T \beta \sim_T \gamma$. Then we can write

$$\alpha =_\epsilon \alpha' \; c(T) \; \beta' =_\epsilon \beta =_\epsilon \beta'' \; c(T) \; \gamma' =_\epsilon \gamma$$

Let $\beta' = f(\boldsymbol{x})$ and $\beta'' = g(\boldsymbol{y})$. Then the above relations become:

$$\alpha =_\epsilon f(\boldsymbol{x'}) \; c(T) \; f(\boldsymbol{x}) =_\epsilon g(\boldsymbol{y}) \; c(T) \; g(\boldsymbol{y'}) =_\epsilon \gamma$$

Since we have a dominated presentation, we get h, u, and v such that $h(u) =_\epsilon f(id_n)$, $h(v) =_\epsilon g(id_m)$ and $\boldsymbol{x} \circ u = \boldsymbol{y} \circ v$. The first two statements imply $f(\boldsymbol{x'}) =_\epsilon h(\boldsymbol{x'} \circ u)$ and $g(\boldsymbol{y'}) =_\epsilon h(\boldsymbol{y'} \circ v)$.

We also know $\boldsymbol{x'}(i) \; T \; \boldsymbol{x}(i)$ for all $i \in [1, n]$ and $\boldsymbol{y}(i) \; T \; \boldsymbol{y'}(i)$ for $i \in [1, m]$. Therefore, $(\boldsymbol{x'} \circ u)(i) \; T \; (\boldsymbol{x} \circ u)(i) = (\boldsymbol{y} \circ v)(i) \; T \; (\boldsymbol{y'} \circ v)(i)$ for $i \in [1, k]$, where the middle equality is by the last condition guaranteed by the dominated presentation. Then since T is transitive we know $(\boldsymbol{x'} \circ u)(i) \; T \; (\boldsymbol{y'} \circ v)(i)$ for $i \in [1, k]$.

Therefore we have produced terms such that

$$\alpha =_\epsilon h(\boldsymbol{x'} \circ u) \; c(T) \; h(\boldsymbol{y'} \circ v) =_\epsilon \gamma$$

and so $\sim_T$ is transitive and hence is an equivalence relation.

Lemma 4. *Suppose (X, d) is a specification for a finitary functor preserving weak pullbacks with the dominated presentation (Σ, ϵ). Let T be any equivalence relation on X and R be any relation on X such that xRy implies $d(x) \sim_T d(y)$. Then $e(R)$, the equivalence closure of R, has the property $x \; e(R) \; y$ implies $d(x) \sim_T d(y)$.*

Proof. By Lemmas 1 and 2, we know immediately that $x \; sr(R) \; y$ implies $d(x) \sim_T d(y)$, where $sr(R)$ is the symmetric reflexive closure of R. Hence we only have to consider $(x, y) \in e(R) \smallsetminus sr(R)$. Therefore, suppose we have $x \; sr(R) \; z \; sr(R) \; y$. Then by the noted property of $sr(R)$ we get $d(x) \sim_T d(z) \sim_T d(y)$. By the previous Lemma, since T is an equivalence relation and we have a dominated presentation, $\sim_T$ is an equivalence relation and hence $d(x) \sim_T d(y)$.

Corollary 4. *Suppose F preserves weak pullbacks and its presentation (Σ, ϵ) is dominated. If R is a relation such that xRy implies $d(x) \sim_{e(R)} d(y)$, then $R \subseteq \sim$.*

The following corollary follows directly from the results above, but is of critical importance to our proof of soundness.

Corollary 5. *Suppose F preserves weak pullbacks and its presentation (Σ, ϵ) is dominated. If R is a relation such that xRy implies $d(x) \sim_{e(R \cup \sim)} d(y)$, then $R \subseteq \sim$.*

4   A Proof System for Bisimulation Up to Presentation

In this section, we outline a formal proof system to capture the notion of bisimulation up to presentation. For this whole section, we assume F preserves weak pullbacks. We then prove this system to be sound and complete.

Our system has judgements of the form $R \vdash \sigma = \tau$ where $R \subseteq X \times X$ and $(\sigma, \tau) \in X \times X + H_\Sigma X \times H_\Sigma X$. The inference rules are as follows:

$$\frac{}{R \vdash \sigma = \sigma} \; r \qquad\qquad\qquad \frac{R \vdash \tau = \sigma}{R \vdash \sigma = \tau} \; s$$

$$\frac{\alpha =_\epsilon \beta}{R \vdash \alpha = \beta} \; \epsilon \qquad\qquad \frac{R \vdash \sigma = \tau \quad R \vdash \tau = \rho}{R \vdash \sigma = \rho} \; t$$

$$\frac{}{\{\varphi\} \cup R \vdash \varphi} \; a \qquad \frac{R \vdash x_1 = y_1 \quad \cdots \quad R \vdash x_{ar(f)} = y_{ar(f)}}{R \vdash f(x_1, \ldots, x_{ar(f)}) = f(y_1, \ldots, y_{ar(f)})} \; c$$

$$\frac{R \vdash \varphi \quad \forall (x, y) \in R.R \vdash d(x) = d(y)}{\vdash \varphi} \; b$$

As usual, we say $R \vdash \varphi$ when there is a proof tree using the above rules with the judgement $R \vdash \varphi$ as the root. The notation $\vdash \varphi$ is shorthand for $\varnothing \vdash \varphi$. Recall that $\vDash x = y$ means that x and y are behaviourally equivalent (have the same image in the final coalgebra).

We should point out that R on the left side of the turnstile does not have the usual force of a full assumption. Rather, this R should be thought of as tracking unverified bisimulation hypotheses. In most rules this unverified bisimulation remains on both sides, except the axioms and the b rule, which essentially discharges a verified bisimulation from the left hand side. This b rule is probably the least intuitive, but is really just the coinductive proof principle. We also note its similarity to the Recursion Inference Rule from FLR_0, which was in mind as the system was constructed. [15]

Before we prove soundness and completeness, we give two example proofs using the system. This first example is based on Example 4.2 in Moss et al. [3].

Example 10. Consider the specification on $X = \{x, y, r, s\}$ for $\mathcal{P}_3$ defined by

$$d(x) = \sigma_2(x, y) \qquad d(y) = \sigma_0 \qquad d(r) = \sigma_2(r, s) \qquad d(s) = \sigma_0$$

Let $R = \{(x, r), (y, s)\}$. The proof tree below witnesses $\vdash x = r$.

$$\frac{\dfrac{}{R \vdash x = r} \; a \quad \dfrac{\dfrac{}{R \vdash x = r} \; a \quad \dfrac{}{R \vdash y = s} \; a}{R \vdash \sigma_2(x, y) = \sigma_2(r, s)} \; c \quad \dfrac{}{R \vdash \sigma_0 = \sigma_0} \; c}{\vdash x = r} \; b$$

The next example is adapted from Example 9 in this paper and showcases how some rules allow for shorter proofs.

Example 11. Consider the specification on $X = \{x, y\}$ for $\mathcal{P}_3$ defined by

$$d(x) = \sigma_2(x, y) \qquad\qquad d(y) = \sigma_1(x)$$

Let $R = \{(x, y)\}$. The proof tree below witnesses $\vdash x = y$.

$$
\cfrac{
\cfrac{
\cfrac{\overline{R \vdash x = x}\ r \quad \cfrac{\cfrac{R \vdash x = y}{R \vdash y = x}\ s}{}}{R \vdash \sigma_2(x, y) = \sigma_2(x, x)}\ c \quad \cfrac{R \vdash \sigma_2(x, x) = \sigma_1(x)}{}\ \epsilon
}{
\overline{R \vdash x = y}\ a \qquad \cfrac{R \vdash \sigma_2(x, y) = \sigma_1(x)}{}\ t
}{}
}{\vdash x = y}\ b
$$

Note that R contains a single pair though an unenhanced bisimulation proof would require three: the r rule allows us to omit (x, x), and the s rule elides the mirror-image proof that $d(y) = d(x)$ thereby allowing us to omit (y, x).

4.1   Soundness

To help with our soundness proof, we define a new relational closure which we call the *presentational closure* of a relation $R \subseteq X \times X$ on the carrier of a specification (X, d). Recall from Sect. 3 that $\sim_R$ is defined to be $=_\epsilon \bullet c(R) \bullet =_\epsilon$. That is, $\alpha \sim_R \beta$ if we can rewrite α and β using ϵ laws so they have the same function symbol and all corresponding variables are related by R. The presentational closure of R, denoted $pr(R)$, is defined to be $pr(R) \triangleq e(R \cup \sim) + \sim_{e(R \cup \sim)}$, a relation on $X + H_\Sigma X$.

We note that $e(R \cup \sim)$ is an equivalence relation on X, and $\sim_{e(R \cup \sim)}$ is an equivalence relation on $H_\Sigma X$ as a consequence. Since X and $H_\Sigma X$ are disjoint, $pr(R)$ is also an equivalence relation.

Proposition 1. *If $R \vdash \varphi$, then $\varphi \in pr(R)$.*

Proof. By induction on the proof tree. The base cases are r, ϵ, and a. We know $pr(R)$ is an equivalence relation, hence r. The relation $e(R \cup \sim)$ is reflexive, hence $\sim_{e(R \cup \sim)}$ contains the relation $=_\epsilon$, hence ϵ. All pairs in R are included in $e(R \cup \sim)$, hence a.

The induction steps are s, t, c, and b. s and t follow easily from the fact that $pr(R)$ is an equivalence relation.

Suppose for all $1 \leq i \leq ar(f)$, we know $(x_i, y_i) \in pr(R)$ and hence $(x_i, y_i) \in e(R \cup \sim)$. By definition of the flat contextual closure, $f(x_1, \ldots, x_{ar(f)})$ $c(e(R \cup \sim))$ $f(y_1, \ldots, y_{ar(f)})$. Since $=_\epsilon$ is a reflexive relation, $pr(R)$ contains $c(R)$ and hence these two terms are related by $pr(R)$. Therefore the induction holds across the c rule.

Finally we consider the b rule, which allows one to discharge bisimulations from the left hand side. The induction hypothesis gives $\varphi \in pr(R)$ and $(d(x), d(y)) \in pr(R)$ for each $(x, y) \in R$. Since $d(x), d(y) \in H_\Sigma X$ we know $d(x) \sim_{e(R \cup \sim)} d(y)$. Then by Corollary 5, we know $R \subseteq \sim$. Therefore, $pr(R) = e(R \cup \sim) + \sim_{e(R \cup \sim)} = e(\sim) + \sim_{e(\sim)} = pr(\varnothing)$. Then $\varphi \in pr(\varnothing)$ since $\varphi \in pr(R)$.

Corollary 6 (Soundness). *If $\vdash x = y$, then $\vDash x = y$.*

Proof. If $\vdash x = y$, then $(x, y) \in e(\sim \cup \varnothing)$ by the previous proposition. However, clearly $e(\sim \cup \varnothing) = e(\sim) = \sim$, so $x \sim y$. A standard fact about functors preserving weak pullbacks is that two states in a coalgebra are bisimilar if and only if they are behaviourally equivalent [9], so $x \sim y$ implies $\vDash x = y$.

4.2 Completeness

Lemma 5. *If $\alpha \sim_R \beta$, then $R \vdash \alpha = \beta$.*

Proof. $\alpha \sim_R \beta$ means there are $f \in \Sigma_n$ and $(x_1, y_1), \ldots, (x_n, y_n) \in R$ such that

$$\alpha =_\epsilon f(x_1, \ldots, x_n) \; c(R) \; f(y_1, \ldots, y_n) =_\epsilon \beta.$$

then

$$
\cfrac{
  \cfrac{
    \overline{R \vdash \alpha = f(x_1, \ldots, x_n)}\;\epsilon
    \qquad
    \cfrac{\overline{R \vdash x_1 = y_1}\;a \;\cdots\; \overline{R \vdash x_n = y_n}\;a}{R \vdash f(x_1, \ldots, x_n) = f(y_1, \ldots, y_n)}\;c
  }{R \vdash \alpha = f(y_1, \ldots, y_n)}\;t
  \qquad
  \cfrac{}{R \vdash f(y_1, \ldots, y_n) = \beta}\;\epsilon
}{R \vdash \alpha = \beta}\;t
$$

is a witness for $R \vdash \alpha = \beta$.

Corollary 7 (Completeness). *If $\vDash x = y$, then $\vdash x = y$.*

Proof. Recall that $\vDash x = y$ iff $x \sim y$. Since x and y are bisimilar, they are related by a bisimulation up to presentation, which we call R. By Theorem 2, $uRv \to d(u) \sim_R d(v)$. Syllogizing with the previous lemma yields $uRv \to (R \vdash d(u) = d(v))$. Trivially, $R \vdash x = y$ by the a rule. Therefore, $\vdash x = y$ by the b rule.

5 Compositionality of Presentations

Bisimulations up to presentation give a uniform way to reason about behavioural equivalence of variables in specifications for finitary functors, but there is a potential disadvantage. Certain inductive classes of functors have sound, complete, and **compositional** proof systems. That is, the rules for reasoning about coalgebras of a functor are built inductively in a manner corresponding to the definition of the functor. The prime example of this situation is that of the polynomial functors and the Kripke polynomial functors.

A functor is called *polynomial* if it is generated by the following BNF grammar:

$$P ::= A \mid Id \mid P + P \mid P \times P \mid P^B$$

where A is the constant functor having value $A \in Set$ and B is a finite set. The *Kripke polynomial* class of functors adds the finite powerset functor:

$$K:: = A \mid Id \mid P_\omega(K) \mid K + K \mid K \times K \mid K^B$$

Bonsangue et al. build a sound, complete and compositional expression calculus to represent coalgebras of Kripke polynomial functors in [5]. We show presentations are similarly compositional, in that both the signature and the ϵ transformation can be built inductively to parallel the construction of the functor. For the following three constructions, suppose F and G are finitary functors with presentations (Σ, ϵ) and (Σ', ϵ').

5.1 Products

Let $J = F \times G$. Then we claim J has a presentation (Σ'', ϵ'') where Σ'' has all pairs of symbols, $\Sigma''_n = \{(f,g) : f \in \Sigma, g \in \Sigma', ar(f) + ar(g) = n\}$, and $\epsilon'' : H_{\Sigma''} \to J$ has components

$$\epsilon''_X : (f,g)(x_1, \ldots, x_n) \mapsto (\epsilon_X(f(x_1, \ldots, x_{ar(f)})), \epsilon'_X(g(x_{ar(f)+1}, \ldots, x_n))).$$

Then ϵ'' is an epic natural transformation as a consequence of ϵ and ϵ' being epic natural transformations.

We single out this particular presentation because it allows us to state the ϵ'' laws in terms of ϵ and ϵ' laws. By definition $\epsilon''((f,g)(x_1, \ldots, x_n)) = \epsilon''((f',g')(y_1, \ldots y_m))$ means

$$\epsilon(f(x_1, \ldots, x_{ar(f)})) = \epsilon(f'(y_1, \ldots, y_{ar(f')})) \text{ and}$$
$$\epsilon'(g(x_{ar(f)+1}, \ldots, x_n)) = \epsilon'(g'(y_{ar(f')+1}, \ldots, y_m)).$$

Therefore, ϵ'' laws in this presentation are pairs of ϵ and ϵ' laws.

We note here that we could represent finite powers with a similar construction. If B is a finite set, a signature for F^B has symbols $|B|$ tuples of symbols from Σ with arity the sum of the arities through the tuple. Then the ϵ'' laws are $|B|$ tuples of ϵ laws.

5.2 Coproducts

Let $J = F + G$. We write $S + T = \{inl s : s \in S\} \cup \{inr t : t \in T\}$. Then J has a presentation (Σ'', ϵ'') where $\Sigma''_n = \Sigma_n + \Sigma'_n$ and ϵ'' has components ϵ''_X such that

$$\begin{cases} \epsilon''_X(inl f(\boldsymbol{x})) = inl \epsilon_X(f(\boldsymbol{x})) \\ \epsilon''_X(inr g(\boldsymbol{x})) = inr \epsilon'_X(g(\boldsymbol{x})) \end{cases}.$$ Since ϵ and ϵ' are epic natural transformations, ϵ'' is also an epic natural transformation.

Again, we can state the ϵ'' laws in terms of the ϵ and ϵ' laws. By definition $\epsilon''\alpha = \epsilon''\beta$ means α and β are both labelled inl or are both labelled inr. In the former case, we have $inl \epsilon(f(\boldsymbol{x})) = inl \epsilon(g(\boldsymbol{y}))$, which is an inl-labelled instance of an ϵ-law. Similarly, the latter case gives an inr-labelled instance of an ϵ'-law.

5.3   Compositions

Let $J = G \circ F$. Then J has a presentation with symbols from the set $\Sigma'' = \{(\sigma', (\sigma_1, \ldots, \sigma_{ar(\sigma')})) : \sigma' \in \Sigma' \text{ and } \sigma_i \in \Sigma\}$. For each symbol $\sigma'' \in \Sigma''$ define $w_{\sigma''}(i) = \sum_{j=1}^{i} ar(\sigma_i)$ for $0 \leq i \leq ar(\sigma')$ and define σ'' to have arity $w_{\sigma''} = w_{\sigma''}(ar(\sigma'))$. Given an $ar(\sigma'')$ tuple from X, we let $\boldsymbol{x}_i = (x_{w_{\sigma''}(i-1)+1}, \ldots, x_{w_{\sigma''}(i)})$ for $1 \leq i \leq ar(\sigma')$, the slice of the variables corresponding to σ_i. The natural transformation ϵ'' has components given by

$$\epsilon'' : \sigma''(x_1, \ldots, x_{ar(\sigma'')}) \mapsto \epsilon'(\sigma'(\epsilon(\sigma_1(\boldsymbol{x}_1)), \ldots, \epsilon(\sigma_{ar(\sigma')}(\boldsymbol{x}_{ar(\sigma'')})))).$$

This is an epic natural transformation and the ϵ'' laws can be stated again in terms of the ϵ and ϵ' laws.

5.4   Kripke Polynomial Functors and Other Polynomial-Like Classes of Functors

We have presentations for constant functors (Example 1), the identity functor (Example 2) and the finite power set functor (Example 3), which means the above constructions give a compositional presentation for each of the Kripke polynomial functors. Due to the previous section, we know bisimulation up to those presentations is a sound and complete proof system.

Example 12. Consider the functor $F = A \times \mathrm{Id}^B$ where B is finite. Following the constructions above, F has a presentation with function symbols $\Sigma = \Sigma_{|B|} = \{(a, \beta) : a \in A, \beta : B \to \{*\}\}$ and equations $\epsilon((a, \beta)(\boldsymbol{x})) = \epsilon((a', \beta')(\boldsymbol{x}'))$ iff $a = a'$ and $\boldsymbol{x} = \boldsymbol{x}'$. Since there is only one function $\beta : B \to \{*\}$, we might as well omit that part of the function symbol and abbreviate (a, β) by just a, still with arity $|B|$.

Suppose we had a specification in this signature, like

$$d(x) = a(\overbrace{x, \ldots, x}^{|B|}) \qquad d(y) = a(\overbrace{z, \ldots, z}^{|B|}) \qquad d(z) = a(\overbrace{y, \ldots, y}^{|B|})$$

The complete relation on $\{x, y, z\}$ is a bisimulation up to presentation for this specification, since for each pair of variables, all corresponding variable pairs in the definitions are in the relation. Hence we can conclude these variables all have the same image in the final coalgebra.

The constructions above generalize previous results about inductive classes of functors since they assure us a compositional proof system exists for any inductive class using any of those formation rules.

6   Conclusion and Future Directions

In this paper, we presented a sequent-style deduction system for reasoning about behavioural equivalence of points in coalgebras and specifications of finitary functors in Set. This system was based on a relaxed version of H_Σ-bisimulations which nicely coincide with F-bisimulations called bisimulations up to presentation. We demonstrated this proof system was sound and complete for finitary functors preserving weak pullbacks. We also demonstrated that three common operations on functors have uniform effects on both the signature and equations in a presentation.

One restriction in our setting we would like to remove is the totality restriction. Since our specifications are defined with a (total) function, there is exactly one related coalgebra and each variable has exactly one interpretation in a final coalgebra. This makes strong soundness and completeness results decidedly less satisfying—either the assumptions are true or false of the single model of the specification. By removing the totality restriction, we could get more meaningful strong soundness and completeness.

Another more practical advantage of partial specifications is the ability to detect equality even in circumstances where values of certain variables are not known or are irrelevant. For example, from

$$d(x) = 1 : \mathrm{zip}(y, z) \qquad\qquad d(y) = 1 : \mathrm{zip}(x, z)$$

we would like to be able to conclude $\vDash x = y$ even if the value of z is not specified.

We implicitly used the fact that finitary functors have *flat* presentations. That is, each point in FX is an image of a point in $H_\Sigma X$ and all equations necessary for the presentation are between flat terms. Flat signatures are not always the most natural though, sometimes one would like to use zip terms like $0 : \mathrm{zip}(1 : x, \mathrm{zip}(y, y))$, where we have three function symbols: $\Sigma_1 = \{0 :, 1 :\}$ and $\Sigma_2 = \{\mathrm{zip}\}$. Then some non-flat equations are necessary to capture all the truths of the system. How many more modifications are necessary to deal successfully with specifications in $X \to T_\Sigma X$ instead of $X \to H_\Sigma X$?

We are also interested in contexts beyond Set, particularly Vect. Milius [6] extended the expression calculi of Bonsangue et al. [4,5] to vector space coalgebras of the functor $FX = \mathbb{R} \times X$, providing a sound and complete system for reasoning about stream circuits. We have some hope that by combining our approach with the general definition of signature from Kelly and Power [16] we might be able to devise a system usable in more categories than Set.

There are also a couple of generic questions suggested by this work. Here we utilized the quotient relationship $\epsilon : H_\Sigma \twoheadrightarrow F$ to relate H_Σ- and F-bisimulations. Do other relationships between functors yield interesting interplay between their coalgebras? Additionally, many of the results here suggest a connection to bisimulation up to literature. How does bisimulation up to presentation fit into the theory of enhanced bisimulations?

Acknowledgements. I owe thanks to Alexandra Silva for helpful conversations at WoLLIC at the start of this project and to Larry Moss for his expertise and encouragement throughout. Thanks also to the anonymous referees for their careful reading and thoughtful comments.

References

1. Adámek, J., Milius, S., Moss, L.S.: On finitary functors and their presentations. In: Pattinson, D., Schröder, L. (eds.) CMCS 2012. LNCS, vol. 7399, pp. 51–70. Springer, Heidelberg (2012)
2. Moss, L.S.: Recursion and corecursion have the same equational logic. Theoret. Comput. Sci. **294**(1), 233–267 (2003)
3. Moss, L.S., Wennstrom, E., Whitney, G.T.: A complete logical system for the equality of recursive terms for sets. In: Constable, R.L., Silva, A. (eds.) Logic and Program Semantics, Kozen Festschrift. LNCS, vol. 7230, pp. 180–203. Springer, Heidelberg (2012)
4. Bonsangue, M., Rutten, J., Silva, A.: A Kleene theorem for polynomial coalgebras. In: de Alfaro, L. (ed.) FOSSACS 2009. LNCS, vol. 5504, pp. 122–136. Springer, Heidelberg (2009)
5. Bonsangue, M., Rutten, J., Silva, A.: An algebra for Kripke polynomial coalgebras. In: 24th Annual IEEE Symposium on Logic in Computer Science, LICS 2009, pp. 49–58. IEEE (2009)
6. Milius, S.: A sound and complete calculus for finite stream circuits. In: 2010 25th Annual IEEE Symposium on Logic in Computer Science (LICS), pp. 421–430. IEEE (2010)
7. Silva, A.M.: Kleene coalgebra. PhD thesis, CWI (2010)
8. Silva, A.M.: Marcello Maria Bonsangue, and Jan JMM Rutten. Kleene coalgebras. CWI. Software Engineering [SEN] (2010)
9. Rutten, J.J.M.M.: Universal coalgebra: a theory of systems. Theoret. Comput. Sci. **249**(1), 3–80 (2000)
10. Mac Lane, S.: Categories for the Working Mathematician, vol. 5. Science & Business Media, Berlin (1978)
11. Kupke, C., Rutten, J.J.M.M.: On the final coalgebra of automatic sequences (2011)
12. Grabmayer, C., Endrullis, J., Hendriks, D., Klop, J.W. Moss, L.S.: Automatic sequences and zip-specifications. In: Proceedings of the 2012 27th Annual IEEE/ACM Symposium on Logic in Computer Science, pp. 335–344. IEEE Computer Society (2012)
13. Rot, J., Bonchi, F., Bonsangue, M., Pous, D., Rutten, J.J.M.M., Silva, A.: Enhanced coalgebraic bisimulation. Math. Struct. Comput. Sci. (to appear, 2014)
14. Adámek, J., Gumm, H.P., Trnková, V.: Presentation of set functors: a coalgebraic perspective. J. Logic Comput. **20**(5), 991–1015 (2010)
15. Hurkens, A.J.C., McArthur, M., Moschovakis, Y.N., Moss, L.S., Whitney, G.T.: The logic of recursive equations. J. Symbolic Logic **63**(02), 451–478 (1998)
16. Kelly, G.M., Power, A.J.: Adjunctions whose counits are coequalizers, and presentations of finitary enriched monads. J. Pure Appl. Algebra **89**(1), 163–179 (1993)

Coalgebraic Completeness-via-Canonicity

Principles and Applications

Fredrik Dahlqvist[(✉)]

University College London, London, UK
f.p.h.dahlqvist@gmail.com

Abstract. We present the technique of completeness-via-canonicity in a coalgebraic setting and apply it to both positive and boolean coalgebraic logics with relational semantics.

1 Introduction

Coalgebraic logic has been very successful at unifying the multitude of modal logics used to describe and specify state-based systems, both semantically and syntactically (see e.g. [CKP+09,KP11]). One of the great insights of coalgebraic logics is that there exists a close correspondence between the coalgebraic semantics and *rank 1 axiomatizations*, i.e. axioms with nesting depth of modal operators uniformly equal to 1. For a **Set**-endofunctor T the class of *all* T-coalgebras can be characterised logically in rank 1 (see [Sch06]). Conversely, given a modal logic axiomatized in rank 1, there exist a **Set**-endofunctor T such that the logic is strongly complete with respect to the class of all T-coalgebras (see [SP10]).

However, one is often interested in providing a sound and complete semantics to modal logics which are known to include axioms of rank greater than one. Most temporal logics for example (see [Gol92]) contain such axioms. Alternatively, one may have a rank 1 axiomatization of the class of T-coalgebras for a functor T of particular interest, and be interested in logically carving out important proper sub-classes of T-coalgebras, which may very well require axioms with nested modalities, for example the axiomatization of transitive Kripke frames by the axiom $\Diamond\Diamond p \to \Diamond p$.

Very little is known about the question of completeness for coalgebraic logics with axioms of arbitrary rank. To our knowledge, the only results in this direction are the work of Pattinson and Schröder in [PS08] as well as our previous work in [DP13] which dealt with the ∇ formalism of coalgebraic logic and [DP15a] which focused on a coalgebraic account of distributive substructural logics. In what follows we will present the general principles of *coalgebraic completeness-via-canonicity*, a method for proving strong completeness of coalgebraic logics with axioms of arbitrary rank, in as much abstraction and generality as possible. To this end we will use the abstract presentation of coalgebraic logic (see

© IFIP International Federation for Information Processing 2016
Published by Springer International Publishing Switzerland 2016. All Rights Reserved
I. Hasuo (Ed.): CMCS 2016, LNCS 9608, pp. 174–194, 2016.
DOI: 10.1007/978-3-319-40370-0_11

e.g. [KKP04, KKP05, KP11, JS10]) which can be summarized by the following *fundamental diagram*:

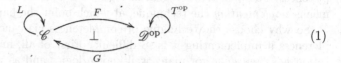

$$(1)$$

where $\mathscr{C}$ is the category in which 'modal formulas' are built from the functor L and interpreted in T-coalgebras over 'carriers' in $\mathscr{D}$. We must however be careful: in this abstract formulation coalgebraic logic is extremely general indeed, and the notion of *canonical extension* (and thus of *canonicity*) does not in general make sense in the base category $\mathscr{C}$ on which the logic is defined. We therefore need to restrict our attention to base categories whose objects have a notion of canonical extension. This presents us with a conceptual restriction which in practice is harmless since all examples of coalgebraic logics are based on a category with a good notion of canonical extension, viz. the categories **BA** of boolean algebras, **DL** or distributive lattice, or **MSL** of meet semilattices (see [JS10] for an example of meet semilattice-based coalgebraic logic, and see [GP14] for a discussion of canonical extensions in **MSL**). A second restriction comes from the fact that canonicity must spontaneously appear from the diagram above, in the sense that for any $\mathscr{C}$-object A, the canonical extension of A must be representable as GFA. This condition is more restrictive. In the case of boolean logics, this poses no problem: if we take $\mathscr{D} = \mathbf{Set}$ and the usual adjunction $F = \mathsf{Uf} \dashv \mathcal{P} = G$ between the ultrafilter and powerset functors then $\mathcal{P}\mathsf{Uf}A$ is indeed the canonical extension of A. However, in the case of positive coalgebraic logics this requirement precludes the use of **Set**-based models for positive coalgebraic logics, and we have to take $F = \mathsf{Pf} : \mathbf{DL} \to \mathbf{Pos}^{\mathrm{op}}$ the prime filter functor and $G = \mathcal{U} : \mathbf{Pos}^{\mathrm{op}} \to \mathbf{DL}$ the upsets functor to represent the canonical extension of a distributive lattice A as $\mathcal{U}\mathsf{Pf}A$. The situation for **MSL**-based logics is much more involved. The canonical extension of general semilattices is described in [GP14], however no adjunction $F \dashv G$ between **MSL** and a category $\mathscr{D}^{\mathrm{op}}$ emerges as a 'natural' way of building it. A duality theory for *distributive* meet-semilattices is given in [BJ11]. It consists in building a distributive lattice $D(A)$ from a distributive semilattice A and then applying the usual functor Pf. It follows from the construction of [BJ11] that $F = \mathsf{Pf} \circ D \dashv U \circ \mathcal{U} = G$ (where U is the obvious forgetful functor), but we have not investigated if GFA is then the canonical extension of A as described in [GP14]. We will therefore restrict our attention to logics based on the category **DL** of which boolean logics (based on **BA**) are a special case.

Having established the scope of logics which coalgebraic completeness-via-cano-nicity can hope to tackle *a priori*, we must make the following remark about what can be achieved *in practice*. Questions of canonicity are very hard; in general it is undecidable whether a given formula is canonical, and establishing that a particular class of formulas is canonical is almost always highly non-trivial. What we will present in this paper is a general coalgebraic template which avoids these hard questions altogether, a conceptual roadmap of how the technique of completeness-via-canonicity works in coalgebraic logic. To *actually* prove completeness of a *particular* coalgebraic logic with a *particular* coalgebraic semantic means *implementing* the technique, at which point the hard work begins.

So why choose the technique of completeness-via-canonicity to prove completeness if implementing it is so difficult? First of all, much of the implementation *has been done* for many well-known logics and as we will show, we now have a complete theory for all positive or boolean logics with a relational semantics. Secondly, because of all the methods for proving completeness in modal logic, it is probably the best suited to being generalized to coalgebraic logic since it has a very clean and abstract algebraic formulation which connects in a very generic fashion to the coalgebraic semantics via the well-established coalgebraic Jónsson-Tarski theorem (see [KKP05,KR12,SP09]). Moreover, we believe that generalising completeness-via-canonicity to coalgebraic logics also greatly clarifies the technique itself. The connection between the syntactic/algebraic part of the method on one side, viz. canonical extensions and canonical equations, and the semantics/coalgebraic part of the method on the other side, viz. the construction of 'canonical models', is greatly clarified by the abstracting power of coalgebraic logics and its semantics. Another advantage of coalgebraic completeness-via-canonicity is that it applies equally well to *positive* coalgebraic logics (see [KKV12]). In fact, since the traditional boolean setting is a special case of the more general setting of positive coalgebraic logics, we will formulate most results in terms positive coalgebraic logics. A final advantage of the technique its *modularity*: we can combine strongly complete logics to create new strongly complete logics in a completely mechanical way (in the spirit of [CP07,DP11]). This work is in many ways a continuation and generalisation of the author's previous work with his PhD supervisor Dirk Pattinson [DP13]. The paper will be structured as follows. We start by presenting coalgebraic logics in its 'abstract' flavour. In Sect. 3 we describe the semantics/coalgebraic side of completeness-via-canonicity, whilst Sect. 4 will deal with the syntactic/algebraic side of the technique. Section 5 will show how and when the algebraic and coalgebraic halves of the method can be combined, and strong completeness proved. We will use the example of (positive) modal logic to illustrate every important concept, and conclude with an application to 'positive separation logics'.

2   Preliminaries

Coalgebraic logics require seven mathematical entities, six of which we introduced in the *fundamental diagram* (1). These six entities are:

(1) a *'minimal reasoning structure'* in the form of a category $\mathscr{C}$ whose objects are endowed with the fundamental logical operations we wish to take for granted. Due to the algebraic nature of $\mathscr{C}$-objects, we will assume throughout that there exist a free-forgetful adjunction $\mathsf{F} \dashv \mathsf{U}$

between **Set** and $\mathscr{C}$ (note the sans-serif font for the free functor). We will take $\mathscr{C}$ to be **DL, BDL** or **BA**, the categories of distributive lattices, bounded distributive lattices or boolean algebras with the obvious morphisms.

(2) a *'minimal modelling structure'* in the form of a category $\mathscr{D}$ whose objects have the structure we wish the carriers of models to have. In our examples we will take $\mathscr{D}$ to be either **Pos** or **Set**, the category of posets and monotone functions or sets and functions.

(3)-(4) two functors $F : \mathscr{C} \to \mathscr{D}^{\mathrm{op}}$ and $G : \mathscr{D}^{\mathrm{op}} \to \mathscr{C}$ forming a dual adjunction $F \dashv G$ relating the world of syntax to the world of semantics. In the examples we will take $F = \mathsf{Pf} : \mathbf{DL} \to \mathbf{Pos}^{\mathrm{op}}$ the functor sending a distributive lattice to the poset of its prime filters and **DL**-morphisms to their inverse images, and $G = \mathcal{U} : \mathbf{Pos}^{\mathrm{op}} \to \mathbf{DL}$ the functor sending a poset to the distributive lattice of its upsets and monotone maps to their inverse images. An important special case of the adjunction $\mathsf{Pf} \dashv \mathcal{U}$ is its restriction $\mathsf{Uf} \dashv \mathcal{P}$ to boolean algebras and sets: since prime filters are maximal in boolean algebras – i.e. ultrafilters (hence the ultrafilter functor Uf), their posets are trivial and thus simply form sets; upsets of trivial posets are simply subsets (hence the powerset functor $\mathcal{P}$).

(5) a *syntax building functor* $L : \mathscr{C} \to \mathscr{C}$ which specifies how to build 'modal algebras', and in particular how to build modal formulas.

(6) a *model building functor* $T : \mathscr{D} \to \mathscr{D}$ which specifies the kind of transition structure we want our models to have.

Languages, Logics and Free Algebras. As will be illustrated in the examples, L can specify much more than a grammar, it can also enforce axioms. What is included in L is a matter of convenience, but as we shall see, including *some* axioms – specifically *distribution laws* – is a good idea. The distinction between *language* and *logic* therefore becomes blurred, and indeed may not be terribly useful in this presentation of coalgebraic logics. The relevant notion is that of a *free L-algebra*. For our purpose it will be enough to say that the free L-algebra over a $\mathscr{C}$-object A, written as $\mathsf{F}_L A$, is the initial $L(-) + A$ algebra. We will assume throughout that these algebras exist, i.e. that L is a *varietor*, and focus on a particular choice of objects in $\mathscr{C}$, namely those which are themselves free objects (recall that we assume a free-forgetful adjunction between **Set** and $\mathscr{C}$). For example, if $\mathscr{C} = \mathbf{DL}$ and V is a set of propositional variables, we will consider the free L-algebra over the free distributive lattice over V, i.e. we will consider L-algebras of the type $\mathsf{F}_L \mathsf{F} V$. These are the entities which play the role of *language* since their carriers contain terms freely built from propositional variables, modulo the axioms of $\mathscr{C}$ and those encoded in L. In particular, it is the elements of these algebras which we will want to *interpret*.

Coalgebraic Semantics. Terms in a free L-algebra are interpreted as 'predicates' on the carriers of T-coalgebras. The exact meaning of the word 'predicate' is specified by the functor G which maps the carrier of a T-coalgebra to a $\mathscr{C}$-structure whose elements are by definition the predicates. An interpretation is thus a map from terms over V, viz. elements $\mathsf{F}_L \mathsf{F} V$, to predicates on

X, viz. elements of GX. We produce such an map by equipping GX with an $L(-)+\mathsf{F}V$-algebra structure and using the *initiality* of $\mathsf{F}_L\mathsf{F}V$ amongst $L(-)+\mathsf{F}V$-algebras. By definition of the coproduct, to define a morphism $LGX+\mathsf{F}V \to GX$ we need:

1. a morphism of the type $\mathsf{F}V \to GX$, and
2. a morphism of the type $LGX \to GX$

By adjunction any morphism of the type $\mathsf{F}V \to GX$ is equivalent to a map $V \to \mathsf{U}GX$ interpreting each propositional variable as a predicate, i.e. a *valuation* $v : V \to \mathsf{U}GX$. The second morphism deals with modal terms whose interpretation should depend on the transition structure $\gamma : X \to TX$. To encode this dependency we make the second morphism factor through $G\gamma : GTX \to GX$. What we therefore need is a morphism $\delta_X : LGX \to GTX$ for *any* $\mathscr{D}$-object X. Moreover, if $\beta : Y \to TY$ is another T-coalgebra and $f : Y \to X$ is a T-coalgebra morphism, it is not hard to check that the unicity of catamorphisms enforces $G\beta \circ GTf \circ \delta_X = G\beta \circ \delta_Y \circ LGf$. In fact we assume the somewhat stronger condition that the maps δ_X in fact define a *natural transformation* $\delta : LG \to GT$. This natural transformation will be called the *semantic transformation* and is the final necessary ingredient of coalgebraic logics. Given a semantics transformation and a valuation we define the interpretation of 'formulas' of $\mathsf{F}_L\mathsf{F}V$ in $\gamma : X \to TX$ as the catamorphism $[\![-]\!]_{(\gamma,v)}$ given by:

$$
\begin{array}{ccc}
L\mathsf{F}_L\mathsf{F}V + \mathsf{F}V & \xrightarrow{\;\;L[\![-]\!]_{(\gamma,v)}+\mathrm{Id}_{\mathsf{F}V}\;\;} & LGX + \mathsf{F}V \\
\downarrow & & \downarrow{\scriptstyle \delta_X+\mathrm{Id}_{\mathsf{F}V}} \\
& & GTX + \mathsf{F}V \\
& & \downarrow{\scriptstyle G\gamma+\hat{v}} \\
\mathsf{F}_L\mathsf{F}V & \xrightarrow[\quad [\![-]\!]_{(\gamma,v)} \quad]{} & GX
\end{array}
$$

Modularity. Coalgebraic logics defined on a common minimal reasoning structure can be freely combined to form new logics combining the modalities of their constituents in a process called the *fusion* of modal logics (see [CP07,DP11]). Formally, if $L_1, L_2 : \mathscr{C} \to \mathscr{C}$ are two syntax constructors, then the *fusion* of $\mathsf{F}_{L_1}\mathsf{F}V$ and $\mathsf{F}_{L_2}\mathsf{F}V$ is the language defined by the (point-wise) coproduct of these functors, i.e. $\mathsf{F}_{L_1+L_2}\mathsf{F}V$. Assuming that free L_i-algebras are interpreted in T_i-coalgebras via a semantics transformation δ_i for $i = 1, 2$, we can combine the semantics in a dual way to the syntax by interpreting free $L_1 + L_2$-algebras in $T_1 \times T_2$-coalgebras via the semantics transformation $G\pi_1 \circ \delta_1 + G\pi_2 \circ \delta_2$ where π_1, π_2 are the usual projections from a product.

Example 1 ((Positive) Modal Logic). Standard Modal Logic, henceforth ML, is boolean and we therefore choose **BA** as our minimal reasoning structure. The syntax building functor is $L^{\mathrm{ML}} : \mathbf{BA} \to \mathbf{BA}$ defined by:

$$
L^{\mathrm{ML}}A = \mathsf{F}\{\Diamond a \mid a \in \mathsf{U}A\}/\{\Diamond(a \vee b) = \Diamond a \vee \Diamond b, \Diamond\bot = \bot\}
$$

i.e. L^{ML} builds the free boolean algebra over the formal expressions $\Diamond a$ with $a \in A$, and then quotients this object by the fully invariant equivalence relation (in **BA**!) generated by the distribution laws above. We will show how $L^{\mathrm{ML}}A$ can be defined categorically in Sect. 4. An L^{ML}-algebra is a *boolean algebra with operator*, i.e. a boolean algebra together with a unary operation which distributes over joins. Given a set V of propositional variables, the object representing the *language* of ML will be $\mathsf{F}_{L\mathrm{ML}}FV$, the colimit of the diagram

$$2 \xrightarrow{\ c_0\ } L^{\mathrm{ML}}2 + FV \xrightarrow{\ c_1 = L^{\mathrm{ML}}(c_0) + \mathrm{Id}_{FV}\ } L^{\mathrm{ML}}(L^{\mathrm{ML}}2 + FV) + FV \ldots$$

where $2 = \{\bot, \top\}$ is the initial object in **BA**. The L^{ML}-algebra $\mathsf{F}_{L\mathrm{ML}}FV$ thus contains all terms which can be built from elements of $V, \top, \bot, \neg, \vee, \wedge$ and $\Diamond$ modulo the axioms of **BA** and the distribution laws encoded in L^{ML}. For the semantics we take $T = \mathsf{P}$, the *covariant* powerset functor on **Set**, and the transformation $\delta : L\mathcal{P} \to \mathcal{P}\mathsf{P}$ given at any set X and generator $\Diamond U \in L^{\mathrm{ML}}\mathcal{P}X$ by:

$$\delta_X^{\mathrm{ML}}(\Diamond U) = \{V \subseteq X \mid U \cap V \neq \emptyset\}$$

It is clear that $\delta_X(\Diamond(U_1 \cup U_2)) = \delta_X(\Diamond U_1 \cup \Diamond U_2)$ and $\delta_X(\Diamond \emptyset) = \delta_X(\emptyset)$, and δ_X is thus well-defined. P and δ give the standard Kripke semantics of ML, the only difference being that here we interpret equivalence classes of formulas.

Mutantis mutandis we can perform the exact same construction for positive ML. The minimal reasoning structure becomes either **DL** or **BDL** depending on whether we want $\top$ and $\bot$ or not, and due to the lack of negation one needs to introduce the dual operator $\Box$ explicitly. The functor becomes $L^{\mathrm{ML}} : \mathbf{DL} \to \mathbf{DL}$:

$$L^{\mathrm{ML}}A = \mathsf{F}\{\Diamond a, \Box a \mid a \in \mathsf{U}A\}/\{\Diamond(a \vee b) = \Diamond a \vee \Diamond b, \Box(a \wedge b) = \Box a \wedge \Box b\}$$

In the case of **BDL** one also adds $\Diamond \bot = \bot$ and $\Box \top = \top$ to the equations defining the quotient. The construction of the language $\mathsf{F}_{L\mathrm{ML}}FV$ is exactly the same as in the boolean case, with the caveat that the initial object in **DL** is the empty distributive lattice $\emptyset$. On the semantics side we need to find an equivalent of the covariant powerset for **Pos**. This is not entirely straightforward, but as was persuasively argued in Example 5.3 of [VK11] and in [BKV13], the **Pos** equivalent of P is the *convex powerset functor* $\mathsf{P}_c : \mathbf{Pos} \to \mathbf{Pos}$ sending a poset to the set of its convex subsets ordered by the Egli-Milner order. Since we are not yet enforcing any relation between $\Diamond$ and $\Box$, we interpret positive ML in T^{ML}-coalgebras for the functor $T^{\mathrm{ML}} = \mathsf{P}_c \times \mathsf{P}_c$ (one copy of P_c per modality). The semantics transformation $\delta^{\mathrm{ML}} : L^{\mathrm{ML}}\mathcal{U} \to \mathcal{U}(\mathsf{P}_c \times \mathsf{P}_c)$ is then defined as:

$$\delta_X^{\mathrm{ML}}(\Diamond U) = \{(V_1, V_2) \mid U \cap V_1 \neq \emptyset\} \qquad \delta_X^{\mathrm{ML}}(\Box U) = \{(V_1, V_2) \mid V_2 \subseteq U\}$$

and it is not hard to check that δ_X is well-defined, although interestingly this relies heavily on the definition of the Egli-Milner order, confirming the choice of P_c as the 'correct' generalization of P.

3 Strong Completeness and Jónsson-Tarski Extensions

We have seen how important the 'predicate' functor $G : \mathscr{D}^{\mathrm{op}} \to \mathscr{C}$ is to define the coalgebraic semantics but have so far ignored its left-adjoint $F : \mathscr{C} \to \mathscr{D}^{\mathrm{op}}$.

The intuition behind F is that it sends a reasoning structure to 'states' on this structure, where a 'state' is a collection of elements structured in such a way that it may be understood as a consistent set of logical terms which can simultaneously hold at some point in a model. A coarse description of the semantic half of completeness-via-canonicity, which we present in this section, is that it consists in equipping a (po)set of such states with the target coalgebraic structure, i.e. in building models on collections of algebraic terms. Formally, starting from a $\mathscr{C}$-object A with an L-algebra structure, we want to place a T-coalgebra structure on its set of 'states' FA. When A is of the shape $F_L FV$, such a T-coalgebra is often referred to as a 'canonical' model, although it is usually far from canonical. In fact such a model almost always requires a non-constructive principle such as the axiom of choice or the Prime Ideal Theorem (henceforth PIT). In this sense 'canonical' models are deeply non-canonical, which is why we will settle for an alternative terminology.

The Coalgebraic Jónsson-Tarski Theorem. To formulate this important result we need the following natural transformation: by using the adjunction $F \dashv G$, we can associate with each semantics transformation $\delta : LG \to GT$ its *adjoint semantic transformation* $\hat{\delta} : TF \to FL$ given by $\hat{\delta} = FL\eta \circ F\delta_F \circ \epsilon_{TF}$ where η, ϵ are the unit and counit of $F \dashv G$.

Theorem 1 (Coalgebraic Jónsson-Tarski Theorem). *Consider the fundamental situation of diagram (1) and let $\delta : LG \to GT$ be a semantic transformation. For any $\mathbf{Alg}_{\mathscr{C}}(L)$-object (A, α), if $\hat{\delta}_A$ has a right-inverse ζ_A then the morphism $\eta_A : A \to GFA$ lifts to an L-algebra morphism.*

Proof [KKP05]. We show that the following diagram commutes

$$ (2) $$

The right-hand-side trapezium commutes by naturality of η, so we need only show that the left-hand-side triangle commute.

$$
\begin{aligned}
\eta_{LA} &= G\zeta_A \circ G\hat{\delta}_A \circ \eta_{LA} & \zeta_A \text{ right-inverse} \\
&= G\zeta_A \circ G\epsilon_{TFA} \circ GF(\delta_{FA} \circ L\eta_A) \circ \eta_{LA} & \text{Definition of } \hat{\delta} \\
&= G\zeta_A \circ G\epsilon_{TFA} \circ \eta_{GTFA} \circ \delta_{FA} \circ L\eta_A & \text{Naturality of } \eta \\
&= G\zeta_A \circ \delta_{FA} \circ L\eta_A & F \dashv G
\end{aligned}
$$

Jónsson-Tarski Extensions. For $\mathscr{C} = \mathbf{DL}$, $F = \mathsf{Pf}$ and $G = \mathcal{U}$ if we assume the PIT or the axiom of choice (the latter being strictly stronger than the former), then the unit of $\mathsf{Pf} \dashv \mathcal{U}$ is a monomorphism, i.e. η_A is injective at every stage A. This means that in the conditions of Theorem 1, GFA is an *extension* of A as an L-algebra. We call such an extension a *Jónsson-Tarski extension* of

(A, α) and denote it by $\alpha^\varsigma : LGFA \to GFA$. As this notation implies, we use the terminology *an* extension, rather than *the* extension because in general, different right-inverses ζ_A will lead to different extensions, although we will encounter nice situations in the last section when there exists a unique Jónsson-Tarski extension. We will however refer to *the* Jónsson-Tarski extension when a particular choice of right-inverse ζ_A has been made and no ambiguity is possible. Note that $\zeta_A \circ \mathrm{Pf}\alpha : \mathrm{Pf}A \to T\mathrm{Pf}A$ is a T-coalgebra – i.e. a model – on 'states'. When A is of the shape $\mathsf{F}_L FV$ this coalgebra is commonly known as a 'canonical model', although in practice the construction of right inverses to the adjoint semantic transformation also requires the PIT or the AC, which makes these models deeply non-canonical. It follows from the unicity of catamorphisms that if $\hat{\delta}_{\mathsf{F}_L FV}$ has a right-inverse $\zeta_{\mathsf{F}_L FV}$ then the interpretation map in

$$\mathrm{Pf}\mathsf{F}_L FV \to \mathrm{Pf}L\mathsf{F}_L FV \xrightarrow{\zeta_{\mathsf{F}_L FV}} T\mathrm{Pf}\mathsf{F}_L FV$$

is given by Diagram (2), in other words $[\![-]\!]_{\mathrm{Pf}\mathsf{F}_L FV} = \eta_{\mathsf{F}_L FV}$. Modal logicians refer to this as the *truth lemma*: a formula a holds at a prime filter w in a 'canonical model' iff $a \in w$, by definition of η.

The Case of Boolean Coalgebraic Logics. In practice, when $\mathscr{C} = \mathbf{DL}$, right-inverses to adjoint semantic transformations must be built explicitly, and in fact this is also done in the construction of the standard 'canonical' model of ML. However, when the minimal reasoning structure is $\mathbf{BA}$ the criterion for the existence of Jónsson-Tarski extensions can be simplified somewhat, at the cost of being even less constructive. Assuming the axiom of choice all epimorphisms in $\mathbf{Set}$ are split, i.e. all surjections have a right inverse. For boolean coalgebraic logics, it is therefore sufficient to require that $\hat{\delta}$ be a pointwise epimorphism, and useful criteria for this to happen have been developed in [Dah15, KR12, SP09].

Strong Completeness. The main application of Jónsson-Tarski extensions is to prove *strong completeness*. Let us first define precisely what we mean by strong completeness. Let $\mathscr{C}$ be $\mathbf{DL}, \mathbf{BDL}$ or $\mathbf{BA}$, let V be a set of propositional variables, let $q : \mathsf{F}_L FV \twoheadrightarrow \mathcal{L}$ be a regular epi, and let $\Phi, \Psi \subseteq \mathcal{L}$ be two families of 'formulas' such that $\Phi \not\vdash \Psi$, i.e. such that no finite set Φ_0 of elements of Φ and no finite set Ψ_0 of elements of Ψ can be found such that $\bigwedge \Phi_0 \leq \bigvee \Psi_0$. The statement that $\mathcal{L}$ is strongly complete w.r.t. to a class $\mathfrak{T}$ of T-coalgebras means that for any such choice of Φ, Ψ there exists a T-coalgebra $\gamma : X \to TX$ in $\mathfrak{T}$, a valuation $v : FV \to GX$, and a point $x \in X$ such that $x \in [\![a]\!]_{(\gamma,v)}$ for all $a \in \Phi$ and $x \notin [\![b]\!]_{(\gamma,v)}$ for all $b \in \Psi$.

Theorem 2 (Strong Completeness). *If the adjoint semantic transformation $\hat{\delta}$ has a right-inverse $\zeta_{\mathsf{F}_L FV}$ at $\mathsf{F}_L FV$, then $\mathsf{F}_L FV$ is strongly complete w.r.t. to the class $\mathbf{Coalg}_{\mathscr{D}}(T)$ of T-coalgebras.*

Proof. Let $\Phi, \Psi \subseteq \mathsf{F}_L FV$ and $\Phi \not\vdash \Psi$. Then the filter $\langle \Phi \rangle^\uparrow$ generated by Φ and the ideal $\langle \Psi \rangle^\downarrow$ generated by Ψ obey $\langle \Phi \rangle^\uparrow \cap \langle \Psi \rangle^\downarrow = \emptyset$. By the PIT there exists a prime filter w_Φ extending $\langle \Phi \rangle^\uparrow$ such that $w_\Phi \cap \langle \Psi \rangle^\downarrow = \emptyset$. By Theorem 1, the L-algebra

$\mathsf{F}_L FV$ has a Jónsson-Tarski extension which provides an interpretation of $\mathsf{F}_L FV$ in the T-coalgebra $\mathsf{PfF}_L FV \to \mathsf{Pf}L\mathsf{F}_L FV \xrightarrow{\varsigma_{\mathsf{F}_L FV}} T\mathsf{PfF}_L FV$, which coincides with $\eta_{\mathsf{F}_L FV}$. In this interpretation $w_\Phi \in \llbracket a \rrbracket$ for all $a \in \Phi$ and $w_\Phi \notin \llbracket b \rrbracket$ for all $b \in \Psi$.

Jónsson-Tarski extensions, and thus strong completeness, are modular:

Theorem 3 [DP15b]. *Let $L_i : \mathscr{C} \to \mathscr{C}, T_i : \mathscr{D} \to \mathscr{D}, \delta^i : L_i G \to GT_i, i = 1,2$. For any $\mathbf{Alg}_\mathscr{C}(L_1 + L_2)$-object (A, α), if $\hat{\delta}^i_A$ has a right-inverse $\zeta^i_A, i = 1,2$, then $\eta_A : A \to GFA$ lifts to an $L_1 + L_2$-algebra morphism.*

In a nutshell, the purpose of coalgebraic completeness-via-canonicity is to determine how and when Theorem 2 resists to quotienting $\mathsf{F}_L FV$.

Example 2. The PIT-based technique of [DP15a] becomes particularly simple for unary operators and shows that the adjoint transformation $\hat{\delta}^{\mathrm{ML}} : T^{\mathrm{ML}}\mathsf{Pf} \to \mathsf{Pf}L^{\mathrm{ML}}$ of the semantic transformation defined in Example 1 has right-inverses $\zeta^{\mathrm{ML}}_A : \mathsf{Pf}L^{\mathrm{ML}}A \to T^{\mathrm{ML}}\mathsf{Pf}A$ at every A in **DL** given by:

$$\zeta^{\mathrm{ML}}_A(F) = (\{F_1 \mid a \in F_1 \Rightarrow \Diamond a \in F\}, \{F_2 \mid \Box a \in F \Rightarrow a \in F_2\})$$

The (positive) language for ML defined by the functor L^{ML} is thus strongly complete w.r.t. $\mathsf{P}_c \times \mathsf{P}_c$-coalgebras. We will show later that quotienting this language by the axioms relating $\Diamond$ and $\Box$ defines a variety closed under Jónsson-Tarski extension. Strong completeness with respect to P_c-coalgebras interpreting both modalities by the same relation will then follow (modulo two lemmas).

4 Canonical Equations and Canonical Extensions of L-algebras

In the previous section we have shown how to construct coalgebraic models whose carriers are the 'states' FA of an L-algebra A in a way that provides an L-algebra embedding of A into GFA. When $\mathscr{C} = \mathbf{DL}$ (or $\mathbf{BDL}$ or $\mathbf{BA}$) and $F \dashv G$ is the adjunction $\mathsf{Pf} \dashv \mathcal{U}$, objects of the form GFA are very well-known to algebraists studying boolean algebras and distributive lattices, and are known as *canonical extensions* and denoted A^σ. Motivated by the algebraic semantics of modal logic, this notion was extended to boolean algebra with operators (BAOs) [JT51] and distributive lattice expansions (DLEs) [GJ94, GJ04]. One of the key areas of research in this domain is to find conditions under which the validity of an equation in an BAO or a DLE can be transferred to its canonical extension, i.e. conditions under which $A \models s = t$ implies $A^\sigma \models s = t$. Such equations are called *canonical*. In this section we will review the basic facts about canonical equations and about a topological technique for establishing the canonicity of equations. As a by-product of this theory we will give a theoretically partial but practically complete answer to the following question:

For which functors $L : \mathbf{DL} \to \mathbf{DL}$ does the canonical extension construction in
$\mathbf{DL}$ lift to $\mathbf{Alg}_{\mathbf{DL}}(L)$?

Canonical Extensions of DLs. For any A in **DL**, $\mathcal{U}\mathsf{Pf}A$ is known as the *canonical extension* of A and denoted A^σ. It can be characterised uniquely up to isomorphism through purely algebraic properties, namely that A is *dense* and *compact* in A^σ. In this sense the adjective 'canonical' is fully justified, in contrast with its usage in the expression 'canonical model'. For our purpose however, *defining* the canonical extension of A as $\mathcal{U}\mathsf{Pf}A$ will be sufficient. The canonical extension A^σ of a distributive lattice A is always *completely distributive* (see [GJ04]). The following terminology will be important: A^σ is a completion of A and all joins of elements of A therefore exist in A^σ, such elements are called *open* and their set is denoted by $O(A)$. Dually, meets in A^σ of elements of A will be called *closed* and their set denoted $K(A)$. Elements of $A = K(A) \cap O(A)$ are therefore called *clopens*.

Canonical Extension of DLEs. It was shown in [JT51] that if A is a BA with a map $f : \mathsf{U}A \to \mathsf{U}A$ preserving joins then $A^\sigma = \mathcal{P}\mathsf{Uf}A$ can be equipped with a map $f^\sigma : \mathsf{U}A^\sigma \to \mathsf{U}A^\sigma$ which extends f and preserves all non-empty joins. This construction was later extended to DLs with n-ary maps and no particular preservation properties in [GJ94, GJ04]. Formally, given a signature Σ with arity map $\mathrm{ar} : \Sigma \to \mathbb{N}$, define the syntax building functor $L_\Sigma : \mathbf{DL} \to \mathbf{DL}$ by:

$$L_\Sigma A = \mathsf{F}\left(\coprod_{s \in \Sigma} \mathsf{U}A^{\mathrm{ar}(s)}\right) \tag{3}$$

An L_Σ-algebra is a distributive lattice with n-ary maps defined by the signature, i.e. a *Distributive Lattice Expansion*, or DLE for short. We now sketch the theory of their canonical extensions. Each map $f : \mathsf{U}A^n \to \mathsf{U}A$ can be extended to a map $(\mathsf{U}A^\sigma)^n \to \mathsf{U}A^\sigma$ in two canonical ways:

$$f^\sigma(x) = \bigvee\{\bigwedge f[d,u] \mid K(A)^n \ni d \leq x \leq u \in O(A)^n\}$$
$$f^\pi(x) = \bigwedge\{\bigvee f[d,u] \mid K(A)^n \ni d \leq x \leq u \in O(A)^n\}$$

where $f[d,u] = \{f(a) \mid a \in A^n, d \leq a \leq u\}$. In many important cases, the two extensions (viz. f^σ and f^π) agree, in which case f is said to be *smooth*. We *define* the *canonical extension* of an L_Σ-algebra A as the L_Σ-algebra A^σ defined by $(A^\sigma, (f_s^\sigma : (\mathsf{U}A^\sigma)^{\mathrm{ar}(s)} \to \mathsf{U}A^\sigma)_{s \in \Sigma})$. This gives us a first class of functors L which answers the question above: for any finitary signature Σ, the **DL**-endofunctor L_Σ defined by Eq. (3) lifts canonical extensions from **DL** to $\mathbf{Alg_{DL}}(L_\Sigma)$.

Topological Methods where introduced in [GJ04, Ven06] to study the canonical extension of maps. These methods are useful because they (a) uniquely characterize canonical extensions, (b) reflect interesting algebraic properties of maps (e.g. the preservation of meets) and, crucially (c) provide a very effective way of studying the *composition of canonical extensions* which is essential to establishing canonicity. We need six topologies on A^σ. First, we define $\sigma^\uparrow, \sigma^\downarrow$ and σ as the topologies defined by the bases $\{\uparrow p \mid p \in K\}, \{\downarrow u \mid u \in O\}$ and $\{\uparrow p \cap \downarrow u, K \ni p, u \in O\}$. The next set of topologies is well-known to domain theorists: a *Scott open* in A^σ is a subset $U \subseteq A^\sigma$ such that (1) U is an upset and (2) for any up-directed set D such that $\bigvee D \in U$, $D \cap U \neq \emptyset$. The collection of Scott opens forms a topology called the *Scott topology*, which we denote $\gamma^\uparrow$. The dual topology will be denoted by $\gamma^\downarrow$, and their join by γ. Since for every $x \in A^\sigma$, $x = \bigvee \downarrow x \cap K = \bigwedge \uparrow x \cap O$, it is easy to see that $\gamma^\uparrow \subseteq \sigma^\uparrow$, $\gamma^\downarrow \subseteq \sigma^\downarrow$, and $\gamma \subseteq \sigma$. We denote the product of topologies by $\times$, and the n-fold product by $(-)^n$.

Proposition 1 [GJ04]. *For any DL A and map $f : \mathsf{U}A^n \to \mathsf{U}A$,*

1. *f^σ is the largest $(\sigma^n, \gamma^\uparrow)$-continuous extension of f,*
2. *f^π is the smallest $(\sigma^n, \gamma^\downarrow)$-continuous extension of f*
3. *f is smooth iff it has a unique (σ^n, γ)-continuous extension.*

The following result which relates algebraic and topological properties, is a straightforward generalization of results from [GH01, GJ94, GJ04, Ven06].

Proposition 2. *Let A be a distributive lattice, and let $f : \mathsf{U}A^n \to \mathsf{U}A$ be a map. For any $(n-1)$-tuple $a = (a_i)_{1 \leq i \leq n-1}$, we denote by $f_a^k : \mathsf{U}A \to \mathsf{U}A$ the map defined by $x \mapsto f(a_1, \ldots, a_{k-1}, x, a_k, \ldots, a_{n-1})$.*

1. *If f_a^k preserves binary joins, $(f^\sigma)_a^k$ preserves all non-empty joins.*
2. *If f_a^k preserves binary meets, $(f^\sigma)_a^k$ preserves all non-empty meets.*
3. *If f_a^k anti-preserves binary joins, $(f^\sigma)_a^k$ anti-preserves all non-empty joins.*
4. *If f_a^k anti-preserves binary meets, $(f^\sigma)_a^k$ anti-preserves all non-empty meets.*
5. *If $(f^\sigma)_a^k$ preserves all non-empty joins, it is $(\sigma^\downarrow, \sigma^\downarrow)$-continuous.*
6. *If $(f^\sigma)_a^k$ preserves all non-empty meets, it is $(\sigma^\uparrow, \sigma^\uparrow)$-continuous.*
7. *If $(f^\sigma)_a^k$ anti-preserves all non-empty joins, it is $(\sigma^\downarrow, \sigma^\uparrow)$-continuous.*
8. *If $(f^\sigma)_a^k$ anti-preserves all non-empty meets, it is $(\sigma^\uparrow, \sigma^\downarrow)$-continuous.*
9. *In each case f_a^k is is smooth.*

Function composition and canonical extension interact in a non-trivial way, but the following consequence of Proposition 1 greatly clarifies their interaction. This result is our main tool for proving canonicity.

Theorem 4 (Principle of Matching Topologies, [GH01, Ven06]). *Let A be a DL, and $f : \mathsf{U}A^n \to \mathsf{U}A$ and $g_i : \mathsf{U}A^{m_i} \to \mathsf{U}A, 1 \leq i \leq n$ be arbitrary maps. Assume that there exist topologies τ_i on A, $1 \leq i \leq n$ such that each g_i^σ is (σ^{m_i}, τ_i)-continuous. If f^σ is*

1. *$(\tau_1 \times \ldots \times \tau_n, \gamma^\uparrow)$-continuous, then $f^\sigma(g_1^\sigma, \ldots, g_n^\sigma) \leq (f(g_1, \ldots, g_n))^\sigma$,*
2. *$(\tau_1 \times \ldots \times \tau_n, \gamma^\downarrow)$-continuous, then $f^\sigma(g_1^\sigma, \ldots, g_n^\sigma) \geq (f(g_1, \ldots, g_n))^\sigma$,*
3. *$(\tau_1 \times \ldots \times \tau_n, \gamma)$-continuous, then $f^\sigma(g_1^\sigma, \ldots, g_n^\sigma) = (f(g_1, \ldots, g_n))^\sigma$.*

Monotone (i.e. isotone *or* antitone) maps have a nice property which complements the Principle of Matching Topologies very effectively. The proof of this property can already be found for isotone maps in [Rib52], and generalizes tediously but straightforwardly to monotone maps (i.e. either isotone *or* antitone).

Proposition 3. *Let* $g_i : (\mathsf{U}A)^{n_i} \to \mathsf{U}A$, $1 \leq i \leq m$ *and* $f : (\mathsf{U}A)^m \to \mathsf{U}A$ *be monotone maps, then* $(f(g_1, \ldots, g_m))^\sigma \leq f^\sigma(g_1^\sigma, \ldots, g_m^\sigma)$.

Canonicity. Before we consider the canonical extension of more general L-algebras, we need to talk about canonicity. Let us fix a signature Σ. Recall that an equation $s = t$ in the language of L_Σ-algebras (i.e. DLEs with signature Σ) is *canonical* if it has the property that $A^\sigma \models s = t$ whenever $A \models s = t$. To say anything about the canonicity of equations, we therefore need to compare interpretations in A with those in A^σ. It is natural to try to use the extension $(\cdot)^\sigma$ to mediate between these interpretations, but $(\cdot)^\sigma$ is defined on maps, not on terms. Moreover, not every valuation on A^σ originates from valuation on A. We therefore want to recast the problem in such a way that (1) terms are viewed as maps, and (2) we do not need to worry about valuations. The solution is to adopt the language of *term functions* (as first suggested in [Jón94]). Let Σ be a signature and t be a term in the language $\mathsf{F}_{L_\Sigma}\mathsf{F}V$, there exist a finite set $V_0 = \{p_1, \ldots, p_n\} \subseteq V$ containing the propositional variables of t. For any L_Σ-algebra A, we can put an $L_\Sigma(-) + \mathsf{F}V_0$-algebra structure on the distributive lattice $\mathcal{A}_n$ of n-ary maps on A (with pointwise meets and joins) as follows:

- Define $v : V_0 \to \mathsf{U}\mathcal{A}_n, p_i \mapsto \pi_i$, the ith projection $(\mathsf{U}A)^n \to \mathsf{U}A$.
- For each $f \in \Sigma$, we overload and define $f : (\mathsf{U}\mathcal{A}_n)^{\mathrm{ar}(f)} \to \mathsf{U}\mathcal{A}_n$ by $(g_1, \ldots, g_{\mathrm{ar}(f)}) \mapsto f \circ \langle g_1, \ldots, g_{\mathrm{ar}(f)} \rangle$, where $\langle \rangle$ denotes the product.

By taking the adjoint transpose of these maps (i.e. freely extending) we equip $\mathcal{A}_n$ with the desired algebraic structure. We can now interpret a term t as the *term function* $t^A : \mathsf{U}A^n \to \mathsf{U}A$ given by the catamorphism $(\cdot)^A$:

$$
\begin{array}{ccc}
L_\Sigma \mathsf{F}_{L_\Sigma}\mathsf{F}V_0 + \mathsf{F}V_0 & \xdashrightarrow{\; L_\Sigma(\cdot)^A + \mathrm{Id}_{\mathsf{F}V_0} \;} & L_\Sigma \mathcal{A}_n + \mathsf{F}V_0 \\
\downarrow & & \downarrow{\scriptstyle \Sigma_{f \in \Sigma}\, \hat{f} + \hat{v}} \\
\mathsf{F}_{L_\Sigma}\mathsf{F}V_0 & \xdashrightarrow[\quad (\cdot)^A \quad]{} & \mathcal{A}_n
\end{array}
$$

For any two terms $s, t \in \mathsf{F}_{L_\Sigma}\mathsf{F}V$ we can take V_0 to be the set of propositional variables required to build both s and t and thus get a common catamorphism $(\cdot)^A$ interpreting both terms as maps $(\mathsf{U}A)^n \to \mathsf{U}A$. It is then well-known and easy to check that $A \models s = t$ iff $s^A = t^A$. Following [Jón94], we say that $t \in \mathsf{F}_{L_\Sigma}\mathsf{F}V$ is *stable* if $(t^A)^\sigma = t^{A^\sigma}$, that t is *expanding* if $(t^A)^\sigma \leq t^{A^\sigma}$, and that t is *contracting* if $(t^A)^\sigma \geq t^{A^\sigma}$, for any A. The inequality between maps is taken pointwise. The following proposition illustrates the usefulness of these notions:

Proposition 4 [Jón94]. *If* $s, t \in \mathsf{F}_{L_\Sigma}\mathsf{F}V$ *are stable then* $s = t$ *is canonical. If* s *is contracting and* t *is expanding, then* $s \leq t$ *is canonical.*

In practice, we use the Principle of Matching Topologies (Theorem 4) to determine when a term is stable, expanding or contracting, and thus when equations or inequations are canonical.

Canonical Extension of L-Algebras. We now show that L^{ML} defined in Example 1 belongs to a general class of functors of the form $L_\Sigma/\{E\}$ for which the canonical extension construction always lifts to $\mathbf{Alg_{DL}}(L)$. To categorically formalize functors of the type $L_\Sigma/\{E\}$ for a set E of equations, we need to capture the notion of fully invariant equivalence relation generated by a set of equations. We will only sketch the construction which can be performed in great generality in any well-powered cocomplete regular category (see [Dah15], Chap. 1). Consider a set E of equations in $\mathsf{F}_{L_\Sigma}FV$ (e.g. $E = \{\Diamond(a \vee b) = \Diamond a \vee \Diamond b, \Diamond\bot = \bot\}$), it is equivalent to a pair of jointly monic functions $e_1, e_2 : E \rightrightarrows U\mathcal{L}$ where $\mathcal{L}$ is the distributive lattice underlying the L_Σ-algebra $\mathsf{F}_{L_\Sigma}FV$. By adjunction we can re-write this as a pair of morphisms $\hat{e}_1, \hat{e}_2 : FE \rightrightarrows \mathcal{L}$ where FE is the 'free DL of equations'. Consider the coequalizer $FE \rightrightarrows \mathcal{L} \xrightarrow{q} Q$ of $\hat{e}_1, \hat{e}_2$ and all the terms $s, t \in \mathcal{L}$ such that $q(s) = q(t)$. They form an equivalence relation in **DL** (the kernel pair of q) containing E, but not a fully invariant one (e.g. $\Diamond(c \vee d) = \Diamond c \vee \Diamond d$ for $c \neq a, d \neq b$ does not in general belong to this relation). To capture substitution instances we must consider a 'bigger' coequalizer, namely

$$\coprod_{f\in\text{hom}(\mathcal{L},\mathcal{L})} FE \xrightarrow[\coprod_{f\in\text{hom}(\mathcal{L},\mathcal{L})} L_\Sigma f\circ\phi^{-1}\circ\hat{e}_2]{\coprod_{f\in\text{hom}(\mathcal{L},\mathcal{L})} L_\Sigma f\circ\phi^{-1}\circ\hat{e}_1} L_\Sigma\mathcal{L} \xrightarrow{q_{\mathcal{L}}} L\mathcal{L} := L_\Sigma/\{E\}(\mathcal{L}) \quad (4)$$

where $\phi : L_\Sigma\mathcal{L} \to \mathcal{L}$ is the iso structure map of the free L_Σ-algebra. The pairs of terms s, t such that $q(s) = q(t)$ now form a fully invariant equivalence relation in **DL**. We nearly have a rigorous definition of functors of the shape $L_\Sigma/\{E\}$, the final step is to notice that (4) can to some extent be made parametric in the choice of the middle object. We define for any A the coequalizer q_A:

$$\coprod_{f\in\text{hom}(\mathcal{L},A)} FE \xrightarrow[\coprod_{f\in\text{hom}(\mathcal{L},A)} L_\Sigma f\circ\phi^{-1}\circ\hat{e}_2]{\coprod_{f\in\text{hom}(\mathcal{L},A)} L_\Sigma f\circ\phi^{-1}\circ\hat{e}_1} L_\Sigma A \xrightarrow{q_A} LA := L_\Sigma/\{E\}(A) \quad (5)$$

It is easy to see that (5) defines a functor: for any $f : A \to B$, amongst the morphisms $\mathcal{L} \to L_\Sigma B$ are all the ones which factor through $L_\Sigma A$ via $L_\Sigma f$, and thus LB is a co-cone for the diagram defining LA and so there must exist a unique $Lf : LA \to LB$. For the same reason $L\mathcal{L}$ is initial amongst all objects of the form LA with A in **DL**. Any L-algebra $\alpha : LA \to A$ defines an L_Σ-algebra, i.e. a Σ-DLE, $\alpha \circ q_A : L_\Sigma A \to LA \to A$ which will call the *associated Σ-DLE*.

Theorem 5 (Canonical Extension Lifting). *Let Σ be a finitary signature, let E be a set of equations between terms in $\mathsf{F}_{L_\Sigma}FV$ of modal depth at most one and let $L : \mathbf{DL} \to \mathbf{DL}$ be defined by $LA = L_\Sigma/\{E\}(A)$ as in (5). If for any L-algebra $\alpha : LA \to A$, the n-ary maps of the associated Σ-DLE are monotone, then the canonical extension construction lifts from $\mathbf{DL}$ to $\mathbf{Alg_{DL}}(L)$.*

Proof. For any $\alpha : LA \to A$, the map $\phi = \alpha \circ q_A : L_\Sigma A \to LA \to A$ defines an L_Σ-algebra, and we know how to build the canonical extension of L_Σ-algebras. Let $\phi^\sigma : L_\Sigma A^\sigma \to A^\sigma$ be this canonical extension and let us define $\alpha^\sigma : LA^\sigma \to A^\sigma$ by $\alpha^\sigma(x) = \phi^\sigma(y)$ for $y \in q_{A^\sigma}^{-1}(x)$. We need to show that α^σ is well-defined, i.e. that if $q_{A^\sigma}(y) = q_{A^\sigma}(y') = x$ then $\phi^\sigma(y) = \phi^\sigma(y')$. Note that if it is the case, then it is immediate to check that α^σ is a **DL**-morphism. If $q_{A^\sigma}(y) = q_{A^\sigma}(y')$ then there must exist a $z \in FE$ and $f \in \hom(\mathcal{L}, L_\Sigma A^\sigma)$ such that $y = f \circ \hat{e}_1(z), y' = f \circ \hat{e}_1(z)$. But we know that $\phi(g \circ \hat{e}_1(z)) = \phi(g \circ \hat{e}_2(z))$ for any $g \in \hom(\mathcal{L}, L_\Sigma A)$ by definition of ϕ. This means that $(A, \phi) \models \hat{e}_1(z) = \hat{e}_1(z)$, and therefore if the equation is canonical we are done, for then $(A^\sigma, \phi^\sigma) \models \hat{e}_1(z) = \hat{e}_2(z)$, i.e. $\phi^\sigma(g \circ \hat{e}_1(z)) = \phi^\sigma(g \circ \hat{e}_2(z))$ for any $g \in \hom(\mathcal{L}, L_\Sigma A^\sigma)$, and thus for $g = f$.

The result thus amounts to showing that equations involving terms of modal depth at most one are canonical, and this will follow immediately from Proposition 4 if we can show that terms of modal depth at most one are stable. Let A be in **DL** and $t \in F_{L_\Sigma} FV$ be a term of modal depth 0 built from propositional variables in $V_0 = \{p_1, \ldots, p_n\}$. By distributivity, we can assume $t = \bigvee_{i=1}^{l} \bigwedge_{j=1}^{m_i} p_{k(i,j)}$ where k picks for each (i, j) the index of a variable in V. By definition

$$t^A = \vee^A \circ \langle \wedge^A \circ \langle \pi_{k(1,1)}, \ldots, \pi_{k(1,m_1)} \rangle, \ldots, \wedge^A \circ \langle \pi_{k(l,1)}, \ldots, \pi_{k(l,m_l)} \rangle \rangle$$

where $\vee^A : (UA)^l \to UA$ is the l-ary join in A, and similarly for every $\wedge^A$. Each $\pi_i^\sigma : (UA^\sigma)^n \to UA^\sigma$ is (σ^n, σ)-continuous by definition of σ^n. Moreover, $\vee^{A^\sigma}$ and $\wedge^{A^\sigma}$ preserve meets and joins in every argument by distributivity, and are thus (σ^l, σ)- and (σ^{m_i}, σ)-continuous respectively by Proposition 2. It follows that $t^{A^\sigma} = (t^A)^\sigma$ by the Principle of Matching Topologies. Assume now that t is of modal depth 1, i.e. $t = \bigvee_{i=1}^{n} \bigwedge_{j=1}^{m_i} f_{ij}(a_{ij1}, \ldots, a_{ij\mathrm{ar}(f_{ij})})$, where each a_{ijk} is of modal depth 0. Since every extension f_{ij} is assumed to be monotone, Proposition 3 implies that $(t^A)^\sigma \leq t^{A^\sigma}$. So we need only show the reverse inequality. We have established above that each $a_{ijk}^{A^\sigma}$ is (σ^n, σ)-continuous, and by Proposition 1 $f_{ij}^{A^\sigma}$ is $(\sigma^{\mathrm{ar}(f_{ij})}, \gamma^\uparrow)$-continuous. Finally since $\vee^{A^\sigma}$ and $\wedge^{A^\sigma}$ preserve all joins they preserve up-directed ones and are thus $((\gamma^\uparrow)^k, \gamma^\uparrow)$-continuous. The result then follows from the Principle of Matching Topologies. A completely analogous proof can be shown to hold in boolean algebras by using the de Morgan laws and the antitone preservation properties of Proposition 2.

Remark. Not all sets E of equations satisfying the conditions of Theorem 5 make sense, take for example $\Sigma = \{\Diamond\}$ and $E = \{a = b, \Diamond(c \vee d) = \Diamond c \vee \Diamond d\}$ with $a, b, c, d \in V$ all distinct. The equation $a = b$ is canonical: choose $V_0 = \{a, b, c, d\}$, then a^A is simply the projection $\pi_1^A : A^4 \to A$ and b^A is simply $\pi_2^A : A^4 \to A$ for any L_Σ-algebra A. It is easy to check from the definition that $(\pi_1^A)^\sigma = (\pi_1)^{A^\sigma}$, and similarly for π_2^A, so the terms are stable, and the equation canonical. But it is vacuously canonical: π_1^A and π_2^A are not equal.

Remark. As was mentioned earlier, [Sch06] shows that for a **Set**-endofunctor T, the class of all T-coalgebras can be characterized by axioms of modal depth one. From the point of view of coalgebraic logic, the only restrictive requirement of Theorem 5 is therefore that the expansions defined by such an axiomatization should be monotone. This however covers most coalgebraic logics, for example graded modal logic, probability logic, conditional logic, etc.

Example 3. It is clear from the definition of L^{ML} in Example 1 that L^{ML} satisfies the conditions of Theorem 5, i.e. canonical extensions lift to $\mathbf{Alg_{DL}}(L^{\mathrm{ML}})$. In order to axiomatize the duality between $\Diamond$ and $\Box$ in positive ML, one must enforce Dunn's *Interaction Axioms* on L^{ML}-algebras [Dun95]:

$$\Diamond a \wedge \Box b \leq \Diamond(a \wedge b), \quad \Box(a \vee b) \leq \Box a \vee \Diamond b$$

It follows from the proof of Theorem 5 that these inequations are also canonical.

5   Jónsson-Tarski vs Canonical Extensions.

Combining the results of Sects. 2 and 3, we know that for logics defined by an endofunctor $L : \mathscr{C} \to \mathscr{C}$ satisfying the conditions of Theorem 5 and a semantic transformation $\delta : LG \to GT$ satisfying the conditions of Theorem 1, any L-algebra $\alpha : LA \to A$ has two extensions with a common carrier: the Jónsson-Tarski extension α^{ς} and the canonical extension α^{σ}. There is no reason *a priori* for them to be isomorphic L-algebras, but it turns out that this is frequently the case in practice. It is in these instances that coalgebraic completeness-via-canonicity applies. For now though the situation is the following:

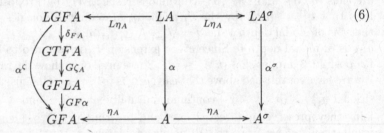

$$\hspace{12cm}(6)$$

The left-hand side of Diagram (6) deals with the model-building part of coalgebraic completeness-via-canonicity, whilst the right-hand side of Diagram (6) deals with the algebraic part the method.

Theorem 6 (Coalgebraic Completeness-via-Canonicity). *Let* $L : \mathbf{DL} \to \mathbf{DL}$ *satisfy the conditions of Theorem 5 and* $\delta : L\mathcal{U} \to \mathcal{U}T$ *the conditions of Theorem 1. If* $q : \mathsf{F}_L \mathsf{F} V \twoheadrightarrow \mathcal{L}$ *is the quotient in* $\mathbf{Alg_{DL}}(L)$ *of a fully invariant equivalence relation defining a variety closed under canonical extensions, and if the Jónsson-Tarski and canonical extensions coincide, then* $\mathcal{L}$ *is strongly complete w.r.t. the class of* T-*coalgebras validating all equations* $s = t$ *s.th.* $q(s) = q(t)$.

Proof. For any $\Phi, \Psi \subseteq \mathcal{L}$ such that $\Phi \not\vdash \Psi$ we can find a point in the coalgebra $\gamma : \mathsf{Pf}\mathcal{L} \to T\mathsf{Pf}\mathcal{A}$ satisfying every formula in Φ and none of Ψ exactly as in

Theorem 2, but we must also check that this T-coalgebra validates all the equations $s = t$ where $s, t \in \mathsf{F}_L \mathsf{F} V$ and $q(s) = q(t)$. If $q(s) = q(t)$, then $(\mathcal{L}, \alpha) \models s = t$ by construction of $\mathcal{L}$. Since $\mathcal{L}$ belongs to the variety it defines (in fact it is its initial object), and since this variety is assumed to be closed under canonical extensions, it follows that $(\mathcal{L}^\sigma, \alpha^\sigma) \models s = t$. Finally, since the Jónsson-Tarski and canonical extensions coincide we have $(\mathcal{U}\mathsf{Pf}\mathcal{L}, \alpha^\varsigma) \models s = t$ which means that $[\![s]\!]_{(\gamma, v)} = [\![t]\!]_{(\gamma, v)}$ for any valuation $v : \mathsf{F} V \to \mathcal{U}\mathsf{Pf}\mathcal{L}$.

Remark. If E is a set of equations between terms of $\mathsf{F}_L \mathsf{F} V$, then the quotient $q : \mathsf{F}_L \mathsf{F} V \twoheadrightarrow \mathcal{L}$ defined from E in the fashion of Diagram (4) is usually called the *Lindenbaum-Tarski algebra of the logic defined by L and E*.

Remark. We choose to require that a variety *defined by a regular quotient* of the free L-algebra should be canonical, i.e. closed under canonical extensions. This is strictly more general than requiring that a variety be *defined by canonical equations* (in which case it is also canonical). Indeed, as was shown in [HV05], there exist canonical varieties of BAOs with *no* canonical axiomatization.

Remark. In fact we could require less than a full isomorphism of L-algebras, what is really needed is the implication $(A^\sigma, \alpha^\sigma) \models s = t \Rightarrow (\mathcal{U}\mathsf{Pf}A, \alpha^\varsigma) \models s = t$.

We now give a useful criterion for the Jónsson-Tarski and canonical extensions to be equal. Consider for a finitary signature Σ a set E of equations of the shape

$$\{ f(a_1, \ldots, a_{i-1}, \bigwedge^{f,i} X, a_{i+1}, \ldots, a_n) = \bigvee_{b \in X}^{f,i} f(a_1, \ldots, a_{i-1}, b, a_{i+1}, \ldots, a_n) \mid \quad (7)$$

$$f \in \Sigma, 1 \le i \le \mathrm{ar}(f), \bigwedge^{f,i}, \bigvee^{f,i} \in \{\bigwedge, \bigvee\}, X \in \mathcal{P}_f(\mathsf{F}_{L_\Sigma} \mathsf{F} V)\}$$

and let $L : \mathbf{DL} \to \mathbf{DL}$ be defined as $LA = L_\Sigma / \{E\}(A)$ as in Eq. (5). It is easy to see that for any $\alpha : LA \to A$, the associated Σ-DLE has n-ary expansions which (anti)-preserve meets or joins in each argument.

Theorem 7. *Let E be as in Eq. (7) and $LA = L_\Sigma / \{E\}(A)$. Let E^∞ denote the set equations defined as (7) but with $\mathcal{P}_f(\mathsf{F}_{L_\Sigma} \mathsf{F} V)$ replaced by $\mathcal{P}(\mathsf{F}_{L_\Sigma} \mathsf{F} V)$. If at every A in $\mathbf{DL}$ the adjoint transformation $\hat{\delta}$ of $\delta : L\mathcal{U} \to \mathcal{U}T$ has a right-inverse ζ_A and $\delta_{\mathsf{Pf}A} \circ q_A$ coequalizes the pair of morphisms defined by plugging E^∞ in (5), then the Jónsson-Tarski and canonical extensions coincide.*

Proof. Let (A, α) be an L-algebra. Note first that L is of the shape required by Theorem 5: E is a set of equations of modal depth at most one, and if the n-ary expansions of the Σ-DLE associated with (A, α) (anti)-preserve meets or joins in each argument, they are in particular monotone in each argument.

The structure map of the Jónsson-Tarski extension of (A, α) is denoted by α^ς and that of the canonical extension by α^σ. The situation can be summarised in the following diagram whose innermost and outermost triangles commute:

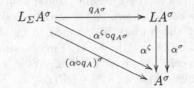

We need to show $\alpha^\varsigma = \alpha^\sigma$. It follows from the definition of E and Proposition 2 that the n-ary expansions of the Σ-DLE associated with (A, α) are smooth, and therefore have *unique* (σ^n, γ)-continuous extensions given by $(\alpha \circ q_A)^\sigma$, moreover α^σ is defined in such a way that $\alpha^\sigma \circ q_{A^\sigma} = (\alpha \circ q_A)^\sigma$ (see Theorem 5).

By definition $\alpha^\varsigma = \mathcal{U}\mathrm{Pf}\alpha \circ \mathcal{U}\zeta_A \circ \delta_{\mathrm{Pf}A}$. Since $\delta_{\mathrm{Pf}A} \circ q_{A^\sigma}$ is assumed to coequalize the maps defined by E^∞ in the way of Diagram (5), so does $\alpha^\varsigma \circ q_{A^\sigma}$, and it follows that that the n-ary expansions of the Σ-DLE associated with α^ς satisfy one of the conditions 5,6,7,8 of Proposition 2, and are in particular (σ^n, γ)-continuous. They therefore define the same L_Σ-algebra structure on A^σ as $(\alpha \circ q_A)^\sigma$. It follows that $\alpha^\varsigma \circ q_{A^\sigma} = \alpha^\sigma \circ q_{A^\sigma}$, i.e. that $\alpha^\varsigma = \alpha^\sigma$ since q_{A^σ} is (regular) epi.

Example 4. As was shown in Example 2, L^{ML}-algebras have Jónsson-Tarski extensions. Moreover, it is not difficult to see directly from the definition of δ^{ML} that for any A in **DL** its composition with the quotient $q_A : L_{\{\Diamond,\Box\}}A \twoheadrightarrow L^{\mathrm{ML}}A$ determines two unary maps on A preserving respectively all non-empty joins and all non-empty meets, i.e. δ^{ML} meets the criterion of Theorem 7.

Moreover as was shown in Example 3, L^{ML}-algebras also have canonical extensions, and it is clear from the definition that the equations defining L^{ML} are of the general shape of (7). It follows that L^{ML} satisfies the conditions of Theorems 7 and coalgebraic completeness-via-canonicity, i.e. Theorem 6, can therefore be used. Consider for example Dunn's Interaction axioms:

$$I = \{\Diamond a \wedge \Box b \leq \Diamond(a \wedge b), \Box(a \vee b) \leq \Box a \vee \Diamond b\}$$

Since they are canonical (see Example 3), it follows from Theorem 6 that the quotient of $\mathsf{F}_{L_{\mathrm{ML}}}FV$ under the fully invariant equivalence relation in **DL** defined by I, is strongly complete w.r.t. to $P_c \times P_c$-coalgebras validating I. We will denote this logic $\mathbf{K}_+$. These axioms *do not* collapse the relations interpreting $\Diamond$ and $\Box$ as might be expected and as is the case in standard Kripke frames (see [Dun95] and 6.1 of [GNV05] for a discussion on models with one or two relations). However, we can always find such a model if we accept to have a relation closed upward and downward. The following lemma is very useful in practice and greatly clarifies correspondence theory for positive ML (see [CJ97]). We denote by $\downarrow \gamma$ (resp. $\uparrow \gamma$) the pointwise downward (resp. upward) closure of a map $\gamma : W \to P_cW$.

Lemma 1. *Let* $\gamma_\Diamond \times \gamma_\Box : W \to P_c(W) \times P_c(W), w \in W$ *and* $a \in \mathsf{F}_{L_{\mathrm{ML}}}FV$. *then* $(w, \gamma_\Diamond \times \gamma_\Box, v) \models a$ *iff* $(w, \downarrow \gamma_\Diamond \times \uparrow \gamma_\Box, v) \models a$ *for any valuation* v.

Proof. Immediate from the fact that denotations are upsets.

Lemma 2. *Let $\gamma_\Diamond \times \gamma_\Box : W \to \mathsf{P}_c(W) \times \mathsf{P}_c(W)$ be a coalgebra validating the Interaction axioms, and let $w \in W$ and $a \in \mathsf{F}_{L\mathrm{ML}}\mathsf{F}V$, then $(w, \gamma_\Diamond \times \gamma_\Box, v) \models a$ iff $(w, (\gamma_\Diamond \cap \dot\gamma_\Box) \times (\gamma_\Diamond \cap \gamma_\Box), v) \models a$ for any valuation v.*

Proof. By induction on a, the interesting cases being $a = \Diamond b$ and $a = \Box b$. We show the $\Diamond b$ case, the other is dual. From Lemma 1, we can assume w.l.o.g. that $\gamma_\Box$ is upward-closed. We fix a valuation $v : V \to \mathcal{U}(W)$ and show the non-trivial direction: assume $(w, \gamma_\Diamond \times \gamma_\Box, v) \models \Diamond b$. Since $\Diamond c \wedge \Box d \leq \Diamond(c \wedge d)$ is valid, it must hold at w for any valuation. Consider for instance the following valuation: let q be a free variable, i.e. not occurring in b, and let us define $v'(p) = v(p)$ on $V \setminus \{q\}$ and $v'(q) = \gamma_\Box(w)$, which is an upset. The denotations of b under v and v' are equal. By construction we have $(w, \gamma_\Diamond \times \gamma_\Box, v') \models \Diamond b \wedge \Box q$, and thus $(w, \gamma_\Diamond \times \gamma_\Box, v') \models \Diamond(b \wedge q)$, and therefore there exist $x \in \gamma_\Diamond(w) \cap \gamma_\Box(w) \cap [\![b]\!]_{v'}$ but since $[\![b]\!]_{v'} = [\![b]\!]_v$ this means that there exists $x \in \gamma_\Diamond(w) \cap \gamma_\Box(w) \cap [\![b]\!]_v$, i.e. $(w, (\gamma_\Diamond \cap \gamma_\Box) \times (\gamma_\Diamond \cap \gamma_\Box), v) \models \Diamond b$ as desired.

The choice of which type of model to consider, viz. models with one or two relations, will *in fine* depend on what the models represent. In the next example we will present models where states are memory resources and accessibility relations correspond to the action of programs. A single relation then interprets each pair of existential and universal modalities, and I is then trivially satisfied.

Example 5 (Modal separation logics). We conclude with a more elaborate family of examples. In [DP15a] we introduced the functor $L^{\mathrm{SL}} : \mathbf{DL} \to \mathbf{DL}$ defined by

$$L^{\mathrm{SL}}A = \mathsf{F}\{I, a * b, a \mathbin{-\!\!*} b, a *\!\!- b \mid a, b \in \mathsf{U}A\}/$$
$$\{(a \vee b) * c = (a * c) \vee (b * c), a * (b \vee c) = (a * b) \vee (a * c)$$
$$a \mathbin{-\!\!*}(b \wedge c) = (a \mathbin{-\!\!*}b) \wedge (a \mathbin{-\!\!*}c), (a \vee b) \mathbin{-\!\!*}c = (a \mathbin{-\!\!*}c) \wedge (b \mathbin{-\!\!*}c)$$
$$(a \wedge b) *\!\!- c = (a *\!\!- c) \wedge (a *\!\!- c), a *\!\!-(b \vee c) = (a *\!\!- b) \wedge (a *\!\!- c)\}$$

We interpret L^{SL}-formulas in T^{SL}-coalgebras for $T^{\mathrm{SL}} : \mathbf{Pos} \to \mathbf{Pos}$ defined by:

$$T^{\mathrm{SL}}W = 2 \times \mathsf{P}_c(W \times W) \times \mathsf{P}_c(W^{\mathrm{op}} \times W) \times \mathsf{P}_c(W \times W^{\mathrm{op}})$$

via the semantic transformation $\delta^{\mathrm{SL}} : L^{\mathrm{SL}}\mathcal{U} \to \mathcal{U}T^{\mathrm{SL}}$ defined on generators at each poset W by $\delta_W^{\mathrm{SL}}(I) = \{t \in T^{\mathrm{SL}}W \mid \pi_1(t) = 0\}$ and

$$\delta_W^{\mathrm{SL}}(U * V) = \{t \in T^{\mathrm{SL}}W \mid \exists (x, y) \in \pi_2(t), x \in U, y \in V\}$$
$$\delta_W^{\mathrm{SL}}(U \mathbin{-\!\!*}V) = \{t \in T^{\mathrm{SL}}W \mid \forall (x, y) \in \pi_3(t), x \in U \Rightarrow y \in V\}$$
$$\delta^{\mathrm{SL}}W(U *\!\!- V) = \{t \in T^{\mathrm{SL}}W \mid \forall (x, y) \in \pi_4(t), y \in V \Rightarrow x \in U\}$$

The intended interpretation of this language is that worlds represent *resources* which can be split and $w \models p * q$ means that the resource w can be split into two resource s, t such that $s \models p$ and $t \models q$. The operations $\mathbin{-\!\!*}$ and $*\!\!-$ are left and right residuals to $*$, and I acts as a unit. This is encoded by the (in)equations:

FC1 $a * 1 = a, 1 * a = a$ FC4 $(c *\!\!-b) * a \leq c *\!\!-(a * b)$

FC2 $1 \leq a -\!\!*a, 1 \leq a *\!\!-a$ FC5 $(a -\!\!*b) * b \leq a$

FC3 $a * (b -\!\!*c) \leq (a * b) -\!\!*c$ FC6 $b * (b *\!\!-a) \leq a$

The logic defined by L^{SL} and these (in)equalities is known as *separation logic* or *the logic of bunched implication* or *the distributive Lambek calculus* depending on the context, and we shall denote it as **SL**. These (in)equations are canonical (residuated maps and their residuals are very well-behaved under canonical extension, even in posets see [Mor14]). As was shown in [DP15a], the adjoint transformation $\hat{\delta}^{SL}$ has right-inverses at every A in **DL**, and L^{SL}-algebras therefore have Jónsson-Tarski extensions. Moreover, as can be seen from the definition of δ^{SL} the criterion of (anti)-preservation of arbitrary joins or meets of Theorem 7 is also satisfied. Finally, the equations defining L^{SL} satisfy the conditions of Theorem 5 and canonical extensions thus lift to $\mathbf{Alg_{DL}}(L^{SL})$. All the conditions of Theorem 7 are thus satisfied, and we can use Theorem 6 on the regular quotient defined by FC1-FC6. The logic **SL** is thus strongly complete w.r.t. the T^{SL}-coalgebras validating these axioms, viz. T^{SL}-coalgebras such that

$$(x, y) \in \gamma_*(w) \text{ iff } (x, w) \in \gamma_{-*}(y) \text{ iff } (w, y) \in \gamma_{*-}(x) \qquad (8)$$

and for every $w \in W$ there exists $(w, x), (y, w) \in \gamma_*(w)$ with $x, y \models I$. A typical example of such coalgebra is given by memory states represented by the set H of *heaps*, i.e. partial maps $f : \mathbb{N}_+ \rightharpoonup_f \mathbb{N}$ with finite domain. These are ordered by $f \leq g$ if $g \restriction \text{dom} f = f$, the empty heap is the unit and $\gamma_* : H \to \mathsf{P}_c(H \times H), f \mapsto \{(g, h) \mid \text{dom} g \cap \text{dom} h = \emptyset, g, h \leq f\}$ interprets the *separation conjunction* $*$ and its residuals via (8). In this context, it is reasonable to combine modal logics for program specification with separation logic to describe heaps evolving under the action of programs. Various fragments of PDL (see [Gol92]) are good candidates. For example, consider the simple program syntax $\alpha :: = \pi \mid \alpha; \alpha$ with $\pi \in \Pi$ a set of atomic programs. By Theorem 3, the results for **SL** and $\mathbf{K_+}$, and Lemma 2, it follows that the *fusion* $\bigoplus_{\Pi^*} \mathbf{K_+} \oplus \mathbf{SL}$ is strongly complete w.r.t. to $\prod_{\Pi^*} \mathsf{P}_c(-) \times 2 \times \mathsf{P}_c((-) \times (-))$-coalgebras. If we want to encode the sequential composition of the grammar in the interpretation we need the axioms

$$\mathsf{Comp} = \{\langle \alpha_1; \alpha_2 \rangle a = \langle \alpha_1 \rangle \langle \alpha_2 \rangle a, [\alpha_1; \alpha_2]a = [\alpha_1][\alpha_2]a \mid \alpha_1, \alpha_2 \in \Pi^*\}$$

It is easy to check from Theorem 4 that these axioms are canonical. They are valid in a model with a single relation R_α interpreting each pair $\langle \alpha \rangle, [\alpha]$ if $R_{\alpha_1}^\downarrow \circ R_{\alpha_2}^\downarrow = R_{\alpha_1;\alpha_2}^\downarrow$ and $R_{\alpha_1}^\uparrow \circ R_{\alpha_2}^\uparrow = R_{\alpha_1;\alpha_2}^\uparrow$, where $R_\alpha^\downarrow$ and $R_\alpha^\uparrow$ are the downward and upward closure of R_α respectively. Theorem 6 gives us strong completeness of $\bigoplus_{\Pi^*} \mathbf{K_+}/\{\mathsf{Comp}\}$ with respect to such coalgebras, and modularity then gives us strong completeness of $\bigoplus_{\Pi^*} \mathbf{K_+}/\{\mathsf{Comp}\} \oplus \mathbf{SL}$. Note how the use of *positive logics* allows us to talk about existential access to all resources smaller than certain *upper bounds*, and universal access to resources larger than certain *lower bounds*. More interestingly perhaps, we could consider $\alpha :: = \pi \mid \alpha; \alpha \mid \alpha \parallel \alpha$ with a parallel composition operation and interaction axioms of the shape

$$\langle \alpha_1 \parallel \alpha_2 \rangle a = \langle \alpha_1 \rangle a * \langle \alpha_2 \rangle a$$

This time, strong completeness will not simply transfer by modularity since we are making the languages interact, however since such equations are canonical, we can still apply completeness-via-canonicity, only this time to the entire logic.

6   Conclusion

We have described the key steps of the coalgebraic version of completeness-via-canonicity, and in particular the key role played by the Jónsson-Tarski and canonical extensions. We have sketched an implementation of the method for all positive or boolean logics with modalities satisfying a set of equations in the shape of (7) and a relational semantics. Much work remains to be done. We have a complete implementation for boolean graded logics, but not yet in the positive case, and no implementation at all for probability logic. For this we would like to explore whether the method can be applied to **MSL**-based logics, since an expressive logics for Markov chains can be formulated over **MSL** [JS10].

References

[BJ11]   Bezhanishvili, G., Jansana, R.: Priestley style duality for distributive meet-semilattices. Stud. Logica **98**(1–2), 83–122 (2011)

[BKV13]   Balan, A., Kurz, A., Velebil, J.: Positive fragments of coalgebraic logics. In: Heckel, R., Milius, S. (eds.) CALCO 2013. LNCS, vol. 8089, pp. 51–65. Springer, Heidelberg (2013)

[CJ97]   Celani, S., Jansana, R.: A new semantics for positive modal logic. Notre Dame J. Formal Logic **38**(1), 1–18 (1997)

[CKP+09]   Cîrstea, C., Kurz, A., Pattinson, D., Schröder, L., Venema, Y.: Modal logics are coalgebraic. Comput. J. (2009)

[CP07]   Cîrstea, C., Pattinson, D.: Modular construction of complete coalgebraic logics. Theor. Comput. Sci. **388**(1–3), 83–108 (2007)

[Dah15]   Dahlqvist, F.: Completeness-via-canonicity for coalgebraic logics, Ph.D. thesis, Imperial College London (2015)

[DP11]   Dahlqvist, F., Pattinson, D.: On the fusion of coalgebraic logics. In: Corradini, A., Klin, B., Cîrstea, C. (eds.) CALCO 2011. LNCS, vol. 6859, pp. 161–175. Springer, Heidelberg (2011)

[DP13]   Dahlqvist, F., Pattinson, D.: Some Sahlqvist completeness results for coalgebraic logics. In: Pfenning, F. (ed.) FOSSACS 2013 (ETAPS 2013). LNCS, vol. 7794, pp. 193–208. Springer, Heidelberg (2013)

[DP15a]   Dahlqvist, F., Pym, D.: Completeness via canonicity for distributive substructural logics: a coalgebraic perspective. In: Kahl, W., Winter, M., Oliveira, J. (eds.) RAMiCS 2015. LNCS, vol. 9348, pp. 119–135. Springer, Heidelberg (2015)

[DP15b]   Dahlqvist, F., Pym, D.: Completeness-via-canonicity for distributive substructural logics, a coalgebraic perspective, Technical report RN/15/04, UCL (2015)

[Dun95]   Dunn, J.M.: Positive modal logic. Stud. Logica **55**(2), 301–317 (1995)

[GH01]   Gehrke, M., Harding, J.: Bounded lattice expansions. J. Algebra **238**(1), 345–371 (2001)

[GJ94]   Gehrke, M., Jónsson, B.: Bounded distributive lattices with operators. Math. Japon. **40**(2), 207–215 (1994)

[GJ04]   Gehrke, M., Jónsson, B.: Bounded distributive lattice expansions. Math. Scand. **94**, 13–45 (2004)

[GNV05] Gehrke, M., Nagahashi, H., Venema, Y.: A Sahlqvist theorem for distributive modal logic. Ann. Pure Appl. Logic **131**(1–3), 65–102 (2005)

[Gol92] Goldblatt, R.: Logics of Time and Computation. CSLI Lecture Notes. Center for the Study of Language and Information, Stanford (1992)

[GP14]   Gouveia, M.J., Priestley, H.A.: Canonical extensions and profinite completions of semilattices and lattices. Order **31**(2), 189–216 (2014)

[HV05]   Hodkinson, I., Venema, Y.: Canonical varieties with no canonical axiomatisation. Trans. AMS **357**(11), 4579–4605 (2005)

[Jón94] Jónsson, B.: On the canonicity of Sahlqvist identities. Stud. Logica **53**(4), 473–492 (1994)

[JS10]   Jacobs, B., Sokolova, A.: Exemplaric expressivity of modal logics. J. Log. Comput. **20**, 1041–1068 (2010)

[JT51]   Jónsson, B., Tarski, A.: Boolean algebras with operators. Part 1. Amer. J. Math. **33**, 891–937 (1951)

[KKP04] Kupke, C., Kurz, A., Pattinson, D.: Algebraic semantics for coalgebraic logics. Electr. Notes in Th. Comp. Sc. **106**, 219–241 (2004). CMCS

[KKP05] Kupke, C., Kurz, A., Pattinson, D.: Ultrafilter Extensions for coalgebras. In: Fiadeiro, J.L., Harman, N.A., Roggenbach, M., Rutten, J. (eds.) CALCO 2005. LNCS, vol. 3629, pp. 263–277. Springer, Heidelberg (2005)

[KKV12] Kapulkin, K., Kurz, A., Velebil, J.: Expressiveness of positive coalgebraic logic. AiML **9**, 368–385 (2012)

[KP11]   Kupke, C., Pattinson, D.: Coalgebraic semantics of modal logics: an overview. Th. Comp. Sc. **412**(38), 5070–5094 (2011). CMCS 2010

[KR12]   Kurz, A., Rosický, J.: Strongly complete logics for coalgebras. Logical Methods Comput. Sci. **8**(3), 1–32 (2012)

[Mor14] Morton, W.: Canonical extensions of posets. Algebra Univers. **72**(2), 167–200 (2014)

[PS08]   Pattinson, D., Schröder, L.: Beyond rank 1: algebraic semantics and finite models for coalgebraic logics. In: Amadio, R.M. (ed.) FOSSACS 2008. LNCS, vol. 4962, pp. 66–80. Springer, Heidelberg (2008)

[Rib52] Ribeiro, H.: A remark on boolean algebras with operators. Amer. J. Math. **74**, 162–167 (1952)

[Sch06] Schröder, L.: A finite model construction for coalgebraic modal logic. In: Aceto, L., Ingólfsdóttir, A. (eds.) FOSSACS 2006. LNCS, vol. 3921, pp. 157–171. Springer, Heidelberg (2006)

[SP09]   Schröder, L., Pattinson, D.: Strong completeness of coalgebraic modal logics. STACS, Dagstuhl Seminar Proceedings, vol. 09001, pp. 673–684 (2009)

[SP10]   Schröder, L., Pattinson, D.: Rank-1 logics are coalgebraic. J. Log. Comput. **20**(5), 1113–1147 (2010)

[Ven06] Venema, Y.: Algebras and coalgebras. In: van Benthem, J., Blackburn, P., Wolter, F. (eds.) Handbook of Modal Logic. Elsevier, Amsterdam (2006)

[VK11]   Velebil, J., Kurz, A.: Equational presentations of functors and monads. Math. Struct. in Comp. Sc. **21**(2), 363–381 (2011)

Relational Lattices via Duality

Luigi Santocanale[✉]

LIF, CNRS UMR 7279, Aix-Marseille Université, Marseille, France
luigi.santocanale@lif.univ-mrs.fr

Abstract. The *natural join* and the *inner union* combine in different ways tables of a relational database. Tropashko [18] observed that these two operations are the meet and join in a class of lattices—called the *relational lattices*—and proposed lattice theory as an alternative algebraic approach to databases. Aiming at query optimization, Litak et al. [12] initiated the study of the equational theory of these lattices. We carry on with this project, making use of the duality theory developed in [16]. The contributions of this paper are as follows. Let A be a set of column's names and D be a set of cell values; we characterize the dual space of the relational lattice $\mathsf{R}(D, A)$ by means of a generalized ultrametric space, whose elements are the functions from A to D, with the $P(A)$-valued distance being the Hamming one but lifted to subsets of A. We use the dual space to present an equational axiomatization of these lattices that reflects the combinatorial properties of these generalized ultrametric spaces: symmetry and pairwise completeness. Finally, we argue that these equations correspond to combinatorial properties of the dual spaces of lattices, in a technical sense analogous of correspondence theory in modal logic. In particular, this leads to an exact characterization of the finite lattices satisfying these equations.

1 Introduction

Tropashko [18] has recently observed that the *natural join* and the *inner union* , two fundamental operations of the relational algebra initiated by Codd [2]—the algebra by which we construct queries—can be considered as the meet and join operations in a class of lattices, known by now as the class of *relational lattices*. Elements of the relational lattice $\mathsf{R}(D, A)$ are the relations whose variables are listed by a subset of a total set A of attributes, and whose tuples' entries are taken from a set D. Roughly speaking, we can consider a relation as a table of a database, its variables as the columns' names, its tuples being the rows.

Let us illustrate these operations with examples. The natural join takes two tables and constructs a new one whose columns are indexed by the union of the headers, and whose rows are the glueings of the rows along identical values in common columns. As we emphasize in this paper the lattice theoretic aspects of the natural join operation, we shall depart from the standard practice of denoting it by the symbol $\bowtie$ and use instead the meet symbol $\wedge$.

© IFIP International Federation for Information Processing 2016
Published by Springer International Publishing Switzerland 2016. All Rights Reserved
I. Hasuo (Ed.): CMCS 2016, LNCS 9608, pp. 195–215, 2016.
DOI: 10.1007/978-3-319-40370-0_12

Author	Area
Santocanale	Logic
Santocanale	CS

$\wedge$

Area	Reviewer
CS	Turing
Logic	Gödel

$=$

Author	Area	Reviewer
Santocanale	Logic	Gödel
Santocanale	CS	Turing

The inner union restricts two tables to the common columns and lists all the possible rows. The following example suggests how to construct, using this operation, a table of users given two (or more) tables of people having different roles.

Authors		
Name	Surname	Conf
Luigi	Santocanale	CMCS

$\vee$

Reviewers		
Name	Surname	Area
Alan	Turing	CS
Kurt	Gödel	Logic

$=$

Users	
Name	Surname
Luigi	Santocanale
Alan	Turing
Kurt	Gödel

Considering the lattice signature as a subsignature of the relational algebra, Litak, Mikulás and Hidders [12] proposed to study the equational theory of the relational lattices. The capability to recognize when two queries are equivalent— that is, a solution to the word problem of such a theory—is of course an important step towards query optimization.

Spight and Tropashko [17] exhibited equational principles in a signature strictly larger than the one of lattice theory. A main contribution of Litak et al. [12]—a work to which we are indebted in many respects—was to show that the quasiequational theory of relational lattices *with the header constant* is undecidable. The authors also proposed a base of equations for the theory in the signature extended with the header constant, and exhibited two non-trivial pure lattice equations holding on relational lattices. It was argued there that the lattice $R(D, A)$ arises via a closure operator on the powerset $P(A \sqcup D^A)$ and, at the same time, as the Grothendieck construction for the functor $P(D^{(-)})$, from $P(A)^{op}$ to $SL_\vee$ (the category of complete lattices and join-preserving mappings), sending $X \subseteq A$ to D^X and then D^X covariantly to $P(D^X)$.

The focus of this paper is on the pure lattice signature. We tackle the study of the equational theory of relational lattices in a coalgebraic fashion, that is, by using the duality theory developed in [16] for finite lattices and here partially extended to infinite lattices. Let us recall some key ideas from the theory, which in turn relies on Nation's representation Theorem [14, Sect. 2]. For a complete lattice L, a join-cover of $x \in L$ is a subset $Y \subseteq L$ such that $x \leq \bigvee Y$. A lattice is *pluperfect* if it is a complete spatial lattice and every join-cover of a completely join-irreducible element refines to a minimal one—see Sect. 2 for a complete definition. Every finite lattice is pluperfect; moreover, relational lattices are pluperfect, even when they are infinite. This property, i.e. pluperfectness, allows to define the dual structure of a lattice L, named the OD-graph in [14]. This is the triple $\langle \mathcal{J}(L), \leq, \vartriangleleft_m \rangle$ with $\mathcal{J}(L)$ the set of completely join-irreducible elements, $\leq$ the restriction of the order to $\mathcal{J}(L)$, and the relation $j \vartriangleleft_m C$ holds when $j \in \mathcal{J}(L)$, $C \subseteq \mathcal{J}(L)$, and C is a minimal join-cover of j. The original lattice L can be recovered up to isomorphism from its OD-graph as the lattice of closed downsets of $\mathcal{J}(L)$—where a downset $X \subseteq \mathcal{J}(L)$ is closed if $j \vartriangleleft_m C \subseteq X$ implies $j \in X$.

We characterize the OD-graph of the lattice $R(D, A)$ as follows. Firstly recall from [12] that we can identify completely join-irreducible elements of $R(D, A)$ with elements of the disjoint sum $A \sqcup D^A$. The order on completely join-irreducible elements is trivial, i.e. it is the equality. All the elements of A are join-prime, whence the only minimal join-cover of some $a \in A$ is the singleton $\{a\}$. The minimal join-covers of elements in D^A are described via an ultrametric distance valued in the join-semilattice $P(A)$; this is, morally, the Hamming distance, $\delta(f, g) = \{x \in A \mid f(x) \neq g(x)\}$. Whenever $f, g \in D^A$ we have $f \lhd_{\mathrm{m}} \delta(f, g) \cup \{g\}$ and these are all the minimal join-covers of f.

As in correspondence theory for modal logic, the combinatorial structure of the dual spaces is an important source for discovering axioms/equations that uniformly hold in a class of models. For relational lattices, most of these combinatorial properties stem from the structure of the ultrametric space (D^A, δ). When we firstly attempted to show that equations (RL1) and (RL2) from [12] hold in relational lattices using duality, we realized that the properties necessary to enforce these equations were the following:

P1. Every non-trivial minimal join-cover contains at most one join-irreducible element which is not join-prime.

Moreover, the generalized ultrametric space (D^A, δ) is

P2. *symmetric*, i.e. $\delta(f, g) = \delta(g, f)$, for each $f, g \in D^A$,
P3. *pairwise complete* : if $\delta(f, g) \subseteq X \cup Y$, then $\delta(f, h) \subseteq X$ and $\delta(h, g) \subseteq Y$ for some $h \in D^A$.

Various notions of completeness for generalized ultrametric spaces are discussed in [1]. At first we called pairwise completeness the Beck-Chevalley-Malcev property of (D^A, δ). Indeed, it is equivalent to saying that the functor $P(D^{(-)}) : P(A)^{op} \longrightarrow \mathsf{SL}_\vee$ mentioned above sends a pullback square (i.e., a square of inclusions with objects $X \cap Y, X, Y, Z$) to a square satisfying the Beck-Chevalley condition. As the property implies that a collection of congruences of join-semilattices commute, it is also a sort of Malcev condition.

We show with Theorem 4 that property P1 of an OD-graph is definable by an equation that we name (Unjp). We investigate the deductive strength of this equation and show in particular that (RL2) is derivable from (Unjp), but not the converse.

In presence of P1, symmetry and pairwise completeness can also be understood as properties of an OD-graph. Symmetry is the following property: if $k_0 \lhd_{\mathrm{m}} C \cup \{k_1\}$ with k_1 not join-prime, then $k_1 \lhd_{\mathrm{m}} C \cup \{k_0\}$. Pairwise completeness can be read as follows: if $k_0 \lhd_{\mathrm{m}} C_0 \cup C_1 \cup \{k_2\}$ with k_2 not join-prime and $C_0, C_1, \{k_2\}$ pairwise disjoint, then $k_0 \lhd_{\mathrm{m}} C_0 \cup \{k_1\}$ and $k_1 \lhd_{\mathrm{m}} C_1 \cup \{k_2\}$ for some completely join-irreducible element k_1.

We exhibit in Sect. 6 three equations valid on relational lattices and characterize, via a set of properties of their OD-graphs, the pluperfect lattices satisfying (Unjp) and these equations. We propose these four equations as an axiomatization of the theory of relational lattices that we call [[AxRel]]. The main result of this paper, Theorem 7, sounds as follows. If we restrict to *finite* lattices that are *atomistic*—that is, lattices in which any element is the join of the atoms below it, so the order on join-irreducible elements in the dual space is trivial—then a lattice satisfies [[AxRel]] if and only if its OD-graph is symmetric and pairwise complete, in the sense just explained.

We can build lattices similar to the relational lattices from $P(A)$-valued ultra-metric spaces. It is tempting to look for further equations so to represent the OD-graph of finite atomistic lattices satisfying these equations as $P(A)$-valued ultrametric space. Unfortunately this is not possible, since a key property of the OD-graph of lattices of ultrametric spaces—the ones ensuring that the distance function is well defined—is not definable by lattice equations. Yet Theorem 7 also exhibits a deep connection between the OD-graph of finite atomistic lattices satisfying [[AxRel]] and the frames of the commutator logic $[S5]^A$, see [10]. Considering the complexity of the theory of combination of modal logics, Theorem 7 can be used to foresee and shape future researches. For example, we shall discuss in Sect. 7 how to derive undecidability results from the correspondent ones in multidimensional modal logic. In particular, a refinement of the main result of Litak et al. [12, Corollary 4.8] can be derived.

The paper is structured as follows. We introduce in Sect. 2 the notation as well as the least lattice theoretic tools that shall allow the reader to go through the paper. In Sect. 3 we describe the relational lattices, present some known results in the literature, and give a personal twist to these results. In particular, we shall introduce semidirect products of lattices, ultrametric spaces as a tool for studying relational lattices, emphasize the role of the Beck-Chevalley property in the theory. In Sect. 4 we characterize the OD-graphs of relational lattices. In Sect. 5 we present our results on the equation (Unjp). In Sect. 6, we describe our results relating equations valid on relational lattices to symmetry and pairwise completeness. In the last Section we discuss the results presented as well as ongoing researches, by the author and by other researchers, trace a road-map for future work.

2   Some Elementary Lattice Theory

A *lattice* is a poset L such that every finite non-empty subset $X \subseteq L$ admits a smallest upper bound $\bigvee X$ and a greatest lower bound $\bigwedge X$. We assume a minimal knowledge of lattice theory—otherwise, we invite the reader to consult a standard monograph on the subject, such as [3] or [5]. The technical tools that we use may be found in the monograph [4], that we also invite to explore. A lattice can also be understood as a structure $\mathfrak{A}$ for the functional signature $(\vee, \wedge)$, such that the interpretations of these two binary function symbols both give $\mathfrak{A}$ the structure of an idempotent commutative semigroup, the two semigroup structures being tied up by the absorption laws $x \wedge (y \vee x) = x$ and $x \vee (y \wedge x) = x$. Once a lattice is presented as such structure, the order is recovered by stating that $x \leq y$ holds if and only if $x \wedge y = x$.

A lattice L is *complete* if any subset $X \subseteq L$ admits a smallest upper bound $\bigvee X$. It can be shown that this condition implies that any subset $X \subseteq L$ admits a greatest lower bound $\bigwedge X$. A complete lattice is *bounded*, since $\bot := \bigvee \emptyset$ and $\top := \bigwedge \emptyset$ are respectively the least and greatest elements of the lattice.

A *closure operator* on a complete lattice L is an order-preserving function $j : L \longrightarrow L$ such that $x \leq j(x)$ and $j^2(x) = j(x)$, for each $x \in L$. We shall use $\mathtt{Clop}(L)$ to denote the poset of closure operators on L, under the pointwise ordering. It can be shown that $\mathtt{Clop}(L)$ is itself a complete lattice. If $j \in \mathtt{Clop}(L)$, then the set L/j of fixed points of j is itself a complete lattice, with $\bigwedge_{L/j} X = \bigwedge_L X$ and $\bigvee_{L/j} X = j(\bigvee_L X)$. For the correspondence between closure operators and congruences in the category of complete join-semilattices, see [9].

Let L be a complete lattice. An element $j \in L$ is said to be *completely join-irreducible* if $j = \bigvee X$ implies $j \in X$, for each $X \subseteq L$; the set of completely join-irreducible element of L is denoted here $\mathcal{J}(L)$. A complete lattice is *spatial* if every element is the join of the completely join-irreducible elements below it. An element $j \in \mathcal{J}(L)$ is said to be *join-prime* if $j \leq \bigvee X$ implies $j \leq x$ for some $x \in X$, for each finite subset X of L. We say that $j \in \mathcal{J}(L)$ is *non-join-prime* if it is not join-prime. An *atom* of a lattice L is an element of L such that $\bot$ is the only element strictly below it. A spatial lattice is *atomistic* if every element of $\mathcal{J}(L)$ is an atom.

For $j \in \mathcal{J}(L)$, a *join-cover* of j is a subset $X \subseteq L$ such that $j \leq \bigvee X$. For $X, Y \subseteq L$, we say that X *refines* Y, and write $X \ll Y$, if for all $x \in X$ there exists $y \in Y$ such that $x \leq y$. A join-cover X of j is said to be *minimal* if $j \leq \bigvee Y$ and $Y \ll X$ implies $X \subseteq Y$; we write $j \lhd_m X$ if X is a minimal join-cover of j. In a spatial lattice, if $j \lhd_m X$, then $X \subseteq \mathcal{J}(L)$. If $j \lhd_m X$, then we say that X is a *non-trivial* minimal join-cover of j if $X \neq \{j\}$. It is common to use the word perfect for a lattice which is both spatial and dually spatial. We need here something different:

Definition 1. *A complete lattice is* pluperfect *if it is spatial and for each $j \in \mathcal{J}(L)$ and $X \subseteq L$, if $j \leq \bigvee X$, then $Y \ll X$ for some Y such that $j \lhd_m Y$. The OD-graph of a pluperfect lattice L is the structure $\langle \mathcal{J}(L), \leq, \lhd_m \rangle$.*

That is, in a pluperfect lattice every cover refines to a minimal one. Notice that every finite lattice is pluperfect. If L is a pluperfect lattice, then we say that $X \subseteq \mathcal{J}(L)$ is *closed* if it is a downset and $j \lhd_m C \subseteq X$ implies $j \in X$. As from standard theory, the mapping $X \mapsto \bigcap \{ Y \subseteq \mathcal{J}(L) \mid X \subseteq Y, Y \text{ is closed} \}$ defines a closure operator whose fixed points are exactly the closed subsets of $\mathcal{J}(L)$. The interest of considering pluperfect lattices stems from the following representation Theorem.

Theorem 1 (Nation [14]). *Let L be a pluperfect lattice and let $\mathsf{L}(\mathcal{J}(L), \leq, \lhd_m)$ be the lattice of closed subsets of $\mathcal{J}(L)$. The mapping $l \mapsto \{ j \in \mathcal{J}(L) \mid j \leq l \}$ is a lattice isomorphism from L to $\mathsf{L}(\mathcal{J}(L), \leq, \lhd_m)$.*

It was shown in [16] how to extend this representation theorem to a duality between the category of finite lattices and the category of OD-graphs. The following Lemma shall be repeatedly used in the proofs of our statements.

Lemma 1. *Let L be a pluperfect lattice, let $j \lhd_m C$ and $k \in C$. If $j \leq \bigvee D$ with $D \ll \{ \bigvee (C \setminus \{k\}), k \}$, then $k \in D$. In particular, if $k' < k$, then $\{ \bigvee C \setminus \{k\}, k' \}$ is not a cover of j.*

3   The Relational Lattices $R(D, A)$

In this Section we define relational lattices, recall some known facts, and develop then some tools to be used later, semidirect products of lattices, generalized ultrametric spaces, a precise connection to the theory of combination of modal logics (as well as multidimensional modal logic and relational algebras).

Let A be a collection of attributes (or column names) and let D be a set of cell values. A *relation* (or, more informally, a *table*) on A and D is a pair (X, T) where $X \subseteq A$ and $T \subseteq D^X$; X is the header of the table while T is the collection of rows. Elements of the relational lattice $R(D, A)$ are relations on A and D.

Before we define the natural join, the inner union operations, and the order on $R(D, A)$, let us recall a few key operations. If $X \subseteq Y \subseteq A$ and $f \in D^Y$, then we shall use $f_{\upharpoonright X} \in D^X$ for the restriction of f to X; if $T \subseteq D^Y$, then $T{\upharpoonright}_X$ shall denote projection to X, that is, the direct image of T along restriction, $T{\upharpoonright}_X := \{f_{\upharpoonright X} \mid f \in T\}$; if $T \subseteq D^X$, then $i_Y(T)$ shall denote cylindrification to Y, that is, the inverse image of restriction, $i_Y(T) := \{f \in D^Y \mid f_{\upharpoonright X} \in T\}$. Recall that i_Y is right adjoint to ${\upharpoonright}_X$. With this in mind, the natural join and the inner union of tables are respectively described by the following formulas:

$$(X_1, T_1) \wedge (X_2, T_2) := (X_1 \cup X_2, T)$$
$$\text{where } T = \{f \mid f_{\upharpoonright X_i} \in T_i, i = 1, 2\} = i_{X_1 \cup X_2}(T_1) \cap i_{X_1 \cup X_2}(T_2),$$

$$(X_1, T_1) \vee (X_2, T_2) := (X_1 \cap X_2, T)$$
$$\text{where } T = \{f \mid \exists i \in \{1, 2\}, \exists g \in T_i \text{ s.t. } g_{\upharpoonright X_1 \cap X_2} = f\}$$
$$= T_1{\upharpoonright}_{X_1 \cap X_2} \cup T_2{\upharpoonright}_{X_1 \cap X_2}.$$

The order is then given by

$$(X_1, T_1) \leq (X_2, T_2) \qquad \text{iff} \qquad X_2 \subseteq X_1 \text{ and } T_1{\upharpoonright}_{X_2} \subseteq T_2.$$

It was observed in [12] that $R(D, A)$ arises—as a category with at most one arrow between two objects—via the Grothendieck construction for the functor sending $X \subseteq A$ contravariantly to D^X and then D^X covariantly to $P(D^X)$. Let us record the following important property:

Lemma 2. *The image of a pullback square by the functor $P(D^{(-)}) : P(A)^{op} \longrightarrow$ $SL_\vee$ satisfies the Beck-Chevalley property.*

The above statement means that if we apply the functor to inclusions of the form $X_1 \cap X_2 \subseteq X_i \subseteq X_3$, $i = 1, 2$, then the two possible diagonals in the diagram on the right, ${\upharpoonright}_{X_2} \circ i_{X_3}$ and $i_{X_2} \circ {\upharpoonright}_{X_1 \cap X_2}$, are equal. The Beck-Chevalley property is a consequence of the glueing property of functions:

if $f \in D^{X_2}, g \in D^{X_1}$ *and* $f_{\restriction X_1 \cap X_2} = g_{\restriction X_1 \cap X_2}$, *then there exists* $h \in D^{X_3}$ *such that* $h_{\restriction X_2} = f$ *and* $h_{\restriction X_1} = g$.

We can recast the previous category-theoretic observations in an algebraic framework. An *action* of a complete lattice L over a complete lattice M is a monotonic mapping $\langle \ \rangle : L \longrightarrow \mathsf{Clop}(M)$, thus sending $X \in L$ to a closure operator $\langle X \rangle$ on M. Given such an action, if we define $j(X, T) := (X, \langle X \rangle T)$, then $j(X, T)$ is a closure operator on the product $L \times M$. In particular, the set of j-fixed points, $L \ltimes_j M := \{(X, T) \in L \times M \mid \langle X \rangle T = T\}$, is itself a complete lattice, where the meet coincides with the one from $L \times M$, while the join is given by the formula $(X_1, T_1) \vee_{L \ltimes_j M} (X_2, T_2) := (X_1 \vee X_2, \langle X_1 \vee X_2 \rangle (T_1 \vee T_2))$. We call $L \ltimes_j M$ the *semidirect product* of L and M via j. The naming is chosen here after the semidirect product of groups, which is a similar instance of the Grothendieck construction. Given such an action, the correspondence $X \mapsto M/\langle X \rangle$ gives rise to a covariant functor from L to the category $\mathsf{SL_\vee}$, so that it makes sense to ask when the Beck-Chevalley property holds, as in Lemma 2. This happens—and then we say that an action $\langle \ \rangle$ satisfies the Beck-Chevalley property—exactly when

$$\langle X_1 \vee X_2 \rangle T = \langle X_1 \rangle \langle X_2 \rangle T, \qquad \text{for each } X_1, X_2 \in L \text{ and } T \in M. \tag{1}$$

Notice that the identity $\langle X_1 \rangle \langle X_2 \rangle T = \langle X_2 \rangle \langle X_1 \rangle T$ is a consequence of (1). As these closure operators correspond to congruences of complete join-semilattices, we also think of the Beck-Chevalley property as a form of Malcev property, stating that a collection of congruences (thought as binary relations) pairwise commute (w.r.t composition of relations).

Relational lattices from ultrametric spaces. Let us come back to the lattice $\mathsf{R}(D, A)$. Define on the set D^A the following $P(A)$-valued ultrametric distance:

$$\delta(f, g) := \{x \in A \mid f(x) \neq g(x)\}.$$

Thus $\delta(f, f) \subseteq \emptyset$ and $\delta(f, g) \subseteq \delta(f, h) \cup \delta(h, g)$ for any $f, g, h \in D^A$, making (D^A, δ) into a generalized metric space in the sense of [11].[1] With respect to the latter work—where axioms for the distance are those of a category enriched over $(P(A)^{op}, \emptyset, \cup)$—for $f, g \in D^A$ we also have that $\delta(f, g) = \emptyset$ implies $f = g$ and symmetry, $\delta(f, g) = \delta(g, f)$. We can define then an action of $P(A)$ on $P(D^A)$:

$$\langle X \rangle T = \{f \in D^A \mid \exists g \in T \text{ s.t. } \delta(f, g) \subseteq X\}. \tag{2}$$

We can now restate (and refine) Lemma 2.1 from [12]—which constructs the lattice $\mathsf{R}(D, A)$ via a closure operator on $P(A + D^A)$—as follows:

Theorem 2. *The correspondence sending* (X, T) *to* $(A \setminus X, i_A(T))$ *is an isomorphism bewteen the relational lattice* $\mathsf{R}(D, A)$ *and* $P(A) \ltimes_j P(D^A)$.

---

[1] This is a lifting of the Hamming distance to subsets. Yet, in view of [15] and of their work on generalized ultrametric spaces, such a distance might reasonably tributed to Priess-Crampe and Ribenboim.

The action defined in (2) satisfies the identity (1). As a matter of fact, (1) is equivalent to *pairwise completeness* of (D^A, δ) as an ultrametric space, see [1], namely the following property:

$$if\ \delta(f, g) \subseteq X_1 \cup X_2,$$
$$then\ there\ exists\ h\ such\ that\ \delta(f, h) \subseteq X_1\ and\ \delta(h, g) \subseteq X_2. \tag{3}$$

It is easily verified that (3) is yet another spelling of the glueing property of functions.

Observe that, given any generalized ultrametric space (F, δ) whose distance takes values in $P(A)$, Eq. (2)—with D^A replaced by F—defines an action of $P(A)$ on $P(F)$. The lattice $P(A) \ltimes_j P(F)$ shall have similar properties to those of the lattices $R(D, A)$ and will be useful when studying the variety generated by the relational lattices. As an example, we construct *typed relational lattices*, i.e. lattices of relations where each column has a fixed type. To this goal, fix a surjective mapping $\pi : D \longrightarrow A$. For each $a \in A$, we think of the set $D_a = \pi^{-1}(a)$ as the type of the attribute a. Let $S(\pi)$ be the set of sections of π, that is, $s \in S(\pi)$ if and only if $s(a) \in D_a$, for each $a \in A$. Notice that $(S(\pi), \delta)$ is a pairwise complete sub-metric space of (D^A, δ). The lattice $R(\pi) := P(A) \ltimes_j P(S(\pi))$ is the typed relational lattice. It can be shown that relational lattices and typed relational lattices generate the same variety.

Relational lattices from multidimensional modal logics. In order to illustrate and stress the value of identity (1), i.e. of the Beck-Chevalley-Malcev property, we derive next a useful formula for computing the join of two tables under the $P(A) \ltimes_j P(D^A)$ representation.

$$(X_1, T_1) \vee (X_2, T_2) = (X_1 \cup X_2, \langle X_1 \cup X_2\rangle(T_1 \cup T_2))$$
$$= (X_1 \cup X_2, \langle X_1 \cup X_2\rangle T_1 \cup \langle X_1 \cup X_2\rangle T_2)$$
since the modal operators $\langle X\rangle$ are normal, in the usual sense of modal logic,
$$= (X_1 \cup X_2, \langle X_2\rangle\langle X_1\rangle T_1 \cup \langle X_1\rangle\langle X_2\rangle T_2) \qquad \text{by (1)}$$
$$= (X_1 \cup X_2, \langle X_2\rangle T_1 \cup \langle X_1\rangle T_2)$$
$$\text{since } \langle X_1\rangle T_1 = T_1 \text{ and } \langle X_2\rangle T_2 = T_2.$$

Theorem 2 suggests that a possible way to study the equational theory of the lattices $R(D, A)$ is to interpret the lattice operations in a two sorted modal logic, where the modal operators are indexed by the first sort and act on the second. It is easily recongnized that each modal operator satisfies the S5 axioms, while Eq. (1) implies that, when A is finite, each modal operator $\langle X\rangle$ is determined by the modal operators of the form $\langle a\rangle$ with a an atom below X. That is, the kind of modal logic we need to interpret the lattice theory is the commutator logic $[S5]^n = \underbrace{[S5, \ldots, S5]}_{n\text{-times}}$, with $n = \mathrm{card}\, A$, see [10, Definition 18].

4   Minimal Join-Covers in $R(D, A)$

The lattices $R(D, A)$ are pluperfect, even when A or D is an infinite set. The completely join-irreducible elements were characterized in [12] together with the meet-irreducible elements and the canonical context (see [3, Chapter 3] for the definition of canonical context). If we stick to the representation given in Theorem 2, the completely join-irreducible elements are of the form $\widehat{a} = (\{a\}, \emptyset)$ and $\widehat{f} = (\emptyset, \{f\})$. We can think of $\widehat{a}$ as an empty named column, while $\widehat{f}$ is an everywhere defined row. They are all atoms, so that, in particular, we shall not be concerned with the restriction of the order to $\mathcal{J}(R(D, A))$ (since this order coincides with the equality). In order to characterize the OD-graph of the lattice $R(D, A)$, we only need to characterize the minimal join-covers. Taking into account that if an element $j \in \mathcal{J}(L)$ is join-prime, then it has just one minimal join-cover, the singleton $\{j\}$, the following Theorem achieves this goal.

Theorem 3. *The lattices $R(D, A)$ are atomistic pluperfect lattices. As a matter of fact, every element $\widehat{a}$, $a \in A$, is join-prime; for $f \in D^A$, the minimal join-covers of $\widehat{f}$ are of the form*

$$\widehat{f} \leq \bigvee_{a \in \delta(f,g)} \widehat{a} \vee \widehat{g}, \qquad for \ g \in D^A.$$

The proof of this statement is almost straightforward, given the characterization of $R(D, A)$ as the semidirect product $P(A) \ltimes_j P(D^A)$ and the definition of the closure operators given with Eq. (2). For this reason, we skip it.

In particular, *every minimal join-cover contains at most one non-join-prime element.* In view of Theorem 1, we obtain a more precise description of the closure operator described in [12, Lemma 2.1] that gives rise to relational lattices.

Corollary 1. *The relational lattice $R(D, A)$ is isomorphic to the lattice of closed subsets of $A \sqcup D^A$, where a subset X is closed if $\delta(f, g) \cup \{g\} \subseteq X$ implies $f \in X$.*

In order to ease the reading, we shall use in the rest of this paper the same notation for a completely join-irreducible element of $R(D, A)$ and an element of $A \sqcup D^A$. This is consistent with the above Corollary, as under the isomorphism we have $\widehat{a} = \{a\}$ and $\widehat{f} = \{f\}$. Thus a shall stand for $\widehat{a}$, and f for $\widehat{f}$.

In a spatial lattice L (thus in a relational lattice), an inequation $s \leq t$ holds if and only if $j \leq s$ implies $j \leq t$, for each $j \in \mathcal{J}(L)$—thus we shall often consider inequations of the form $j \leq s$ with $j \in \mathcal{J}(L)$. When $L = R(D, A)$, the characterization of minimal join-cover leads to the following principle that we shall repeatedly use:

$$f \leq x_1 \vee \ldots \vee x_n \ \text{iff} \ \delta(f, g) \cup \{g\} \ll \{x_1, \ldots, x_n\}, \ \text{for some } g \in D^A. \qquad (4)$$

5   Uniqueness of non-join-prime Elements

For an *inclusion* we mean a pair (s, t) of terms (in the signature of lattice theory) such that the equation $t \vee s = s$ (i.e. the inequality $t \leq s$) is derivable from the usual axioms of lattices. Thus, the equality $s = t$ reduces to the inequality $s \leq t$. We write $s \leq t$ for a lattice inclusion and say it holds in a lattice if the identity $s = t$ holds in that lattice. Next, let us set

$$\mathsf{d}_\ell(\boldsymbol{u}) := u_0 \wedge (u_1 \vee u_2), \quad \mathsf{d}_\rho(\boldsymbol{u}) := (u_0 \wedge u_1) \vee (u_0 \wedge u_2),$$

so $\mathsf{d}_\ell(\boldsymbol{u}) \leq \mathsf{d}_\rho(\boldsymbol{u})$ is (an inclusion equivalent to) the usual distributive law. Consider the following inclusion:

$$\begin{aligned}
x \wedge (\mathsf{d}_\ell(\boldsymbol{y}) \vee \mathsf{d}_\ell(\boldsymbol{z}) \vee w) & \hspace{3cm} \text{(Unjp)} \\
\leq (x \wedge (\mathsf{d}_\rho(\boldsymbol{y}) \vee \mathsf{d}_\ell(\boldsymbol{z}) \vee w)) \vee (x \wedge (\mathsf{d}_\ell(\boldsymbol{y}) \vee \mathsf{d}_\rho(\boldsymbol{z}) \vee w)).
\end{aligned}$$

Theorem 4. *The inclusion (Unjp) holds on relational lattices. As a matter of fact, (Unjp) holds in a pluperfect lattice if and only if every minimal join-cover contains at most one non-join-prime element.*

Proof. Let us prove the first statement. To this goal, it will be enough to argue that any join-irreducible element below the left-hand side of the inclusion is below its right-hand side. Let k be such a join-irreducible element. It is not difficult to see that if k is join-prime, then k is also below the right-hand side of the inclusion. Suppose then that k is non-join-prime, whence $k = f$ for some $f \in D^A$. From $f \leq \mathsf{d}_\ell(\boldsymbol{y}) \vee \mathsf{d}_\ell(\boldsymbol{z}) \vee w$ and (4), it follows that there exists $g \in D^A$ such that $\delta(f, g) \cup \{g\} \ll \{\mathsf{d}_\ell(\boldsymbol{y}), \mathsf{d}_\ell(\boldsymbol{z}), w\}$. In particular, $\{g\} \ll \{\mathsf{d}_\ell(\boldsymbol{y}), w\}$ or $\{g\} \ll \{\mathsf{d}_\ell(\boldsymbol{z}), w\}$. We firstly suppose that the last case holds. If $a \in \delta(f, g)$ and $a \leq \mathsf{d}_\ell(\boldsymbol{y}) = y_0 \wedge (y_1 \vee y_2)$, then $a \leq (y_0 \wedge y_1) \vee (y_0 \wedge y_2) = \mathsf{d}_\rho(\boldsymbol{y})$, since a is join-prime. It follows that $\delta(f, g) \cup \{g\} \ll \{\mathsf{d}_\rho(\boldsymbol{y}), \mathsf{d}_\ell(\boldsymbol{z}), w\}$, whence $f \leq x \wedge (\mathsf{d}_\ell(\boldsymbol{y}) \vee \mathsf{d}_\ell(\boldsymbol{z}) \vee w)$. If $\{g\} \ll \{\mathsf{d}_\ell(\boldsymbol{y}), w\}$, then we conclude similarly that $f \leq x \wedge (\mathsf{d}_\ell(\boldsymbol{y}) \vee \mathsf{d}_\rho(\boldsymbol{z}) \vee w)$. Whence k is below the right-hand side of this inclusion, and the inclusion holds since k was arbitrary.

We leave the reader to generalize the argument above so to prove that if a pluperfect lattice is such that every minimal join-cover has at most one non-join-prime element, then (Unjp) holds. For the converse we argue as follows.

Let L be a pluperfect lattice, let $k_x \in \mathcal{J}(L)$, $C_x \subseteq \mathcal{J}(L)$ with $k_x \lhd_{\mathsf{m}} C_x$, and suppose that $k_y, k_z \in C_x$ are distinct and non-join-prime. For $u \in \{y, z\}$, since k_u is non-join-prime, there is a non-trivial minimal join-cover $k_u \lhd_{\mathsf{m}} C_u$; as every non-trivial minimal join-cover has at least two elements, let $C_{u,1}, C_{u,2}$ be a partition of C_u such that $C_{u,i} \neq \emptyset$ for each $i = 1, 2$.

We construct a valuation which fails (Unjp). Let $x := k_x$, $y_0 := k_y$, $z_0 := k_z$, $w := \bigvee(C_x \setminus \{k_y, k_z\})$ and, for $u \in \{y, z\}$ and $i = 1, 2$, let $u_i := \bigvee C_{u,i}$. The left-hand side of the (Unjp) evaluates to k_x. Assume, by the way of contradiction, that (Unjp) holds, so k_x is below the right-hand side of the inclusion. Since the only minimal join-cover D of k_x such that $D \ll \{k_x\}$ is $\{k_x\}$, either $k_x \leq \mathsf{d}_\rho(\boldsymbol{y}) \vee \mathsf{d}_\ell(\boldsymbol{z}) \vee w$ or $k_x \leq \mathsf{d}_\ell(\boldsymbol{y}) \vee \mathsf{d}_\rho(\boldsymbol{z}) \vee w$; let us assume that the first case holds. We have then $k_x \leq \mathsf{d}_\rho(\boldsymbol{y}) \vee k_z \vee \bigvee(C_x \setminus \{k_y, k_z\}) = \mathsf{d}_\rho(\boldsymbol{y}) \vee \bigvee(C_x \setminus \{k_y\})$. Considering that $\mathsf{d}_\rho(\boldsymbol{y}) \leq k_y$, Lemma 1 implies that $k_y = \mathsf{d}_\rho(\boldsymbol{y}) = (y_0 \wedge y_1) \vee (y_0 \wedge y_2)$. Since

k_y is join-irreducible $k_y = y_0 \wedge y_i$ for some $i \in \{1, 2\}$. Yet this is not possible, as such relation implies that $C_{y,i}$ is a join-cover of k_y; considering that $C_{y,i}$ is a proper subset of the minimal join-cover C_y, this contradicts the minimality of C_y. If $k_x \leq \mathsf{d}_\ell(y) \vee \mathsf{d}_\rho(z) \vee w$, then we get to a similar contradiction. Whence, k_x is not below the right-hand side of (Unjp), which therefore fails.   □

It is worth noticing that the statement "*every minimal join-cover contains exactly one non-join-prime element*" is not definable by equations: for $A = D = \{0, 1\}$, there is a sublattice of $\mathsf{R}(D, A)$ which fails this property.

While Theorem 4 gives a semantic characterization of (Unjp), we might also wish to measure its power at the syntactic level. Theorem 5 and Corollary 2 illustrate the deductive strength of (Unjp), by pinpointing an infinite set of its consequences.

Theorem 5. *If $s_\ell = s_\rho$ and $t_\ell = t_\rho$ are equations valid on distributive lattices, then the equation*

$$(x \wedge (s_\ell \vee t_\ell \vee w)) \vee (x \wedge (s_\rho \vee t_\rho \vee w))$$
$$= (x \wedge (s_\rho \vee t_\ell \vee w)) \vee (x \wedge (s_\ell \vee t_\rho \vee w))$$

is derivable from (Unjp) and general lattice axioms.

Proof. For a lattice term s, let $\mathsf{dnf}(s)$ be its disjunctive normal form. Recall that we can obtain $\mathsf{dnf}(s)$ from s by means of a sequence $s = s_0, \ldots, s_n = \mathsf{dnf}(s)$ where, for each $i = 0, \ldots, n-1$, s_{i+1} is obtained from s_i by one application of the distributive law at the toplevel of the term, and by general lattice axioms. Thus, for two lattice terms s^1, s^2, let s_i^1, $i = 0, \ldots, n$, and s_j^2, $j = 0, \ldots, m$ be the sequences leading to the respective normal forms.

For $i = 0, \ldots, n$ and $j = 0, \ldots, m$, let now $t_{i,j} = x \wedge (s_i^1 \vee s_j^2 \vee w)$. Using (Unjp) and general lattice axioms, we can compute as follows:

$$t_{0,0} = t_{1,0} \vee t_{0,1} = t_{2,0} \vee t_{1,1} \vee t_{0,2} = \ldots$$
$$= \bigvee_{j=0,\ldots,m} t_{n,j} \vee \bigvee_{i=0,\ldots,n} t_{i,m} \overset{?}{=} t_{n,0} \vee t_{0,m},$$

where only the last equality needs to be justified. Notice that the relation $s_{i+1}^k \leq s_i^k$ holds in every lattice. Whence we have $s_{i'}^k \leq s_i^k$ when $i < i'$, and both $t_{n,j} \leq t_{n,0}$ and $t_{i,m} \leq t_{i,0}$. It follows that the indexed join at the last line evaluates to $t_{n,0} \vee t_{0,m}$. We have derived, up to now, the identity

$$x \wedge (s^1 \vee s^2 \vee w) = (x \wedge (\mathsf{dnf}(s^1) \vee s^2 \vee w)) \vee (x \wedge (s^1 \vee \mathsf{dnf}(s^2) \vee w))$$

for every pair of lattice terms s^1 and s^2.

Let us call co-clause a conjunction of variables. By using lattice axioms only, we can suppose that, within $\mathsf{dnf}(t)$, there are no repeated literals in co-clauses and that no co-clause subsumes another. Under this assumption, we have that an identity $s_\ell = s_\rho$ holds in all distributive lattices if and only if $\mathsf{dnf}(s_\ell)$ is equal to $\mathrm{dnf}(s_\rho)$. Whence, to derive the statement of the Theorem, we can compute as follows:

$$(x \wedge (s_\ell \vee t_\rho \vee w)) \vee (x \wedge (s_\rho \vee t_\ell \vee w))$$
$$= (x \wedge (\mathrm{dnf}(s_\ell) \vee t_\rho \vee w)) \vee (x \wedge (s_\ell \vee \mathrm{dnf}(t_\rho) \vee w))$$
$$\vee (x \wedge (\mathrm{dnf}(s_\rho) \vee t_\ell \vee w)) \vee (x \wedge (s_\rho \vee \mathrm{dnf}(t_\ell) \vee w))$$
$$= (x \wedge (\mathrm{dnf}(s_\rho) \vee t_\rho \vee w)) \vee (x \wedge (s_\ell \vee \mathrm{dnf}(t_\ell) \vee w))$$
$$\vee (x \wedge (\mathrm{dnf}(s_\ell) \vee t_\ell \vee w)) \vee (x \wedge (s_\rho \vee \mathrm{dnf}(t_\rho) \vee w))$$
$$= (x \wedge (\mathrm{dnf}(s_\ell) \vee t_\ell \vee w)) \vee (x \wedge (s_\ell \vee \mathrm{dnf}(t_\ell) \vee w))$$
$$\vee (x \wedge (\mathrm{dnf}(s_\rho) \vee t_\rho \vee w)) \vee (x \wedge (s_\rho \vee \mathrm{dnf}(t_\rho) \vee w))$$

—where we have permuted the order of the four joinands

$$= (x \wedge (s_\ell \vee t_\ell \vee w)) \vee (x \wedge (s_\rho \vee t_\rho \vee w)). \qquad \square$$

In [12] two equations were shown to hold on relational lattices. One of them is (RL2) that we describe next. Set

$$\mathrm{d}_\ell^o(\boldsymbol{u}) := (u_0 \vee u_1) \wedge (u_0 \vee u_2), \quad \mathrm{d}_\rho^o(\boldsymbol{u}) := u_0 \vee (u_1 \wedge u_2),$$

the equation is

$$x \wedge (\mathrm{d}_\ell^o(\boldsymbol{y}) \vee \mathrm{d}_\ell^o(\boldsymbol{z})) \leq (x \wedge (\mathrm{d}_\rho^o(\boldsymbol{y}) \vee \mathrm{d}_\ell^o(\boldsymbol{z}))) \vee (x \wedge (\mathrm{d}_\ell^o(\boldsymbol{y}) \vee \mathrm{d}_\rho^o(\boldsymbol{z}))) \quad \text{(RL2)}$$

Corollary 2. *If $s_\ell = s_\rho$ and $t_\ell = t_\rho$ are equations valid on distributive lattices, then the inequation*

$$(x \wedge (s_\ell \vee t_\ell \vee w)) \leq (x \wedge (s_\rho \vee t_\ell \vee w)) \vee (x \wedge (s_\ell \vee t_\rho \vee w)) \qquad (5)$$

is derivable from (Unjp). *In particular* (RL2) *is derivable from* (Unjp).

The Corollary follows from the Theorem and from the fact that $x \leq x \vee y$. In order to derive (RL2) from (5) (if we do not include the bottom constant $\perp$ as part of the signature of lattice theory), we instantiate $s_\ell := \mathrm{d}_\ell^o(\boldsymbol{y})$, $s_\rho := \mathrm{d}_\rho^o(\boldsymbol{y})$, $t_\ell := \mathrm{d}_\ell^o(\boldsymbol{z})$, $t_\rho := \mathrm{d}_\rho^o(\boldsymbol{z})$, and $w := \mathrm{d}_\rho^o(\boldsymbol{z})$.

It can be shown that (Unjp) is not derivable from (RL2)—mainly due to the role of the variable w in the (Unjp). The construction of a lattice L satisfying (RL2) but failing (Unjp) proceeds via the construction of its OD-graph $\langle J(L), \leq, \lessdot_m \rangle$. Due to the consistent number of variables in the two equations, an automated tool such as Mace4 [13] could not help finding a countermodel. Similarly, automated tools such as Prover9 and Waldmeister [6,13] were of no help to show that (RL2) is a consequence of (Unjp).

Natural questions—e.g. decidability—may be raised concerning the equational theory of (Unjp). Since we can give an easy semantic proof that an equation of the form (5) holds on finite lattices (or pluperfect) satisfying (Unjp), a reasonable conjecture is that this theory has some sort of finite model property.

Yet, proving this might not be immediate, since the variety of lattices satisfying (Unjp) is not locally finite (i.e., not every finitely generated lattice satisfying (Unjp) is finite). The construction used in [16, Proposition 7.5] may be used to argue that the lattice freely generated in this variety by three generators is infinite.

6 Symmetry and pairwise completeness

Due to its syntactic shape (Unjp) falls in a class of inclusions described in [16, Section 8] that admit a correspondent property in the OD-graph. Here, the meaning of the word correspondent is analogous to its use in modal logic, where some formulas might be uniformly valid in a frame if and only if the frame satisfies a correspondent first order property. Thus Theorem 4 is not completely unexpected. A more surprising result comes from considering the three equations below, that fall outside the syntactic fragment described in [16]; a strengthening of Lemma 1 (Lemma 4 to follow) allows to characterize the OD-graphs of pluperfect lattices satisfying (Unjp) and these equations, see Theorem 8.

$$x \wedge (y \vee z) \leq \qquad\qquad\qquad\qquad \text{(SymPC)}$$
$$(x \wedge (y \vee (z \wedge (x \vee y)))) \vee (x \wedge (z \vee (y \wedge (x \vee z))))$$
$$x \wedge ((y \wedge z) \vee (y \wedge x) \vee (z \wedge x)) \leq (x \wedge y) \vee (x \wedge z) \qquad \text{(VarRL1)}$$
$$x \wedge ((x \wedge y) \vee \mathsf{d}_\ell(z)) \leq (x \wedge ((x \wedge y) \vee \mathsf{d}_\rho(z))) \vee (x \wedge \mathsf{d}_\ell(z)) \quad \text{(RMod)}$$

Let us first illustrate the way in which these equations hold in relational lattices. In particular, the proof shall illustrate the crucial role played by symmetry and pairwise completeness—i.e., condition (3)—of the ultrametric space (D^A, δ).

Theorem 6. *The inclusions* (SymPC), (VarRL1), (RMod) *hold in relational lattices.*

Proof. (SymPC). Let k be a join-irreducible element below $x \wedge (y \vee z)$. If k is join-prime, then k is also below $(x \wedge y) \vee (x \wedge z)$, whence it is below the right-hand side of this inclusion. Therefore, let k be non-join-prime, so $k = f$ for some $f \in D^A$; by (4), let $g \in D^A$ be such that $\delta(f, g) \cup \{g\} \ll \{y, z\}$. Let us suppose first that $g \leq z$. Since $\delta(f, g) \ll \{y, z\}$, using pairwise completeness we can find h such that $\delta(f, h) \ll \{y\}$ and $\delta(h, g) \ll \{z\}$. It follows that $h \leq \bigvee \delta(h, g) \vee g \leq z$; moreover, since $f \leq x$, $\delta(h, f) = \delta(f, h)$, and $\delta(f, h) \ll \{y\}$, then $h \leq \delta(h, f) \vee f \leq x \vee y$. Consequently, we have $h \leq z \wedge (x \vee y)$ and, considering that $\delta(f, h) \ll \{y\}$, we have $f \leq x \wedge (y \vee (z \wedge (x \vee y)))$. If $g \leq y$, then we similarly deduce that $f \leq x \wedge (z \vee (y \wedge (x \vee z)))$. In both cases, f is below the right-hand side of this inclusion.

(VarRL1). Let k be below the left-hand side of this inclusion. If k is join-prime, then it is below the right-hand side of this inclusion as well. Otherwise $k = f$ is non-join-prime and $\delta(f, g) \cup \{g\} \ll \{y \wedge z, y \wedge x, z \wedge x\}$ for some $g \in D^A$.

Since $g \leq r$ for some $r \in \{y \wedge z, y \wedge x, z \wedge x\}$, we consider three cases; by pairwise completeness we can also assume that g is the only element of $\delta(f, g) \cup \{g\}$ below r—since if $\delta(f, g') \ll \{y \wedge z, y \wedge x, z \wedge x\} \setminus \{r\}$ and $\delta(g', g) \ll \{r\}$, then $g' \leq r$. Also, the last two cases, $g \leq y \wedge x$ and $g \leq z \wedge x$, are symmetric in y and z, so that we consider among them the second-to-last only.

Suppose firstly that $g \leq y \wedge z$. Then, from $\delta(g, f) = \delta(f, g) \ll \{x \wedge y, x \wedge z\} \ll \{x\}$ and $f \leq x$, we deduce $g \leq x$; whence $g \leq x \wedge y$ and $f \leq (x \wedge z) \vee (x \wedge y)$.

Suppose next that $g \leq x \wedge y$. By pairwise completeness, let h be such that $\delta(f, h) \ll \{x \wedge z\}$ and $\delta(h, g) \ll \{y \wedge z\}$. We deduce then $h \leq y$ from $\delta(h, g) \ll \{y\}$ and $g \leq y$, and $h \leq x$, from $\delta(h, f) = \delta(f, h) \ll \{x\}$ and $f \leq x$. Thus $h \leq x \wedge y$ and $f \leq (x \wedge z) \vee (x \wedge y)$.

(RMod). Let k be a join-irreducible below the left-hand side of this inclusion. If k is join-prime, then k is below $x \wedge ((x \wedge y) \vee \mathsf{d}_\rho(\boldsymbol{z}))$. Otherwise $k = f$ and, for some $g \in D^A$, $\delta(f, g) \cup \{g\} \ll \{x \wedge y, \mathsf{d}_\ell(\boldsymbol{z})\}$. If $g \leq x \wedge y$, then all the elements that are not below $x \wedge y$ are below $\mathsf{d}_\ell(\boldsymbol{z})$ and join-prime, whence they are below $\mathsf{d}_\rho(\boldsymbol{z})$. It follows that $f \leq x \wedge ((x \wedge y) \vee \mathsf{d}_\rho(\boldsymbol{z}))$. Otherwise $g \leq \mathsf{d}_\ell(\boldsymbol{z})$ and, by pairwise completeness, we can also assume that g is the only element below $\mathsf{d}_\ell(\boldsymbol{z})$, so $\delta(f, g) \ll \{x \wedge y\}$. It follows then that $\delta(g, f) \cup \{f\} = \delta(f, g) \cup \{f\} \ll \{x\}$, $g \leq x$, whence $g \leq x \wedge \mathsf{d}_\ell(\boldsymbol{z})$. Consequently, $f \leq (x \wedge y) \vee (x \wedge \mathsf{d}_\ell(\boldsymbol{z})) \leq (x \wedge ((x \wedge y) \vee \mathsf{d}_\rho(\boldsymbol{z}))) \vee (x \wedge \mathsf{d}_\ell(\boldsymbol{z}))$. $\qquad\square$

In [12] a second inclusion was shown to hold on relational lattices:

$$x \wedge ((y \wedge (z \vee x)) \vee (z \wedge (y \vee x))) \leq (x \wedge y) \vee (x \wedge z) \qquad \text{(RL1)}$$

The same kind of tools used in the proof of Theorem 6 can be used to argue that this inclusion holds on relational lattices. The reader will have noticed the similarity of (VarRL1) with (RL1). As a matter of fact, (VarRL1) was suggested when trying to derive (RL1) from (Unjp) and the other equations as in the following Proposition.

Proposition 1. (RL1) *is a consequence of* (Unjp), (RMod) *and* (VarRL1).

Proof. Using (Unjp) and considering that $y \wedge z \leq z \wedge (y \vee x)$, we have:

$$x \wedge ((y \wedge (z \vee x)) \vee (z \wedge (y \vee x)))$$
$$= (x \wedge ((y \wedge x) \vee (z \wedge (y \vee x)))) \vee (x \wedge ((y \wedge (z \vee x)) \vee (z \wedge x))).$$

Using now (RMod) and considering that $x \wedge z \leq y \vee x$, we compute as follows:

$$x \wedge ((y \wedge x) \vee (z \wedge (y \vee x)))$$
$$= (x \wedge ((y \wedge x) \vee (z \wedge y) \vee (z \wedge x))) \vee (x \wedge z \wedge (y \vee x))$$
$$= (x \wedge ((y \wedge x) \vee (z \wedge y) \vee (z \wedge x))) \vee (x \wedge z)$$
$$= x \wedge ((y \wedge z) \vee (y \wedge x) \vee (z \wedge x)).$$

Considering the symmetric role of y and z, we obtain:

$$x \wedge ((y \wedge (z \vee x)) \vee (z \wedge (y \vee x))) = \quad x \wedge ((y \wedge z) \vee (y \wedge x) \vee (z \wedge x))$$
$$= \quad (x \wedge y) \vee (x \wedge z), \quad \text{by(VarRL1)}. \qquad \square$$

We present now what we consider our strongest result in the study of the equational theory of relational lattices. To this end, let us denote by [[AxRel]] the (set composed of the) four equations (Unjp), (VarRL1), (RMod) and (SymPC). Also, given that we restrict to lattices satisfying (Unjp), and considering the characterization given with Theorem 4, it is convenient to introduce the notation $k_0 \lhd_\mathrm{m}^C k_1$ for the statement $k_0, k_1 \in J(L)$, k_1 *is non-join-prime*, $k_1 \notin C$, *and* $k_0 \lhd_\mathrm{m} C \cup \{k_1\}$.

Theorem 7. *Let L be a finite atomistic lattice. Then $L \models$ [[AxRel]] if and only if every nontrivial minimal join-cover contains exactly one non-join-prime element and, moreover, the following properties hold in the OD-graph:*

- *If $k_0 \lhd_\mathrm{m}^C k_1$, then $k_1 \lhd_\mathrm{m}^C k_0$.* $\qquad\qquad\qquad\qquad\qquad\qquad$ (6)
- *If $k_0 \lhd_\mathrm{m}^{C_0 \sqcup C_1} k_2$, then $k_0 \lhd_\mathrm{m}^{C_0} k_1$ and $k_1 \lhd_\mathrm{m}^{C_1} k_2$, for some $k_1 \in J(L)$.*

Given Theorem 7, it becomes tempting to look for a representation Theorem. Given a pluperfect atomistic lattice satisfying the above four equations, we would like to define an ultrametric space on the set of non-join-prime elements with distance valued on the powerset of the join-prime ones, and then argue that the lattice constructed via the standard action, defined in (2), is isomorphic to the given lattice. Unfortunately, this idea does not work, since if we try to set $\delta(k_0, k_1) = C$ whenever $k_0 \lhd_\mathrm{m}^C k_1$, this might be ill defined since the implication "$k_0 \lhd_\mathrm{m}^C k_1$ and $k_0 \lhd_\mathrm{m}^D k_1$ implies $C = D$" might fail. Moreover, there is no equation nor quasiequation enforcing this, as an immediate consequence of the next Proposition.

Proposition 2. *There is an atomistic sublattice of $\mathsf{R}(\{0, 1\}, \{0, 1\})$ which does not arise from an ultrametric space.*

Theorem 7 is a consequence of a more general Theorem, to be stated next, characterizing the OD-graphs of pluperfect lattices in the variety axiomatized by [[AxRel]]. While the conditions stated next may appear quite complex, they are the ones to retain if we aim at studying further the theories of relational lattices by duality—e.g., a sublattice of a relational lattice need not be atomistic.

Theorem 8. *A pluperfect lattice belongs to the variety axiomatized by [[AxRel]] if and only if every minimal join-cover contains at most one non-join-prime element and, moreover, the following properties hold in its OD-graph:*

- If $k_0 \vartriangleleft_{\mathrm{m}} C$, then there exists at most one $c \in C$ with $c \le k_0$.   (π-VarRL1)
- If $k_0 \vartriangleleft_{\mathrm{m}}^C k_1$, then no element of C is below k_0.   (π-RMod)
- if $k \vartriangleleft_{\mathrm{m}} C_0 \sqcup C_1$ with C_0, C_1 non-empty, then for some $k' \in \mathcal{J}(L)$,

$$\text{either } k \vartriangleleft_{\mathrm{m}} \{k'\} \sqcup C_1, \ k' \vartriangleleft_{\mathrm{m}} C_0, \text{ and } k' \le \bigvee C_1 \vee k,$$

$$\text{or } k \vartriangleleft_{\mathrm{m}} C_0 \sqcup \{k'\}, \ k' \vartriangleleft_{\mathrm{m}} C_1, \text{ and } k' \le \bigvee C_0 \vee k. \quad (\pi\text{-StrongSymPC})$$

Let us notice that the conditions

- If $k_0 \vartriangleleft_{\mathrm{m}}^{C_0 \sqcup C_1} k_2$ then, for some $k_1 \in \mathcal{J}(L)$,

$$k_0 \vartriangleleft_{\mathrm{m}}^{C_0} k_1, \ k_1 \vartriangleleft_{\mathrm{m}}^{C_1} k_2, \text{ and } k_1 \le \bigvee C_0 \vee k_0. \quad (\pi\text{-SymPC})$$

- If $k_0 \vartriangleleft_{\mathrm{m}}^C k_1$, then $k_1 \le \bigvee C \vee k_0$ (π-Sym)

follow from the above properties. On atomistic pluperfect lattices the last condition is equivalent to (6).

Lemma 3. *If L is a pluperfect lattice with $L \models$ (Unjp) and whose OD-graph satisfies (π-StrongSymPC) and (π-RMod), then (π-SymPC) holds as well.*

Proof. Let $k_1 \vartriangleleft_{\mathrm{m}} C_0 \sqcup C_1 \sqcup \{k_2\}$ with $k_2 \in \mathcal{J}(L)$ and non-join-prime, and use (π-StrongSymPC) to find k_1 such that either (i) $k_0 \vartriangleleft_{\mathrm{m}} \{k_1\} \sqcup C_1 \sqcup \{k_2\}$ and $k_1 \le k_0 \vee \bigvee C_1 \vee k_2$, or (ii) $k_0 \vartriangleleft_{\mathrm{m}} C_0 \sqcup \{k_1\}$ and $k_1 \le k_0 \vee \bigvee C_0$. Let us argue, by contradiction, that (i) cannot arise. By (Unjp), k_1 is join-prime, whence the relation $k_1 \le k_0 \vee \bigvee C_1$ yields $k_1 \le k_0$. This, however, contradicts (π-RMod). $\square$

We close this section by proving Theorem 8. To this end, we need a generalization of Lemma 1. As the refinement relation is an extension to subsets of the order relation, the relation $\lessdot\!\!\lessdot_{\mathrm{m}}$, defined next, can be considered as an extension to subsets of the minimal join-covering relation.

Definition 2. *Let L be a pluperfect lattice and let $X, Y \subseteq \mathcal{J}(L)$ be antichains. Put $X \lessdot\!\!\lessdot_{\mathrm{m}} Y$ if $X \ll \{\bigvee Y\}$ and $y \in C_{x_y}$ for some $x_y \in X$, for each $y \in Y$ and whenever $\{C_x \mid x \in X\}$ is a family of coverings of the form $x \vartriangleleft_{\mathrm{m}} C_x \ll Y$.*

Lemma 4. *Let L be a pluperfect lattice and let $j \vartriangleleft_{\mathrm{m}} C_0 \sqcup C_1$. Suppose that $\bigvee X \le \bigvee C_0$ and $j \le \bigvee X \vee \bigvee C_1$. Then there exists a minimal join-cover of the form $j \vartriangleleft_{\mathrm{m}} D_0 \sqcup C_1$ with $D_0 \ll X$ and $D_0 \lessdot\!\!\lessdot_{\mathrm{m}} C_0$.*

While the proof that a pluperfect lattice whose OD-graphs satisfies those properties essentially mimics the proof of Theorem 6, we prove instead the converse direction through a series of Lemmas.

Lemma 5. *If (VarRL1) holds on a pluperfect lattice, then its OD-graph satisfies (π-VarRL1).*

Proof. Suppose $C = \{k_1\} \sqcup \{k_2\} \sqcup D$ with $k_0 \lhd_{\mathrm{m}} C$ and $k_1, k_2 \leq k_0$. Let $x := k_0$, $y := k_1 \vee \bigvee D$, $z := k_2 \vee \bigvee D$. Then $k_1 \leq x \wedge y$, $k_2 \leq x \wedge z$ and $\bigvee D \leq y \wedge z$, whence the left-hand side of (VarRL1) evaluates to k_0, which is therefore below the right-hand side of this inclusion. It follows that either $k_0 \leq y$, or $k_0 \leq z$, in both cases contradicting the fact that C is a minimal join-cover.   □

The inclusion

$$x \wedge (y \vee \mathsf{d}_\ell(z)) \leq (x \wedge (y \vee \mathsf{d}_\rho(z))) \vee (xland(y \vee (\mathsf{d}_\ell(z) \wedge (y \vee x)))),   \text{(Sym)}$$

is derivable from (Unjp), (SymPC) and (RMod). It can also be shown that (RMod) is a consequence of (Sym).

Lemma 6. *If (Sym) holds in a pluperfect lattice, then its OD-graph satisfies* (π-*Sym*).

Proof. Suppose that the inclusion holds and let $k_0 \lhd_{\mathrm{m}}^C k_1$. Since k_1 is non-join-prime, there exists a minimal join-cover $k_1 \lhd_{\mathrm{m}} D$ which we can partition into two non empty subsets D_1 and D_2. Let now $x := k_0$, $y := \bigvee C$, $z_0 := k_1$, $z_1 = \bigvee D_1$, $z_2 = \bigvee D_2$. Then, the left-hand side of the inclusion evaluates to k_0, which therefore is below the right-hand side. Considering that x is k_0 and that the unique minimal join-cover of k_0 whose elements are all below k_0 is $\{k_0\}$, it follows that either $k_0 \leq y \vee \mathsf{d}_\rho(z)$ or $k_0 \leq y \vee (\mathsf{d}_\ell(z) \wedge (y \vee x))$.

Argue that $\mathsf{d}_\rho(z) < \mathsf{d}_\ell(z)$, since k_1 is join-irreducible, whence by Lemma 1, $\{y, \mathsf{d}_\rho(z)\}$ is not a cover of k_0, excluding the first case. Therefore $\{y, \mathsf{d}_\ell(z) \wedge (y \vee x)\}$ is a cover of k_0, whence, by Lemma 1, $k_1 = \mathsf{d}_\ell(z) \wedge (y \vee x)$, showing that $k_1 \leq y \vee x = \bigvee C \vee k_0$ and proving the statement.   □

Lemma 7. *If L is a pluperfect lattice such that $L \models [[AxRel]]$, then* (π-RMod) *holds in its OD-graph.*

Proof. Let $k_0 \lhd_{\mathrm{m}}^C k_1$ and put $C = C_0 \sqcup C_1$ with $C_0 \ll \{k_0\}$ and $c \not\leq k_0$ for each element $c \in C_1$. As (Sym) whence (π-Sym) hold, we have $k_1 \leq \bigvee C_1 \vee \bigvee C_0 \vee k_0$.

We consider next equation (RMod). Put $x := k_0$, $y := \bigvee C_0$, $z_0 := \bigvee C_1 \vee k_1$, $z_1 := \bigvee C_1 \vee \bigvee C_0$, $z_2 := k_0$. From $k_1 \leq \bigvee C_1 \vee \bigvee C_0 \vee k_0$, we get $z_0 \wedge (z_1 \vee z_2) = z_0$. Whence, the left-hand side of (RMod) evaluates to k_0 so k_0 is below the right-hand side of (RMod). Considering that $\{k_0\}$ is the unique minimal join-cover of k_0 whose elements are all below k_0, it follows that either $k_0 \leq z_0 \wedge (z_1 \vee z_2)$ or $k_0 \leq y \vee (z_0 \wedge z_1) \vee (z_0 \wedge z_2)$.

As $k_0 \not\leq \bigvee C_1 \vee k_1 = z_0$, it follows that $k_0 \leq y \vee (z_0 \wedge z_1) \vee (z_0 \wedge z_2)$. We can use then Lemma 4 to deduce that k_0 has a minimal join-cover of the form $k_0 \lhd_{\mathrm{m}} C_0 \sqcup D$, with $D \ll \{z_0 \wedge z_1, z_0 \wedge z_2\} \ll \{z_1, z_2\}$. If all the elements of D are below $z_1 = \bigvee C_0 \vee \bigvee C_1$, then

$$k_0 \leq \bigvee C_0 \vee \bigvee D \leq \bigvee C_0 \vee \bigvee C_1,$$

contradicting the minimality of $k_0 \lhd_{\mathrm{m}} C_0 \sqcup C_1 \sqcup \{k_1\}$. Therefore, at least one element of D is below $z_2 = k_0$. If $C_0 \neq \emptyset$, then in the minimal join-cover $C_0 \sqcup D$

there are at least two elements that are below k_0. This however contradicts (π-VarRL1), whence (VarRL1). We have, therefore, $C_0 = \emptyset$.   □

Lemma 8. *If L is a pluperfect lattice with $L \models [[AxRel]]$, then (π-StrongSymPC) holds in is OD-graph.*

Proof. By Lemmas 5 and 7, (π-RMod) and (π-VarRL1) hold in the OD-graph.

Let $x := k_0$, $y := \bigvee C_0$, $z := \bigvee C_1$. Then the left-hand side of (SymPC) evaluates to k_0 which is therefore below the right-hand side. Thus, by Lemma 4, there is a minimal join-cover of the form $k_0 \lhd_m D_0 \sqcup D_1$ with either (i) $D_0 \lhd\!\!\lhd_m C_0$, $D_0 \ll \{y \wedge (x \vee z)\}$, and $D_1 = C_1$, or (ii) $D_0 = C_0$, $D_1 \lhd\!\!\lhd_m C_1$, and $D_1 \ll \{z \wedge (x \vee y)\}$.

W.l.o.g. we can suppose that (i) holds. From $D_0 \ll \{y \wedge (x \vee z)\} \ll \{x \vee z\} = \{k_0 \vee \bigvee C_1\}$, we can argue as follows. We notice first that if an element of D_0 is join-prime, then it is either below k_0 or below some $c \in C_1$; since $D_0 \sqcup C_1$ is an antichain, this element is below k_0. Therefore, if all the elements of D_0 are join-prime, then, by (π-VarRL1), $D_0 = \{k'\}$. Otherwise, there exists a non-join-prime element k' in D_0 and, by (Unjp), this is the only non-join-prime in D_0. Write $D_0 = \{k'\} \sqcup E$, then every element of E is join-prime and, as seen before, we need to have $E \ll \{k_0\}$. Then (π-RMod) enforces $E_0 = \emptyset$ and $D_0 = \{k'\}$. In both cases, the relation $\{k'\} = D_0 \lhd\!\!\lhd_m C_0$ yields $k' \lhd_m C_0$.   □

Finally, in order to understand the structure of finite lattices in the variety of axiomatized by [[AxRel]], let $\mathcal{J}_p(L)$ denote the set of join-prime elements of L and consider the following property:

- *If $k_0 \lhd_m C$ and $C \subseteq \mathcal{J}_p(L)$, then $c_0 \leq k_0$ for some $c_0 \in C$*   (π-JP)

The next Lemma ensures the existence of a non-join-prime element in a cover in finite atomistic lattices, as stated in Theorem 7.

Lemma 9. *If a finite lattice L satisfies [[AxRel]], then (π-JP) holds in its OD-graph. In particular, if L is atomistic, then $k_0 \lhd_m C$ implies that $k_1 \in C$ for some non-join-prime k_1.*

It can be shown that the finiteness assumption in Lemma 9 is necessary.

7   Conclusions and Further Directions

Some undecidable problems. Our main result, Theorem 7, characterizes the OD-graphs of finite atomistic lattices satisfying [[AxRel]] as structures similar to frames for the commutator logic $[S5]^n$, the multimodal logic with n distinct pairwise commuting S5 modal operators, see [10]. We exemplify next how to take advantage of such similarity and of the existing theory on combination of modal logics, to deduce undecidability results. As this is not the main goal of the paper, we delay a full exposition of these ideas to an upcoming set of notes.

An $[S5]^n$ frame is a structure $\mathfrak{F} = (F, R_1, \ldots, R_n)$ where each R_i is an equivalence relation on F and, moreover, the confluence property holds: *if $i \neq j$, xR_iy and xR_jz, then yR_jw and zR_iw for some $w \in F$*. A particular class of $[S5]^n$ frames are the universal $S5^n$-products, those of the form $\mathfrak{U} = (F, R_1, \ldots, R_n)$ with $F = X_1 \times \ldots \times X_n$ and $(x_1, \ldots, x_n)R_i(y_1, \ldots, y_n)$ if and only if $x_j = y_j$ for each $j \neq i$.

For a frame $\mathfrak{F} = (W, R_1, \ldots, R_n)$ and $X \subseteq \{1, \ldots, n\}$, let us say that $Y \subseteq W$ is X-closed if $w_0 \in Y$, whenever there is a path $w_0R_{i_0}w_1 \ldots w_{k-1}R_{i_k}w_k$ with $\{i_0, \ldots, i_k\} \subseteq X$ and $w_k \in Y$. Then X-closed subsets are closed under intersections, so subsets of $\{1, \ldots, n\}$ give rise to closure operators $\langle X \rangle$ and to an action as defined in Sect. 3. Let $\mathsf{L}(\mathfrak{F}) = P(\{1, \ldots, n\}) \ltimes_j P(W)$ and notice that $\mathsf{L}(\mathfrak{F})$ is atomistic. A frame $\mathfrak{F}$ is initial if there is $f_0 \in F$ such every other $f \in F$ is reachable from f_0; it is full if, for each $i = 1, \ldots, n$, R_i is not included in the identity. If $\mathfrak{F}$ is initial and full, then $\mathsf{L}(\mathfrak{F})$ is subdirectly irreducible. A p-morphism is defined as usual in modal logic. The key observation leading to undecidability is the following statement.

Theorem 9. *There is a surjective p-morphism from a universal $S5^n$-product frame $\mathfrak{U}$ to a full initial frame $\mathfrak{F}$ if and only if $\mathsf{L}(\mathfrak{F})$ embeds in a relational lattice.*

Proof (Sketch). The construction L is extended to a contravariant functor, so if $\psi : \mathfrak{U} \longrightarrow \mathfrak{F}$ is a p-morphism, then we have an embedding $\mathsf{L}(\psi)$ of $L(\mathfrak{F})$ into $\mathsf{L}(\mathfrak{U})$. We can assume that all the components $X_1, \ldots, X_n$ of $\mathfrak{U}$ are equal, so $X_i = X$ for each $i = 1, \ldots, n$; if this is the case, then $\mathsf{L}(\mathfrak{U})$ is isomorphic to the relational lattice $\mathsf{R}(\{1, \ldots, n\}, X)$.

The converse direction is subtler. Let $\chi : \mathsf{L}(\mathfrak{F}) \longrightarrow \mathsf{R}(A, D)$ be a lattice embedding; since $\mathsf{L}(\mathfrak{F})$ is subdirectly-irreducible, we can suppose that χ preserves bounds; its left adjoint $\mu : \mathsf{R}(A, D) \longrightarrow \mathsf{L}(\mathfrak{F})$ is then surjective. Since both $\mathsf{L}(\mathfrak{F})$ and $\mathsf{R}(D, A)$ are generated (under possibly infinite joins) by their atoms, each atom $x \in \mathsf{L}(\mathfrak{F})$ has a preimage $y \in \mathsf{R}(D, A)$ which is an atom. Consider now $S_0 = \{f \in D^A \mid \mu(f)$ is a non-join-prime atom$\}$ and make it into a $P(\{1, \ldots, n\})$-valued ultrametric space by letting $\delta_{S_0}(f, g) = \mu(\delta(f, g)) \subseteq \{1, \ldots, n\}$—we use here the fact that μ sends join-prime elements to join-prime elements. S_0 is shown to be a pairwise complete ultrametric space over $\{1, \ldots, n\}$. We prove that pairwise complete ultrametric spaces over a finite set B are in bijection with universal $S5^n$-product frames, with $n = \operatorname{card} B$. Then the restriction of μ to S_0 is a surjective p-morphism from S_0 to (a frame isomorphic to) $\mathfrak{F}$. $\square$

In view of the following statement, which relies on [8] and can be inferred from [7]: "*for $n \geq 3$, it is undecidable whether, given a finite full initial frame $\mathfrak{F}$, there is a surjective p-morphism from a universal $S5^n$-product $\mathfrak{U}$ to $\mathfrak{F}$*", we deduce the following undecidability results, which partially answer Problem 4.10 in [12].

Corollary 3. *It is undecidable whether a finite subdirectly irreducible atomistic lattice embeds into a relational lattice. Consequently, the quasiequational theory of relational lattices in the pure lattice signature is also undecidable.*

Comparison with Litak et al. [12]. We have presented our first contribution to the study of the equational theory of relational lattices. In [12] two equations in the larger signature with the header constant are presented as a base for the equational theory of relational lattices. As mentioned there, the four equations of [[AxRel]] are derivable from these two equations. Therefore, we can also think of the present work as a contribution towards assessing or disproving completeness of these two axiomatizations. Yet, we wish to mention here and emphasize some of our original motivations. Lattice theoretic equations are quite difficult to grasp, in particular if considered on the purely syntactic side, as done for example in [12]. Duality theory attaches a meaning to equations via the combinatorial properties of the dual spaces. This process is nowadays customary in modal and intuitionistic logic and gives rise to a well defined area of research, correspondence theory. Our aim was to attach meaning to the equations of relational lattices. The answer we provide is, at the present state of research, via the relevant combinatorial properties, symmetry and pairwise completeness. From this perspective, the results presented in Sect. 6 undoubtedly need further understanding. In particular it is worth trying to modularize them, so as to discover equations exactly corresponding to symmetry or, respectively, to pairwise completeness; alternatively, argue that these equations do not exist. Finally, the present work opens new directions and challenges for the duality theory developed in [16]—of which, we hope we have illustrated the fruitfulness—including a better understanding of how to generalize it to the infinite case, new mechanisms by which to devise correspondence results, natural conjectures concerning equations having correspondents in finite lattices.

References

1. Ackerman, N.: Completeness in generalized ultrametric spaces. P-Adic Numbers Ultrametr. Anal. Appl. **5**(2), 89–105 (2013)
2. Codd, E.F.: A relational model of data for large shared data banks. Commun. ACM **13**(6), 377–387 (1970)
3. Davey, B.A., Priestley, H.A.: Introduction to Lattices and Order. Cambridge University Press, New York (2002)
4. Freese, R., Ježek, J., Nation, J.: Free Lattices. American Mathematical Society, Providence, RI (1995)
5. Grätzer, G.: General Lattice Theory. Birkhäuser, Basel, new appendices by the author with Davey, B.A., Freese, R., Ganter, B., Greferath, M., Jipsen, P., Priestley, H.A., Rose, H., Schmidt, E.T., Schmidt, S.E., Wehrung, F., Wille, R. (1998)
6. Hillenbrand, T., Löchner, B.: Waldmeister (1996–2008). http://www.waldmeister.org/
7. Hirsch, R., Hodkinson, I., Kurucz, A.: On modal logics between K × K × K and S5 × S5 × S5. J. Symbol. Log. **67**, 221–234 (2002)
8. Hirsch, R., Hodkinson, I.: Representability is not decidable for finite relation algebras. Trans. Amer. Math. Soc. **353**, 1403–1425 (2001)
9. Joyal, A., Tierney, M.: An extension of the Galois theory of Grothendieck. Mem. Amer. Math. Soc. **51**(309) (1984)

10. Kurucz, A.: Combining modal logics. In: Patrick Blackburn, J.V.B., Wolter, F. (eds.) Handbook of Modal Logic Studies in Logic and Practical Reasoning, vol. 3, pp. 869–924. Elsevier, New York (2007)

11. Lawvere, F.W.: Metric spaces, generalized logic and closed categories. Rendiconti del Seminario Matematico e Fisico di Milano **XLIII**, 135–166 (1973)

12. Litak, T., Mikulás, S., Hidders, J.: Relational lattices: from databases to universal algebra. JLAMP (2015, to appear) doi:10.1016/j.jlamp.2015.11.008

13. McCune, W.: Prover9 and Mace4 (2005–2010). http://www.cs.unm.edu/~mccune/prover9/

14. Nation, J.B.: An approach to lattice varieties of finite height. Algebra Univers. **27**(4), 521–543 (1990)

15. Priess-Crampe, S., Ribemboim, P.: Equivalence relations and spherically complete ultrametric spaces. C. R. Acad. Sci. Paris **320**(1), 1187–1192 (1995)

16. Santocanale, L.: A duality for finite lattices, September 2009, preprint. http://hal.archives-ouvertes.fr/hal-00432113

17. Spight, M., Tropashko, V.: Relational lattice axioms (2008, preprint). http://arxiv.org/abs/0807.3795

18. Tropashko, V.: Relational algebra as non-distributive lattice (2006, preprint). http://arxiv.org/abs/cs/0501053

On Local Characterization of Global Timed Bisimulation for Abstract Continuous-Time Systems

Ievgen Ivanov[(✉)]

Taras Shevchenko National University of Kyiv,
Volodymyrska St, 60, Kyiv 01601, Ukraine
ivanov.eugen@gmail.com

Abstract. We consider two notions of timed bisimulation on states of continuous-time dynamical systems: global and local timed bisimulation. By analogy with the notion of a bisimulation relation on states of a labeled transition system which requires the existence of matching transitions starting from states in such a relation, local timed bisimulation requires the existence of sufficiently short (locally defined) matching trajectories. Global timed bisimulation requires the existence of arbitrarily long matching trajectories. For continuous-time systems the notion of a global bisimulation is stronger than the notion of a local bisimulation and its definition has a non-local character. In this paper we give a local characterization of global timed bisimulation. More specifically, we consider a large class of abstract dynamical systems called Nondeterministic Complete Markovian Systems (NCMS) which covers various concrete continuous and discrete-continuous (hybrid) dynamical models and introduce the notion of an f^+-timed bisimulation, where f^+ is a so called extensibility measure. This notion has a local character. We prove that it is equivalent to global timed bisimulation on states of a NCMS. In this way we give a local characterization of the notion of a global timed bisimulation.

Keywords: Bisimulation · Cyber-physical system · Dynamical system · Continuous time · Local characterization

1 Introduction

The focus of this paper is the notion of bisimulation [1–4] in the domain of continuous-time dynamical systems. A general overview of the history of bisimulation, bisimilarity, coinductive definitions and their relevance to computer science, logic and other fields can be found in [4].

Recall that in the simplest case of labeled transition systems (LTS) [4] bisimulation and bisimilarity can be defined as follows:

© IFIP International Federation for Information Processing 2016
Published by Springer International Publishing Switzerland 2016. All Rights Reserved
I. Hasuo (Ed.): CMCS 2016, LNCS 9608, pp. 216–234, 2016.
DOI: 10.1007/978-3-319-40370-0_13

A binary relation R on states of an LTS is a bisimulation, if $(q_1, q_2) \in R$ implies that for each state q_1' and a label a such that $q_1 \to^a q_1'$ there exists a state q_2' such that $q_2 \to^a q_2'$ and $(q_1', q_2') \in R$, and, conversely, for each state q_2' and a label a such that $q_2 \to^a q_2'$ there exists a state q_1' such that $q_1 \to^a q_1'$ and $(q_1', q_2') \in R$.

Bisimilarity is the union of all bisimulations.

Associated with these notions is the bisimulation proof method [4,5], which, in particular, can be used to show behavioral equivalence of processes.

As was pointed out in [4], the features of the definition of bisimulation which make the bisimulation proof method practically interesting are:

- *locality of the checks* in the sense that only immediate transitions from states of a pair $(q_1, q_2) \in R$ need to be examined to verify the conditions of the definition;
- *the lack of hierarchy* on the pairs of the bisimulation (i.e. checks can be done in any order).

Many modifications and extensions of the mentioned definitions were proposed in different contexts [4].

In this paper we are interested in the notions of bsimulation for continuous-time models which are useful for modeling cyber-physical systems [6–9] and giving semantics to related specification and programming languages [10–14]. In this context various definitions of bisimulation relations were proposed [15–21]. A survey and comparison of different approaches can be found in [21,22].

Most of such approaches consider dynamical system models with an explicit notion of a *global* (continuous) time with respect to which the system's global state evolves and define some notion of bisimulation on states of such systems.

Such definitions of bisimulation for continuous-time systems can be classified in different ways.

Generally, on one hand there are *reduction-like approaches* which associate a model which has a pre-existing notion of bisimulation (e.g. LTS) with a continuous-time model and consider bisimulation relations for the associated model (in the sense of the pre-existing definition) to be bisimulation relations for the continuous-time model. Approaches of this kind were used for timed automata and several classes of hybrid systems [15] for abstracting infinite-state systems by finite systems and establishing decidability results [15], for abstracting continuous-time linear control systems [16], etc.

On the other hand, there are approaches which define new notions of bisimulation *specifically* for the considered classes of continuous-time systems. Approaches of this kind were proposed in [18] for continuous-time linear control systems with disturbances and certain kinds of nonlinear systems, in [19] for dynamical systems in the sense of J.C. Willems behavioral approach [23], in [24,25] for dynamical systems on manifolds and control and hybrid systems, in [26] for general flow systems.

The way in which a particular definition of bisimulation for continuous-time systems takes into account timing information gives another classification of such definitions.

On one hand, there were proposed time-abstracting bisimulations [27], bisimulations of time-abstract transition systems [16], reachability bisimulation [26] for continuous-time systems which *do not take into account the times* required by a system to reach a particular state.

On the other hand, *timed bisimulation* definitions require matching of states along trajectories (executions) of a system starting from states related by bisimulation at exactly same time moments, e.g. [18,26]. Intermediate approaches which take into account time information, but do not require exact matching along trajectories starting from states related by a bisimulation were also proposed, e.g. *progress bisimulation* [26].

The mentioned approaches to formalization of dynamical systems and the associated notions of bisimulation and proof methods are quite heterogeneous and currently lack a uniform treatment (e.g. in terms of coalgebras).

However, comparing various definitions of bisimulation for continuous-time dynamical/control/hybrid systems to the definition of a bisimulation on the states of a LTS, an important aspect of these definitions becomes visible: although these definitions do not impose a hierarchy on the pairs (similarly to bisimulations for LTS [4]), timed bisimulation definitions are *non-local* in the sense that checking that a pair of states is in a bisimulation relation involves checking some *"far future"/global properties* of the trajectories of a system starting in these states (relative to the time moment when these trajectories start).

In particular, this is true for the bisimulation definitions proposed for abstract types of continuous-time systems, e.g. in [26] the following notion of a timed simulation was introduced for highly abstract *general flow systems*:

If Φ_1, Φ_2 are general flow systems over value spaces X_1, X_2 with the same time line, a binary relation R between X_1, X_2 is a timed simulation of Φ_1 by Φ_2, if $dom(\Phi_1) \subseteq dom(R)$ and for all $x_1, x_1' \in X_1$, $x_2 \in X_2$ such that $(x_1, x_2) \in R$ and for all times $t > 0$, if there is a path $\gamma_1 \in \Phi_1(x_1)$ such that $x_1' = \gamma_1(t)$, then there is $x_2' \in X_2$ and $\gamma_2 \in \Phi_2(x_2)$ such that $x_2' = \gamma_2(t)$, $dom(\gamma_2) = dom(\gamma_1)$, and $(\gamma_1(s), \gamma_2(s)) \in R$ for all $s \in dom(\gamma_2) \cap [0, t]$. A relation R is a timed bisimulation between Φ_1, Φ_2, if both R and R^{-1} are timed simulations (details about the notions used in this definition are given in [26]).

In principle, we agree with definitions of this kind (on both abstract and concrete levels), but consider their non-local character undesirable for applications based on the bisimulation proof method.

Our aim in this paper is to give a *necessary and sufficient condition* (criterion) of a *local* (in time) character for checking that a given relation satisfies a timed bisimulation definition of this kind. The novelty of the main result is that local characterization of global timed bisimulation for continuous-time systems is possible in the very general case and can be given in a uniform way (using the notion of a so called f^+-bisimulation defined below). Local characterization also makes the notion of bisimulation for systems with continuous-time evolution close in spirit to the classical notion of bisimulation for LTS (which are most often used for representing systems with discrete-time evolution) and *allows one to use a wide variety of well-known methods of local analysis* (in local in time or in state space) of the behavior of systems defined by differential equations, inclusions, certain hybrid (discrete-continuous) formalisms, etc. (e.g. linearization, various series expansions, approximations, singularity analysis, etc.) *for proving that a given relation is a bisimulation.* Such methods are difficult or impossible to apply if one tries to prove that a relation is a global timed bisimulation directly by the definition (since this definition is given in terms of long-term behaviors of a system instead of short-term behaviors). We also suppose that this result will be useful for further development of uniform treatment of continuous time dynamical system models and proof principles related to them using coalgebraic approach (e.g. definition of bisimulation on continuous-time systems in terms of coalgebras).

Note that as we have mentioned above, many different definitions of bisimulation for continuous-time systems can be found in the literature. However, arguably, once a local characterization is obtained for some reasonable formalization X of bisimulation, it may be translated to other formalizations of bisimulation at least when they agree with X (e.g. bisimulation for general flow systems in the sense of Davoren and Tabuada [26]). In this paper we do not include a detailed comparison of different approaches to the definition of bisimulation for continuous-time systems and local characterization and its limits in each of such cases, but this remains a topic of further investigation.

To obtain the main result we will consider dynamical systems on a high level of abstraction comparable to the level of the mentioned general flow systems, but use a particular formalization of such systems called *Nondeterministic Complete Markovian Systems* (NCMS).

This formalization was proposed in [28–32] and inspired by the notion of a *solution system* from O. Hájek's Theory of processes [33,34]. In this formalization the global non-negative real time scale is assumed and continuous-time systems are modeled as sets of trajectories considered as functions on real time intervals which take values in an arbitrary fixed set of states. These sets must satisfy certain weak assumptions (more details are given in Sect. 2) [29]:

- be *closed under proper restrictions* onto intervals;
- satisfy the *Markovian* property which means that if two trajectories meet at one time in one state, their concatenation is a trajectory (note that this Markovian property is not formally related to the probability theory and stochastic processes);
- satisfy the *completeness* property in the following sense: a non-empty chain of trajectories in the sense of a subtrajectory relation has a supremum in the set of trajectories.

One interpretation of the Markovian property is that at any time moment the set of possible future evolutions of a system depends only on its current state and time and does not depend on the path by which the system reached the current state (which is also true for LTS). The definition of Hájek's solution system is rather similar, but lacks an equivalent of the completeness requirement of NCMS. But for us completeness is necessary to be able to establish reductions of global-in-time properties of systems to local-in-time properties.

NCMS are also close to the notion of a TCTL structure in the sense of Alur et al. [35], but the definition of the latter TCTL structures lack an equivalent of the completeness assumption. Only with it Markovian property is sufficient for establishing local characterization of bisimulation (informally, Markovian property of NCMS allows joining a finite sequence of trajectories; with completeness it allows joining an infinite sequence of trajectories).

The main reasons we use NCMS are:

- NCMS do not impose restrictions on the structure of the set of states and impose weak restrictions on the system behavior, support nondeterminism and partial trajectories. These features make NCMS promising for computer science and cyber-physical systems applications like semantics of real-time and embedded systems specification languages [30]. In contrast, well-known concrete dynamical system models (classical dynamical systems, switched systems [36], hybrid automata [37,38]) impose restrictions on the structure of the state space (e.g. assuming that it is a vector space, a manifold, or a related structure) and stronger restrictions on the behavior of a system.
- Concrete continuous-time models (e.g. described by differential equations, switched systems, etc.) can be represented by NCMS [28], similarly to representing different kinds of systems by Hájek's solution systems [33,34]. Some examples of such representations are given in Subsect. 2.2 below.
- The model of NCMS allows one to reduce some types of global analysis of system behavior to local analysis of system behavior, e.g. prove global properties by checking that certain conditions hold in a neighborhood of each time moment [29]. This is described in more detail in Subsect. 2.3.

In this paper we will define the notion of a *labeled NCMS* which can be considered as a continuous-time analog of LTS and the notion of a *global* timed bisimulation on the states of a labeled NCMS. We will also define an obvious local version of this notion of a global timed bisimulation which we will call a *local* timed bisimulation. Both notions turn out to be inequivalent in the case of NCMS (local timed bisimulation is strictly weaker than global timed bisimulation). Then we will strengthen the local definition of bisimulation using so-called extensibility measures [29] and call the obtained notion a f^+-timed bisimulation. This notion will have a local character. Then we will show the equivalence of f^+-timed bisimulation and global timed bisimulation, obtaining a local characterization of global timed bisimulation.

The paper is organized in the following way. To make the paper self-contained, we give all necessary preliminaries about NCMS in Sect. 2. The reader may skip this section or most of it, but consult it whenever necessary. In Sect. 3 we introduce the notion of a labeled NCMS. In Sect. 4 we introduce global and local timed simulations and bisimulations on states of labeled NCMS. In Sect. 5 we formulate and discuss the main result, i.e. the local characterization of global timed bisimulation on states of a labeled NCMS. In Sect. 6 we give an outline of the proof of the main result. In Sect. 7 we give conclusions.

2   Preliminaries

2.1   Notation

We will use the following notation: $\mathbb{N} = \{1, 2, 3, ...\}$ is the set of natural numbers; $\mathbb{R}$ is the set of real numbers; $\mathbb{R}_+$ is the set of nonnegative real numbers; $f : A \to B$ is a total function from a set A to a set B; $f : A \dashrightarrow B$ is a partial function from a set A to a set B, 2^A is the power set of a set A; $f|_A$ is the restriction of a function f to a set A; B^A is the set of all total functions from a set A to a set B; $^A B$ is the set of all partial function from a set A to a set B.

For any function $f : A \dashrightarrow B$ we will use the symbol $f(x) \downarrow$ ($f(x) \uparrow$) to denote that $f(x)$ is defined, or, respectively, is undefined on the argument x.

We will not distinguish the notions of a function and a functional binary relation. When we write that a function $f : A \dashrightarrow B$ is total or surjective, we mean that f is total on the set A specifically ($f(x)$ is defined for all $x \in A$), or, respectively, is onto B (for each $y \in B$ there exists $x \in A$ such that $y = f(x)$).

For any $f : A \dashrightarrow B$ denote $dom(f) = \{x \mid f(x) \downarrow\}$, i.e. the domain of f (note that in some fields like the category theory the domain of a partial function is defined differently).

For any binary relation R denote $R^{-1} = \{(y, x) \mid (x, y) \in R\}$ (the inverse relation).

For any partial functions f, g the notation $f(x) \cong g(x)$ will mean the strong equality: $f(x) \downarrow$ if and only if $g(x) \downarrow$, and $f(x) \downarrow$ implies $f(x) = g(x)$.

Denote by $f \circ g$ the functional composition: $(f \circ g)(x) \cong f(g(x))$.

Denote by T the non-negative real time scale $[0, +\infty)$. We will assume that T is equipped with a topology induced by the standard topology on $\mathbb{R}$.

We will use the symbols $\neg, \vee, \wedge, \Rightarrow, \Leftrightarrow$ to denote the logical operations of negation, disjunction, conjunction, implication, and equivalence respectively.

2.2   Nondeterministic Complete Markovian Systems

The notion of a Nondeterminisitc Complete Markovian System (NCMS) was introduced in [28] for studying the existence of global trajectories of dynamical systems. It is close to the notion of a solution system by Hájek [33].

Let us denote by $\mathfrak{T}$ the set of all intervals in T (connected subsets) which have the cardinality greater than one.

Let Q be a set (a state space) and Tr be some set of functions of the form $s : A \to Q$, where $A \in \mathfrak{T}$. We will call the elements of Tr *(partial) trajectories*.

Definition 1 [28,32]. *A set of trajectories Tr is closed under proper restrictions (CPR), if $s|_A \in Tr$ for each $s \in Tr$ and $A \in \mathfrak{T}$ such that $A \subseteq dom(s)$.*

Let us introduce the following notation: if f, g are partial functions, $f \sqsubseteq g$ means that the graph of f is a subset of the graph of g, and $f \sqsubset g$ means that the graph of f is a proper subset of g.

Definition 2. *Let $s_1, s_2 \in Tr$ be trajectories. Then:*

(1) s_1 is called a subtrajectory of s_2, if $s_1 \sqsubseteq s_2$;
(2) s_1 is called a proper subtrajectory of $s_2 \in Tr$, if $s_1 \sqsubset s_2$;
(3) s_1, s_2 are called incomparable, if neither $s_1 \sqsubseteq s_2$, nor $s_2 \sqsubseteq s_1$.

The pair $(Tr, \sqsubseteq)$ is a possibly empty partially ordered set.

Definition 3 [28,32]. *A CPR set of trajectories Tr is*

(1) Markovian (Fig. 2), if for each $s_1, s_2 \in Tr$ and $t_0 \in T$ such that $t_0 = \sup dom(s_1) = \inf dom(s_2)$, $s_1(t_0) \downarrow$, $s_2(t_0) \downarrow$, and $s_1(t_0) = s_2(t_0)$, the following function s belongs to Tr: $s(t) = s_1(t)$, if $t \in dom(s_1)$ and $s(t) = s_2(t)$, if $t \in dom(s_2)$.
(2) complete, if each non-empty chain in $(Tr, \sqsubseteq)$ has a supremum.

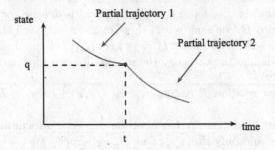

Fig. 1. Markovian property of NCMS. If one (partial) trajectory ends and another begins in a state q at time t, then their concatenation is a (partial) trajectory.

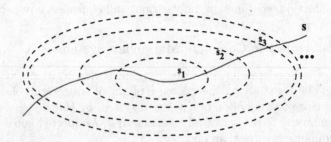

Fig. 2. Illustration of the completeness property of NCMS. The limit s of a $\sqsubseteq$-chain of trajectories (illustrated here as curve fragments bounded by dashed ellipses) of a NCMS is itself a trajectory of this NCMS. The graph of s is the union of graphs of elements of the chain.

Definition 4 [28,32]. *A nondeterministic complete Markovian system (NCMS) is a triple (T, Q, Tr), where Q is a set (state space) and Tr (trajectories) is a set of functions $s : T \rightarrow Q$ such that $dom(s) \in \mathfrak{T}$, which is CPR, complete, and Markovian.*

The notion of an *LR representation* [28,29,32] given below can be used to obtain an overview of the class of all NCMS.

Definition 5 [28,32]. *Let* $s_1, s_2 : T \dashrightarrow Q$. *Then* s_1 *and* s_2 *coincide:*

(1) on a set $A \subseteq T$, *if* $s_1|_A = s_2|_A$ *and* $A \subseteq dom(s_1) \cap dom(s_2)$ *(this is denoted as* $s_1 \doteq_A s_2$);
(2) in a left neighborhood of $t \in T$, *if* $t > 0$ *and there exists* $t' \in [0, t)$ *such that* $s_1 \doteq_{(t',t]} s_2$ *(this is denoted as* $s_1 \doteq_{t-} s_2$);
(3) in a right neighborhood of $t \in T$, *if there exists* $t' > t$, *such that* $s_1 \doteq_{[t,t')} s_2$ *(this is denoted as* $s_1 \doteq_{t+} s_2$).

Let Q be a set and $ST(Q)$ be the set of pairs all (s, t), where $s : A \to Q$ for some $A \in \mathfrak{T}$ and $t \in A$.

Definition 6 [28,32]. *A predicate* $p : ST(Q) \to Bool$ *is*

(1) left-local, if $p(s_1, t) \Leftrightarrow p(s_2, t)$ *whenever* $\{(s_1, t), (s_2, t)\} \subseteq ST(Q)$ *and* $s_1 \doteq_{t-} s_2$ *hold, and, moreover,* $p(s, t)$ *holds whenever* t *is the least element of* $dom(s)$;
(2) right-local, if $p(s_1, t) \Leftrightarrow p(s_2, t)$ *whenever* $\{(s_1, t), (s_2, t)\} \subseteq ST(Q)$ *and* $s_1 \doteq_{t+} s_2$ *hold, and, moreover,* $p(s, t)$ *holds whenever* t *is the greatest element of* $dom(s)$.

Let $LR(Q)$ denote the set of all pairs (l, r), where $l : ST(Q) \to Bool$ is a left-local predicate and $r : ST(Q) \to Bool$ is a right-local predicate.

Definition 7 [32]. *A pair* $(l, r) \in LR(Q)$ *is called a LR representation of a NCMS* $\Sigma = (T, Q, Tr)$, *if*
$$Tr = \{s : A \to Q \,|\, A \in \mathfrak{T} \land (\forall t \in A \; l(s, t) \land r(s, t))\}.$$

The following theorem shows that a NCMS can be represented using predicate pairs.

Theorem 1 [32].

(1) Each pair $(l, r) \in LR(Q)$ *is a LR representation of a NCMS with the set of states* Q.
(2) Each NCMS has a LR representation.

Consider some examples of representation of sets of trajectories of well-known continuous and discrete-continuous dynamical models in the form of NCMS.

1. ***Ordinary differential equations.*** Let $d \in \mathbb{N}$ and $f : \mathbb{R} \times \mathbb{R}^d \to \mathbb{R}^d$ be a continuous function. Let Tr be the set of all $\mathbb{R}^d$-valued functions such that $dom(s) \in \mathfrak{T}$ (i.e. s is defined on a non-degenerate real interval) such that s is differentiable on the interior of $dom(s)$ and
   - $\frac{d}{dt} s(t) = f(t, s(t))$ holds for each t in the interior of $dom(s)$;
   - $\partial_+ s(t) = f(t, s(t))$, if t is the least element of $dom(s)$;
   - $\partial_- s(t) = f(t, s(t))$, if t is the greatest element of $dom(s)$,

where $\partial_- s(t)$ denotes the left derivative at t, and $\partial_+ s(t)$ denotes the right derivative at t. Then $(T, \mathbb{R}^d, Tr)$ is a NCMS.

Indeed, consider predicates $l, r : ST(\mathbb{R}^d) \to Bool$ defined as follows:

- $l(s, t)$ if and only if either $\min dom(s) \downarrow= t$, or $t > \inf dom(s)$ and $\partial_- s(t) \downarrow= f(t, s(t))$;
- $r(s, t)$ if and only if either $\max dom(s) \downarrow= t$, or $t < \sup dom(s)$ and $\partial_+ s(t) \downarrow= f(t, s(t))$.

Obviously, $l(s, t)$ is left-local and $r(s, t)$ is right-local. Moreover, $l(s, t) \wedge r(s, t)$ holds for all $t \in dom(s)$ if and only if $s \in Tr$. Then Theorem 1 implies that $(T, \mathbb{R}^d, Tr)$ is a NCMS. Note that for this result we do not need any assumptions about global existence or uniqueness of solutions of differential equations, because NCMS support partiality and nondeterminism.

2. **Differential inclusions.** Consider a differential inclusion $\dot{x}(t) = F(t, x(t))$, where $F : \mathbb{R} \times \mathbb{R}^d \to 2^{\mathbb{R}^d}$ is a set-valued mapping. Let us introduce an auxiliary variable y and rewrite the inclusion as $\begin{cases} \dot{x}(t) = y(t); \\ y(t) \in F(t, x(t)). \end{cases}$

Let $Q = \mathbb{R}^d \times \mathbb{R}^d$ and Tr be the set of all Q-valued functions s such that $dom(s) \in \mathfrak{T}$ and there exist functions $x : dom(s) \to \mathbb{R}^d$ and $y : dom(s) \to \mathbb{R}^d$ such that $s(t) = (x(t), y(t))$ and $y(t) \in F(t, x(t))$ for all $t \in dom(s)$ and x is absolutely continuous on each compact segment $[a, b] \subseteq dom(s)$ and satisfies $\dot{x}(t) = y(t)$ almost everywhere (a.e.) on $dom(s)$ in the sense of Lebesgue's measure. Then (T, Q, Tr) is a NCMS. Indeed, consider $l, r : ST(Q) \to Bool$:

- $l(s, t)$ if and only if either $\min dom(s) \downarrow= t$, there exists $t' \in [0, t)$, an absolutely continuous function $x : [t', t] \to \mathbb{R}^d$, and a function $y : [t', t] \to \mathbb{R}^d$ such that $[t', t] \subseteq dom(s)$, $s(\tau) = (x(\tau), y(\tau))$ and $y(\tau) \in F(\tau, x(\tau))$ for all $\tau \in [t', t]$ and $\frac{d}{d\tau} x(\tau) = y(\tau)$ a.e. on $[t', t]$.
- $r(s, t)$ if and only if either $\max dom(s) \downarrow= t$, or there exists $t' > t$, an absolutely continuous function $x : [t, t'] \to \mathbb{R}^d$, and a function $y : [t, t'] \to \mathbb{R}^d$ such that $[t, t'] \subseteq dom(s)$, $s(\tau) = (x(\tau), y(\tau))$ and $y(\tau) \in F(\tau, x(\tau))$ for all $\tau \in [t, t']$ and $\frac{d}{d\tau} x(\tau) = y(\tau)$ a.e. on $[t, t']$.

Obviously, $l(s, t)$ is left-local and $r(s, t)$ is right-local. Moreover, it is easy to check that $l(s, t) \wedge r(s, t)$ holds for all $t \in dom(s)$ if and only if $s \in Tr$. Then (T, Q, Tr) is a NCMS by Theorem 1.

3. **Switched dynamical systems.** Let $d \geq 1$ be a natural number, I be a finite non-empty set (modes of a switched system), and $f_i : T \times \mathbb{R}^d \to \mathbb{R}^d$, $i \in I$ be an indexed family of vector fields (behaviors in each mode). Let $\mathfrak{J}$ be the set of all functions $\sigma : T \to I$ (switching signals) which are piecewise-constant on each compact segment $[a, b] \subset T$. Assume that for each $i \in I$, f_i is continuous and bounded on $T \times \mathbb{R}^d$ and there exists a number $L > 0$ such that $||f_i(t, x_1) - f_i(t, x_2)|| \leq L||x_1 - x_2||$ for all $x_1, x_2 \in \mathbb{R}^d$, $t \in T$, and $i \in I$ (Lipschitz-continuity). Consider a (nonlinear) switched system

$$\dot{x}(t) = f_{\sigma(t)}(t, x(t)), \ t \geq 0, \ \sigma \in \mathfrak{J}.$$

Note that by Caratheodory existence theorem, for each $x_0 \in \mathbb{R}^d$, $t_0 \in T$, $\sigma \in \mathfrak{J}$ the initial value problem $\frac{d}{dt} x(t) = f_{\sigma(t)}(t, x(t))$, $x(t_0) = x_0$ has a unique

Caratheodory solution $t \mapsto x(t; t_0; x_0; \sigma)$ defined for all $t \in [t_0, +\infty)$ such that $x(t_0; t_0; x_0; \sigma) = x_0$ (i.e. a function that is absolutely continuous on each compact segment in $[t_0, +\infty)$ and satisfies $\frac{d}{dt}x(t; t_0; x_0; \sigma) = f_{\sigma(t)}(t, x(t; t_0; x_0; \sigma))$ a.e. on $[t_0, +\infty)$).

Let $Q = \mathbb{R}^d \times I$ and Tr be the set of all Q-valued functions s such that $dom(s) \in \mathfrak{T}$ (i.e. $dom(s)$ is a non-degenerate real interval) and there exist $t_0 \in T$, $x_0 \in \mathbb{R}^d$, $\sigma : dom(s) \to I$ that is piecewise constant on each compact segment in $dom(s)$, and $x : dom(s) \to \mathbb{R}^d$ that is absolutely continuous on each compact segment in $dom(s)$ such that $\frac{d}{dt}x(t) = f_{\sigma(t)}(t, x(t))$ almost everywhere (a.e.) on $dom(s)$ in the sense of Lebesgue's measure and $s(t) = (x(t), \sigma(t))$ for $t \in dom(s)$. Then (T, Q, Tr) is a NCMS.

Indeed, consider predicates $l, r : ST(\mathbb{R}^d) \to Bool$ defined as follows:

- $l(s, t)$ if and only if either $\min dom(s) \downarrow = t$, there exists $t' \in [0, t)$, an absolutely continuous function $x : [t', t] \to \mathbb{R}^d$, and a piecewise-constant function $\sigma : [t', t] \to I$ such that $[t', t] \subseteq dom(s)$, $s(\tau) = (x(\tau), \sigma(\tau))$ for all $\tau \in [t', t]$ and $\frac{d}{d\tau}x(\tau) = f_{\sigma(\tau)}(\tau, x(\tau))$ a.e. on $[t', t]$.
- $r(s, t)$ if and only if either $\max dom(s) \downarrow = t$, or there exists $t' > t$, an absolutely continuous function $x : [t, t'] \to \mathbb{R}^d$, and a piecewise-constant function $\sigma : [t, t'] \to I$ such that $[t, t'] \subseteq dom(s)$, $s(\tau) = (x(\tau), \sigma(\tau))$ for all $\tau \in [t, t']$, and $\frac{d}{d\tau}x(\tau) = f_{\sigma(\tau)}(\tau, x(\tau))$ a.e. on $[t, t']$.

Obviously, $l(s, t)$ is left-local and $r(s, t)$ is right-local. Moreover, it is easy to check that $l(s, t) \wedge r(s, t)$ holds for all $t \in dom(s)$ if and only if $s \in Tr$. Then (T, Q, Tr) is a NCMS by Theorem 1.

Sets of trajectories of some more general switched/hybrid systems (possibly with state-dependent switching) can be represented as NCMS analogously.

2.3   Global Trajectories of NCMS

The problem of the existence of trajectories of NCMS defined on the whole time domain (global trajectories) was considered in [28, 29, 32]. In [28, 32] a method for proving the existence of a global trajectory in a NCMS was proposed. This method reduces the problem of proving the existence of a global trajectory to the problem of proving the existence of certain locally defined trajectories and can be informally described as follows: (1) guess a "region" (a subset of trajectories) which presumably contains a global trajectory and has a convenient representation in the form of (another) NCMS; (2) prove that this region indeed contains a global trajectory by finding certain locally defined trajectories independently in a neighborhood of each time moment.

Below we briefly state the results which form the basis of this method (Lemma 1 and Theorem 2 given below) which we will use in this paper.

Let $\Sigma = (T, Q, Tr)$ be a fixed NCMS.

Definition 8 [29]. *Σ satisfies*

(1) local forward extensibility (LFE) property, if for each $s \in Tr$ of the form $s : [a, b] \to Q$ $(a < b)$ there exists a trajectory $s' : [a, b'] \to Q$ such that $s' \in Tr$, $s \sqsubseteq s'$ and $b' > b$.

(2) *global forward extensibility (GFE) property, if for each trajectory s of the form* $s : [a, b] \to Q$ *there is a trajectory* $s' : [a, +\infty) \to Q$ *such that* $s \sqsubseteq s'$.

Definition 9 [29]. *A right dead-end path (in Σ) is a trajectory* $s : [a, b) \to Q$ *$(a, b \in T, a < b)$ such that there is no* $s' : [a, b] \to Q$, $s' \in Tr$ *such that* $s \sqsubset s'$.

Definition 10 [29]. *An escape from a right dead-end path* $s : [a, b) \to Q$ *(in Σ) is a trajectory* $s' : [c, d) \to Q$ *$(d \in T \cup \{+\infty\})$ or* $s' : [c, d] \to Q$ *$(d \in T)$ such that* $c \in (a, b)$, $d > b$, *and* $s(c) = s'(c)$. *An escape* s' *is infinite, if* $d = +\infty$.

Definition 11 [29]. *A right dead-end path* $s : [a, b) \to Q$ *in Σ is called strongly escapable, if there exists an infinite escape from s.*

Definition 12 [29].

(1) *A right extensibility measure is a function* $f^+ : \mathbb{R} \times \mathbb{R} \dot{\to} \mathbb{R}$ *such that* $A = \{(x, y) \in T \times T \mid x \leq y\} \subseteq dom(f^+)$, $f(x, y) \geq 0$ *for all* $(x, y) \in A$, $f^+|_A$ *is strictly decreasing in the first argument and strictly increasing in the second argument, and for each* $x \geq 0$, $f^+(x, x) = x$ *and* $\lim_{y \to +\infty} f^+(x, y) = +\infty$.
(2) *A right extensibility measure* f^+ *is called normal, if* f^+ *is continuous on* $\{(x, y) \in T \times T \mid x \leq y\}$ *and there exists a function* α *of class* K_∞ *(i.e. the function* $\alpha : [0, +\infty) \to [0, +\infty)$ *is continuous, strictly increasing, and* $\alpha(0) = 0$, $\lim_{x \to +\infty} \alpha(x) = +\infty$) *such that* $\alpha(y) < y$ *for all* $y > 0$ *and the function* $y \mapsto f^+(\alpha(y), y)$ *is of class* K_∞.

An example of a right extensibility measure is $f_n^+(x, y) = y + (y - x)^n$ for any $n \in \mathbb{N}$. Let f^+ be a right extensibility measure.

Definition 13 [29]. *A right dead-end path* $s : [a, b) \to Q$ *is called* f^+-*escapable (Fig. 3), if there exists an escape* $s' : [c, d] \to Q$ *from s such that* $d \geq f^+(c, b)$.

Lemma 1 [29]. *Σ satisfies GFE if and only if Σ satisfies LFE and each right dead-end path is strongly escapable.*

Theorem 2 ([29], **About right dead-end path**). *Assume that f^+ is a normal right extensibility measure and Σ satisfies LFE. Then each right dead-end path is strongly escapable if and only if each right dead-end path is f^+-escapable.*

3   Traces on Sets of Trajectories and Labeled NCMS

In the case of labeled transition systems (LTS), labels are some data associated with transitions and traces are sequences of labels along executions of an LTS. We would like to define an analogous notion of a trace for NCMS. The informal idea behind the definition of a trace proposed below is that for continuous-time systems the role of "transitions" play "infinitesimally short trajectories" and "labels" are certain values associated with such trajectories. Thus a trace defines some quantity that evolves in time along a trajectory. This idea of co-evolution of trace and trajectory is formalized in Definition 14 below. Theorem 3 given below shows that this definition implies that at each time moment the value of a trace depends only on the values of the trajectory in vicinity of this time moment supporting the informal ideas of "transitions" and "labels" for NCMS.

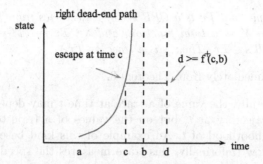

Fig. 3. An f^+-escapable right dead-end path $s : [a, b) \to Q$ (curve) and a corresponding escape $s' : [c, d] \to Q$ (a horizontal segment) such that $d \geq f^+(c, b)$.

Definition 14 (Trace). *Let Tr be a CPR set of trajectories. A function λ on Tr is called a trace on Tr, if the following conditions hold:*

(1) (Preservation of domain) For each $s \in Tr$, $\lambda(s)$ is a function defined on $dom(s)$.

(2) (Monotonicity) If $s_1, s_2 \in Tr$ and $s_1 \sqsubseteq s_2$, then $\lambda(s_1) \sqsubseteq \lambda(s_2)$.

We will define a labeled NCMS as a NCMS with a trace on its trajectories.

Definition 15 (Labeled NCMS). *A labeled NCMS is a pair (Σ, λ), where $\Sigma = (T, Q, Tr)$ is a NCMS and λ is trace on Tr.*

The most important properties of traces are formulated below.

Lemma 2 (Image of trace). *The image of a trace on a CPR set of trajectories is a CPR set of trajectories.*

Lemma 3 (Chain-continuity of a trace). *Let Tr be a CPR set of trajectories and λ be a trace on Tr. Then λ is chain-continuous in the following sense: for any non-empty chain C in the poset $(Tr, \sqsubseteq)$ which has the least upper bound $s^* \in Tr$ the set $\{\lambda(s) \mid s \in C\}$ has the least upper bound $\lambda(s^*)$ in the poset $(\{\lambda(s) \mid s \in Tr\}, \sqsubseteq)$.*

The following theorem gives a convenient criterion for checking if a function is a trace.

Theorem 3 (Criterion of a trace). *Let Tr be a CPR set of trajectories, Y be a set, $\lambda : Tr \to (T \dashrightarrow Y)$ be a total function. Then λ is a trace on Tr if and only if the following conditions hold:*

(1) $dom(\lambda(s)) = dom(s)$ for all $s \in Tr$;

(2) if $s_1, s_2 \in Tr$, $t_0 \in T$, $s_1 \doteq_{t_0+} s_2$, then $\lambda(s_1)(t_0) = \lambda(s_2)(t_0)$;

(3) if $s_1, s_2 \in Tr$, $t_0 \in T$, $s_1 \doteq_{t_0-} s_2$, then $\lambda(s_1)(t_0) = \lambda(s_2)(t_0)$.

The following lemma gives an obvious example of a trace: pointwise application of a total function on the set of states to a trajectory (projection).

Lemma 4. *Assume that Tr is a CPR set of trajectories from T to a set Q, Y is a set, $f : Q \to Y$ is a total function, and $\lambda : Tr \to (T \tilde{\to} Y)$ is such that $\lambda(s) = f \circ s$ for all $s \in Tr$. Then λ is a trace on Tr.*

Proof. Follows immediately from Theorem 3. $\square$

However, generally, the value of a trace at time t may depend not only on the value of a trajectory at t, but on the values of a trajectory in an arbitrarily small neighborhood of t. An example of this kind based on differentiation is given below (informally, this trace measures the speed of change of a trajectory).

Example 1. Assume that $n \in \mathbb{N}$, $Tr \subset T \tilde{\to} \mathbb{R}^n$ is a CPR set of trajectories and each $s \in Tr$ is differentiable on $dom(s) \in \mathfrak{T}$, i.e. s is differentiable at each point of the interior of $dom(s)$, s has the right derivative at the least element of $dom(s)$, if this element exists, and s has the left derivative at the greatest element of $dom(s)$, if this element exists.

Let $\lambda : Tr \to (T \tilde{\to} \mathbb{R}^n)$ be a function such that for each $s \in Tr$:

- $\lambda(s)(t) = \frac{d}{dt} s(t)$, if t is in the interior of $dom(s)$;
- $\lambda(s)(t)$ is the right derivative of s at t, if t is the least element of $dom(s)$;
- $\lambda(s)(t)$ is the left derivative of s at t, if t is the greatest element of $dom(s)$.

Using Theorem 3 it is easy to check that λ is a trace on Tr. $\square$

4   Timed Simulation and Bisimulation on NCMS

For any partial function s on T such that $dom(s) \in \mathfrak{T}$, any $t_0 \in T$, and any element q we will write

$q \overset{s}{\rightsquigarrow}$, if $dom(s)$ has the least element a such that $s(a) = q$;

$q \overset{s}{\rightsquigarrow}_{t_0}$, if t_0 is the least element of $dom(s)$ and $s(t_0) = q$.

Let (Σ, λ) be a fixed labeled NCMS, where $\Sigma = (T, Q, Tr)$.

Definition 16. *Let $s_1, s_2 : T \tilde{\to} Q$ and $R \subseteq Q \times Q$ be a binary relation. Then the functions s_1 and s_2 are:*

(1) pointwise in R, if $dom(s_1) = dom(s_2)$ and $(s_1(t), s_2(t)) \in R$ for $t \in dom(s_1)$;

(2) pointwise in R on a set $A \subseteq T$, if $A \subseteq dom(s_1) \cap dom(s_2)$ and $(s_1(t), s_2(t)) \in R$ for all $t \in A$;

(3) pointwise in R in a right neighborhood of $t \in T$, if there exists $t' > t$, such that s_1, s_2 are pointwise in R on $[t, t')$;

(4) pointwise in R in a deleted left neighborhood of $t \in T$, if $t > 0$ and there is $t' \in [0, t)$ such that s_1, s_2 are pointwise in R on (t', t).

Definition 17 (Global timed simulation). *A relation $R \subseteq Q \times Q$ is a global timed simulation on (Σ, λ), if for each $(q_1, q_2) \in R$ and $s_1 \in Tr$ such that $q_1 \overset{s_1}{\rightsquigarrow}$ there is $s_2 \in Tr$ such that $q_2 \overset{s_2}{\rightsquigarrow}$, $\lambda(s_1) = \lambda(s_2)$, and s_1, s_2 are pointwise in R.*

Definition 18 (Local timed simulation). *A relation $R \subseteq Q \times Q$ is a local timed simulation on (Σ, λ), if for each $(q_1, q_2) \in R$, $s_1 \in Tr$, and $t_0 \in T$ such that $q_1 \overset{s_1}{\leadsto}_{t_0}$ there exists $s_2 \in Tr$ such that $q_2 \overset{s_2}{\leadsto}_{t_0}$, $\lambda(s_1) \doteq_{t_0+} \lambda(s_2)$, and s_1, s_2 are pointwise in R in a right neighborhood of t_0.*

Definition 19 (Timed bisimulation). *A relation $R \subseteq Q \times Q$ is a*

(1) local timed bisimulation on (Σ, λ), if both R and R^{-1} are local timed simulations on (Σ, λ);

(2) global timed bisimulation on (Σ, λ), if both R and R^{-1} are global timed simulations on (Σ, λ).

Lemma 5. *If R is a global timed simulation on (Σ, λ), then R is a local timed simulation on (Σ, λ).*

Lemma 6. *There exists a labeled NCMS (Σ', λ') and a local timed bisimulation R_0 on (Σ', λ') such that R_0 is not a global timed simulation on (Σ', λ').*

Theorem 4 (About global and local timed bisimulation)

(1) If R is a global timed bisimulation on (Σ, λ), then R is a local timed bisimulation on (Σ, λ).

(2) There is a labeled NCMS (Σ', λ') and a local timed bisimulation R_0 on (Σ', λ') such that R_0 is not a global timed bisimulation on (Σ', λ').

Proof. Follows immediately from Lemmas 5, 6, and Definition 19.   □

5 Main Result

As before, let (Σ, λ) be a fixed labeled NCMS, where $\Sigma = (T, Q, Tr)$. Let f^+ be a fixed right extensibility measure.

Definition 20 (f^+-timed simulation). *A relation $R \subseteq Q \times Q$ is a f^+-timed simulation on (Σ, λ), if R is a local timed simulation on (Σ, λ) and for each $s_1, s_2 \in Tr$ and $t_0 \in dom(s_1)$ which satisfy the following conditions:*

- *s_1, s_2 are pointwise in R in a deleted left neighborhood of t_0,*
- *$\lambda(s_1) \doteq_{[t'_0, t_0)} \lambda(s_2)$ for some $t'_0 < t_0$,*

there exist $s'_2 \in Tr$, $t_1 \in dom(s_2) \cap dom(s'_2)$, and $t_2 \in T$ such that

(1) $t_1 < t_0$ and $s_2(t_1) = s'_2(t_1)$;

(2) either $t_2 \geq f^+(t_1, t_0)$, or t_2 is the maximal element of $dom(s_1)$;

(3) $\lambda(s_1) \doteq_{[t_1, t_2]} \lambda(s'_2)$;

(4) s_1 and s'_2 are pointwise in R on $[t_1, t_2]$.

Definition 21 (f^+-timed bisimulation). *A relation $R \subseteq Q \times Q$ is a f^+-timed bisimulation on (Σ, λ), if both R and R^{-1} are f^+-timed simulations on (Σ, λ).*

The main result of this paper is the following theorem:

Theorem 5 (Local characterization of global timed bisimulation). *Let* f^+ *be a normal right extensibility measure. A relation* $R \subseteq Q \times Q$ *is a global timed bisimulation on* (Σ, λ) *if and only if* R *is a* f^+-*timed bisimulation on* (Σ, λ).

This theorem holds for any normal right extensibility measure, for example, $f_1^+(x, y) = y + (y - x) = 2y - x$. The difference between the definition of the f^+-timed simulation and global timed simulation is that the latter definition is non-local, i.e. it requires proving the existence of arbitrarily long trajectories (s_2) for proving that R is a simulation which may be hard, if the dynamics of a system is defined by nonlinear differential equations or in other implicit way. The former definition is local in that for proving that R is a simulation one can show the existence of s_2' that satisfies (1)–(4) on an arbitrarily short interval $[t_1, t_2]$ (the condition (2) imposes a lower bound on its length, but e.g. for $f^+ = f_1^+$ this lower bound can be made arbitrarily small by choosing t_1 close to t_0).

Arguably, the characterization of global timed bisimulation provided by Theorem 5 is non-constructive, because it does not tell how to check the existence of $s_2' \in Tr$, $t_1 \in dom(s_2) \cap dom(s_2')$, and $t_2 \in T$ in Definition 20 (their existence for any s_1, s_2, t_0 that satisfy assumptions of this definition is required for proving that a relation is a global timed bisimulation). But this lack of constructivity is, arguably, a consequence of generality of our model of a system (NCMS). So the role of the local characterization provided by Theorem 5 is logical (to give an alternative view of bisimulation in the general case useful e.g. for proving new theorems about bisimulations) instead of being an executable algorithm.

The question of whether Theorem 5 can be a basis of algorithms for checking properties related to bisimulations and bisimilarity for special types dynamical systems (e.g. described by linear systems, etc.) requires separate investigation.

An informal description of how Theorem 5 can be applied is given below.

Let S be a system that travels through the state space $Q = \mathbb{R}^n$ in accordance with a known law of motion L – an ordinary differential equation with input control. The trace of a trajectory is a pointwise application of some output function to the trajectory (in accordance with Lemma 4). Q contains a (possibly infinite) subset O of isolated point obstacles. If S hits an obstacle, its trajectory ends without possibility of continuation. Trajectories which neither hit nor tend to obstacles can be continued indefinitely.

Suppose that we want to prove that under certain assumptions $R = (Q\backslash O) \times (Q\backslash O)$ is a *global* timed bisimulation.

Proof using Definition 17 involves reasoning about the whole set O. Under assumptions that are close to functional output-controllability of the system one can prove that R is a *local* bisimulation *without reasoning about obstacles at all* (see Definition 18). However, this approach is not directly applicable to the case of global timed bisimulation.

For proving that R is a global timed bisimulation one can use f^+-timed bisimulation which is equivalent to it. In this case one needs to inspect system behavior *near each obstacle individually*, forgetting about others: take $f^+(x, y) = 2y - x$ and consider Definition 20. The main case is when s_2 tends to a some obstacle X as $t \to t_0$ and s_1 avoids all obstacles. Definition 20 requires the existence of a control maneuver s_2' that preserves the trace of s_2, but not for long after t_0 ($t_2 - t_0 \geq t_0 - t_1$ is sufficient). By choosing t_1 such that $t_0 - t_1$ is sufficiently small (informally, "last minute collision avoidance") and taking into account continuity of trajectories of S, proving its existence using L does not require reasoning about obstacles from O other than X.

6   Outline of the Proof of the Main Result

The idea of the proof is to define a family of auxiliary NCMS $\{\Sigma_{s_0,R}(\Sigma,\lambda) \mid s_0 \in Tr\}$ depending on R such that all its members satisfy GFE whenever R is a f^+-timed simulation and show that if they satisfy GFE, then R is a global timed simulation on (Σ,λ). The converse part of the theorem can be proved directly.

We formulate the main steps (milestones) of the proof as a series of lemmas given below (Lemmas 7–13). We assume that their statements are self-describing.

Lemma 7. *Let f^+ be a normal right extensibility measure and $R \subseteq Q \times Q$ be a global timed simulation on (Σ,λ). Then R is a f^+-timed simulation on (Σ,λ).*

For each $s_0 \in Tr$ and a relation $R \subseteq Q \times Q$ let us denote:

- $Tr^0_{s_0,R}(\Sigma,\lambda)$ is the set of all functions $s : T \rightharpoonup Q$ such that $dom(s) \in \mathfrak{T}$ and the following conditions hold:
  - $s|_{dom(s_0)} \in Tr$,
  - $\lambda(s|_{dom(s_0)}) \sqsubseteq \lambda(s_0)$,
  - s_0, s are pointwise in R on $dom(s_0) \cap dom(s)$.
- $Tr_{s_0,R}(\Sigma,\lambda) = \{s : T \rightharpoonup Q \mid dom(s) \in \mathfrak{T} \wedge \exists \hat{s} \in Tr^0_{s_0,R}(\Sigma,\lambda) \ s \sqsubseteq \hat{s}\}$.
- $\Sigma_{s_0,R}(\Sigma,\lambda) = (T, Q, Tr_{s_0,R}(\Sigma,\lambda))$.

Lemma 8. *If $s_0 \in Tr$ and $R \subseteq Q \times Q$, then $\Sigma_{s_0,R}(\Sigma,\lambda)$ is a NCMS.*

Lemma 9. *Let $s_0 \in Tr$ and R be a local timed simulation on (Σ,λ). Then $\Sigma_{s_0,R}(\Sigma,\lambda)$ is a NCMS which satisfies LFE.*

Lemma 10. *Let f^+ be a normal right extensibility measure, $s_0 \in Tr$, and R be a f^+-timed simulation on (Σ,λ). Assume that s_* is a right dead-end path in the NCMS $\Sigma_{s_0,R}(\Sigma,\lambda)$. Then s_* is f^+-escapable.*

Lemma 11. *Let f^+ be a normal right extensibility measure, $s_0 \in Tr$, and R be a f^+-timed simulation on (Σ,λ). Then $\Sigma_{s_0,R}(\Sigma,\lambda)$ satisfies GFE.*

Proof. R is a f^+-timed simulation on (Σ,λ), so R is a local timed simulation on (Σ,λ). Then $\Sigma_{s_0,R}(\Sigma,\lambda)$ is a NCMS which satisfies LFE by Lemma 9. By Lemma 10 each right dead-end path in $\Sigma_{s_0,R}(\Sigma,\lambda)$ is f^+-escapable. By Theorem 2 each right dead-end path in $\Sigma_{s_0,R}(\Sigma,\lambda)$ is strongly escapable. Then by Lemma 1 $\Sigma_{s_0,R}(\Sigma,\lambda)$ satisfies GFE. $\square$

Lemma 12. *Let $R \subseteq Q \times Q$ be a local timed simulation on (Σ, λ). Assume that for each $s_0 \in Tr$, $\Sigma_{s_0, R}(\Sigma, \lambda)$ is a NCMS which satisfies GFE. Then R is a global timed simulation on (Σ, λ).*

Lemma 13. *Let f^+ be a normal right extensibility measure and $R \subseteq Q \times Q$ be a f^+-timed simulation on (Σ, λ). Then R is a global timed simulation on (Σ, λ).*

Proof. By Lemma 11, for each $s_0 \in Tr$, $\Sigma_{s_0, R}(\Sigma, \lambda)$ is a NCMS which satisfies GFE. Because R is a f^+-timed simulation on (Σ, λ), R is a local timed simulation on (Σ, λ). Then by Lemma 12, R is a global timed simulation on (Σ, λ). $\square$

Proof. (Proof of Theorem 5). The "If" part follows from Lemma 13 and the "Only if" part follows from Lemma 7. $\square$

7   Conclusions and Future Work

We have obtained a necessary and sufficient condition (criterion) of a local character for checking that a given relation satisfies the definition of a global timed bisimulation for NCMS.

The obtained results can be useful for applying bisimulation proof method to various continuous-time models for establishing equivalence and constructing abstractions of such systems and for further development of uniform treatment of continuous time dynamical system models and proof principles related to them using coalgebraic approach. We plan to develop bisimulation proof method on the basis of the results obtained in this paper and apply it to cyber-physical system verification problems in the forthcoming papers.

References

1. Milner, R.: Communication and Concurrency. Prentice-Hall Inc., Upper Saddle River (1989)
2. Park, D.: Concurrency and automata on infinite sequences. In: Deussen, P. (ed.) Theoretical Computer Science. LNCS, vol. 104, pp. 167–183. Springer, Heidelberg (1981)
3. Sangiorgi, D.: Introduction to Bisimulation and Coinduction. Cambridge University Press, New York (2011)
4. Sangiorgi, D.: On the origins of bisimulation and coinduction. ACM Trans. Program. Lang. Syst. **31**(4), 15:1–15:41 (2009)
5. Sangiorgi, D.: On the bisimulation proof method. Mathematical. Struct. Comput. Sci. **8**(5), 447–479 (1998)
6. Shi, J., Wan, J., Yan, H., Suo, H.: A survey of cyber-physical systems. In: 2011 International Conference on Wireless Communications and Signal Processing (WCSP), pp. 1–6. IEEE (2011)
7. Baheti, R., Gill, H.: Cyber-physical systems. Impact Control Technol. **12**, 161–166 (2011)
8. Lee, E., Seshia, S.: Introduction to Embedded Systems: A Cyber-physical Systems Approach. Lulu.com, Berkeley (2013)

9. Sifakis, J.: Rigorous design of cyber-physical systems. In: 2012 International Conference on Embedded Computer Systems (SAMOS), p. 319. IEEE (2012)
10. Bouissou, O., Chapoutot, A.: An operational semantics for Simulink's simulation engine. In: Proceedings of 13th ACM SIGPLAN/SIGBED International Conference on Languages, Compilers, Tools and Theory for Embedded Systems, pp. 129–138. ACM (2012)
11. Simulink - Simulation and Model-Based Design. http://www.mathworks.com/products/simulink
12. SCADE Suite. http://www.esterel-technologies.com/products/scade-suite
13. Campbell, S., Chancelier, J.P., Nikoukhah, R.: Modeling and Simulation in Scilab/Scicos with ScicosLab 4.4. Springer, New York (2005)
14. Feiler, P., Gluch, D., Hudak, J.: The architecture analysis and design language (AADL): an introduction. Technical report CMU/SEI-2006-TN-011, Carnegie-Mellon University (2006)
15. Alur, R., Henzinger, T.A., Lafferriere, G., Pappas, G.J.: Discrete abstractions of hybrid systems. Proc. IEEE 88(7), 971–984 (2000)
16. Pappas, G.: Bisimilar linear systems. Automatica 39(12), 2035–2047 (2003)
17. van der Schaft, A.: Equivalence of dynamical systems by bisimulation. IEEE Trans. Autom. Control 49(12), 2160–2172 (2004)
18. van der Schaft, A.: Equivalence of hybrid dynamical systems. In: Proceedings of 16th International Symposium on Mathematical Theory of Networks and Systems, Leuven, Belgium, 5–9 July 2004
19. Julius, A., van der Schaft, A.: Bisimulation as congruence in the behavioral setting. In: Proceedings of 44th IEEE Conference on Decision and Control and the European Control Conference 2005, Seville, Spain, 12–15 December 2005
20. Pola, G., van der Schaft, A., Di Benedetto, M.: Equivalence of switching linear systems by bisimulation. Int. J. Control 79(1), 74–92 (2006)
21. Schmuck, A., Raisch, J.: Simulation and bisimulation over multiple time scales in a behavioral setting (2014). CoRR abs/1402.3484
22. Cuijpers, P.J.L., Reniers, M.A.: Lost in translation: hybrid-time flows vs. real-time transitions. In: Egerstedt, M., Mishra, B. (eds.) HSCC 2008. LNCS, vol. 4981, pp. 116–129. Springer, Heidelberg (2008)
23. Polderman, J., Willems, J.: Introduction to Mathematical Systems Theory: A Behavioral Approach. Springer, Berlin (1997)
24. Haghverdi, E., Tabuada, P., Pappas, G.: Unifying bisimulation relations for discrete and continuous systems. In: Proceedings of International Symposium MTNS2002, South (2002)
25. Haghverdi, E., Tabuada, P., Pappas, G.: Bisimulation relations for dynamical, control, and hybrid systems. Theoret. Comput. Sci. 342(2–3), 229–261 (2005)
26. Davoren, J.M., Tabuada, P.: On simulations and bisimulations of general flow systems. In: Bemporad, A., Bicchi, A., Buttazzo, G. (eds.) HSCC 2007. LNCS, vol. 4416, pp. 145–158. Springer, Heidelberg (2007)
27. Tripakis, S., Yovine, S.: Analysis of timed systems using time-abstracting bisimulations. Form. Methods Syst. Des. 18(1), 25–68 (2001)
28. Ivanov, I.: A criterion for existence of global-in-time trajectories of non-deterministic Markovian systems. Commun. Comput. Inf. Sci. (CCIS) 347, 111–130 (2012)
29. Ivanov, I.: On representations of abstract systems with partial inputs and outputs. In: Gopal, T.V., Agrawal, M., Li, A., Cooper, S.B. (eds.) TAMC 2014. LNCS, vol. 8402, pp. 104–123. Springer, Heidelberg (2014)

30. Ivanov, I., Nikitchenko, M., Abraham, U.: On a decidable formal theory for abstract continuous-time dynamical systems. Commun. Comput. Inf. Sci. (CCIS) **469**, 78–99 (2014)
31. Ivanov, I.: An abstract block formalism for engineering systems. In: CEUR Workshop Proceedings, ICTERI, vol. 1000, pp. 448–463. CEUR-WS.org (2013)
32. Ivanov, I.: On existence of total input-output pairs of abstract time systems. Commun. Comput. Inf. Sci. (CCIS) **412**, 308–331 (2013)
33. Hájek, O.: Theory of processes, I. Czechoslovak Math. J. **17**, 159–199 (1967)
34. Hájek, O.: Theory of processes, II. Czechoslovak Math. J. **17**(3), 372–398 (1967)
35. Alur, R., Courcoubetis, C., Dill, D.: Model checking in dense real-time. Inf. Comput. **104**, 2–34 (1993)
36. Liberzon, D.: Switching in Systems and Control (Systems and Control: Foundations and Applications). Birkhauser Boston Inc., Boston (2003)
37. Alur, R., Courcoubetis, C., Halbwachs, N., Henzinger, T., Ho, P.H., Nicollin, X., Olivero, A., Sifakis, J., Yovine, S.: The algorithmic analysis of hybrid systems. Theoret. Comput. Sci. **138**(1), 3–34 (1995)
38. Goebel, R., Sanfelice, R.G., Teel, A.: Hybrid dynamical systems. IEEE Control Syst. **29**(2), 28–93 (2009)

Author Index

Printed in the United States
By Bookmasters